Advanced Aircraft Systems

TAB
PRACTICAL
FLYING SERIES

Other books in the TAB PRACTICAL FLYING SERIES

The Pilot's Air Traffic Control Handbook, 2nd edition *by Paul E. Illman*

Stalls & Spins *by Paul A. Craig*

The Pilot's Radio Communications Handbook, 4th Edition *by Paul E. Illman*

Night Flying *by Richard F. Haines and Courtney L. Flatau*

Bush Flying *by Steven Levi and Jim O'Meara*

Understanding Aeronautical Charts *by Terry T. Lankford*

Aviator's Guide to Navigation, 2nd Edition *by Donald J. Clausing*

Learning to Fly Helicopters *by R. Randall Padfield*

ABCs of Safe Flying, 3rd Edition *by David Frazier*

Be a Better Pilot: Making the Right Decisions *by Paul A. Craig*

The Art of Instrument Flying, 2nd Edition *by J.R. Williams*

Cross-Country Flying, 3rd Edition
by R. Randall Padfield, previous editions by Paul Garrison and Norval Kennedy

Flying VFR in Marginal Weather, 3rd Edition
by R. Randall Padfield, previous editions by Paul Garrison and Norval Kennedy

Good Takeoffs and Good Landings, 2nd Edition
by Joe Christy Revised and Updated by Ken George

The Aviator's Guide to Flight Planning *by Donald J. Clausing*

Avoiding Common Pilot Errors: An Air Traffic Controller's View *by John Stewart*

Flying In Congested Airspace: A Private Pilot's Guide *by Kevin Garrison*

General Aviation Law *by Jerry A. Eichenberger*

Improve Your Flying Skills: Tips from a Pro *by Donald J. Clausing*

Mountain Flying *by Doug Geeting and Steve Woerner*

The Pilot's Guide to Weather Reports, Forecasts, and Flight Planning *by Terry T. Lankford*

Advanced Aircraft Systems

David Lombardo

TAB Books
Division of McGraw-Hill, Inc.
Blue Ridge Summit, PA 17294-0850

FIRST EDITION
FIRST PRINTING

© 1993 by **TAB Books.**
TAB Books is a division of McGraw-Hill, Inc.

Library of Congress Cataloging-in-Publication Data

Lombardo, David A.
    Advanced aircraft systems : understanding your airplane / David Lombardo.
        p.   cm.
    Includes index.
    ISBN 0-8306-3998-5 (pbk.)   0-8306-3997-7 (Hardbound)
    1. Airplanes. 2. Airplanes—Design and construction.   I. Title.
TL670.L656   1993
629.134—dc20   92-45743
                                                            CIP

Acquisitions Editor: Jeff Worsinger
Editorial team: Steve Bolt, Executive Editor
               Norval G. Kennedy, Editor
               Joann Woy, Indexer
Production team: Katherine G. Brown, Director
               Brenda S. Wilhide, Layout
               Tina M. Sourbier, Typesetting
               Tara Ernst, Proofreader
               N. Nadine McFarland, Quality Control
               Joyce Belella, Computer graphics
Design team: Jaclyn J. Boone, Designer
               Brian Allison, Associate Designer   PFS
Cover photograph courtesy Beech Aircraft Corporation   4170

Contents

Acknowledgments   ix

Introduction   xi

1 Principles of electricity   1
Matter   2
Static electricity   4
Electromotive force   6
Current flow   7
Resistance   8
Power   10
Basic electrical system (dc)   10
Principles of magnetism   14
Electromagnetism   17
Storage batteries   19
Circuit protection   24
Primary causes of electrical circuit failure   25
Alternating current   27
Generating direct current   27
Generating alternating current   29
Miscellaneous electrical components   32

2 Electrical systems   35
Generators (dc)   35
Alternators   44
Combined ac/dc electrical systems   46
Alternator maintenance   47
Motors (dc)   47
Motors (ac)   58
Manufacturer documentation   58

3 Turbine engines   65
Turbine engine history   66
Physics review   67
Newton's laws of motion   74
Factors affecting thrust   76
Aircraft jet engines   78

Turbine engine components   81
Turbine engine ignition systems   94
Turbines   98
Main bearings   102
Exhaust section   102
Thrust reversers   104
Engine noise   105
Accessory section   106
Engine starting systems   106
Water injection   108
Required maintenance   108
Manufacturer documentation   109

4 Lubrication and cooling systems   115
Wet-sump system   116
Dry-sump system   117
Typical dry-sump system   124
Cooling systems   126
Manufacturer documentation   127

5 Aircraft propellers   131
Propeller theory   132
Constant-speed propeller   137
Propeller system operation   142
Synchrophasers   146
Preflight and maintenance   147
Manufacturer documentation   149

6 Hydraulic and pneumatic systems   153
System theory   154
Hydraulic fluid   160
Hydraulic system components   163
Hydraulic pumps   165
Accumulators   168
Fluid lines and fittings   172
Filters   172
Actuating cylinders   173
Pneumatic systems   176
Maintenance   181
Manufacturer documentation   183

7 Fuel systems   187

Turbine engine fuels   188
Fuel system contamination   189
Minimizing contamination   194
Fuel handling safety   195
Fuel system theory   197
Jet fuel controls   198
Fuel system components   204
Fuel system maintenance   211
Manufacturer documentation   211

8 Environmental systems   219

Need for oxygen   219
Atmosphere's composition   220
Atmosphere's pressure   222
Temperature and altitude   223
Pressurization   223
Air-conditioning and pressurization systems   225
Air distribution   239
Air-conditioning systems   240
Heating systems   247
Supplemental oxygen systems   247
Manufacturer documentation   253

9 Landing gear systems   263

Shock struts   264
Main landing gear   266
Emergency extension and safety systems   271
Nosewheels   275
Brake systems   281
Wheels and tires   289
Manufacturer documentation   295

10 Fire protection systems   301

Fires   301
Portable extinguishers   302
Zone classifications   302
Detection systems   303
Detectors   304

Overheat warnings   309
Extinguisher systems   309
Manufacturer documentation   310

11 Aerodynamic control and protection systems   315

Tabs   315
High-lift devices   318
Boundary-layer control devices   319
Vortex generators   321
Flight control systems   323
Precip protection   324
Manufacturer documentation   339

Index   349

About the Author   360

Acknowledgments

THE MAJORITY OF THE MATERIAL FOR THIS BOOK IS EXTRACTED FROM THE three Federal Aviation Administration Airframe and Powerplant Mechanic's Handbooks: *AC 65-9A General Handbook*, *AC 65-15A Airframe Handbook*, and *AC 65-12A Powerplant Handbook*. In almost all instances, the material has been rewritten to more appropriately reflect the depth of knowledge necessary for the pilot, rather than its originally intended readership, the aviation maintenance technician; however, explanations are excerpted as originally published because those descriptions were ideally suited to the task of educating flight crews. Readers interested in a more in-depth knowledge of any of the subject areas covered in this text are encouraged to refer to those publications for greater detail. Similarly, all the illustrations in this text, except as noted, are taken from those advisory circulars. The most notable exception being those illustrations done by professional artist, and sister, Susan M. Lombardo, who created some great illustrations in a very short time.

Special thanks are extended to George D. Rodgers and Mike Potts, of Beech Aircraft Corporation, and Gene Rainville, of Falcon Jet Corporation for their assistance and support in this project. Beech Aircraft and Falcon Jet were willing to provide pilot's operating handbooks for my review and grant permission to duplicate portions of those manuals for inclusion in the text. From the outset, George Rodgers and Gene Rainville said the book made sense and would fill a necessary gap in the education of advancing pilots. To those individuals, and the foresight of their companies, I am deeply indebted.

ACKNOWLEDGMENTS

This book could not have been published without the efforts of an excellent support staff that was able to handle many of the day-to-day matters that keep a writer from writing. For their constant encouragement and support in times of struggle and stress, I especially thank Diane Blazevich, Lisa Collins, and Lynda Moore.

For their willingness to serve as subject matter experts in the final proofreading of the book, and for their encouragement, I would like to thank the following individuals for finding time amid their hectic academic teaching schedules. Phillip Balsamo, Robert Dowse, Raymond Drosz, Robert McAuley, Louis Revisky, Michael Streit, and Nazim Uddin. Also, a special thanks to a gentleman whom I have never even spoken to, Howard Gammon, president of Gammon Technical Products, Inc. I faxed him an author's release form for approval to quote his newsletter. He ended up offering to proofread the entire chapter on fuels while he was on a business trip and get it back to me within a few days. The dedication and support people in aviation have for each other never fails to amaze me.

Special thanks go to Vicki Cohen for proofreading the entire book in fewer than 24 hours and our wacky dogs Brandy, Roscoe, and Molly, for their continued support every day of my life.

Finally, this book probably never would have been written had it not been for my friend Mac. Mac bought and sold corporate turboprop and turbojet equipment and was always willing to give a new commercial pilot, such as myself, experience. He continually impressed me with the fact that in aviation what you don't know will kill you. Both pilot and mechanic, he had an intuitive knowledge of aircraft systems that served him well for more than 40 years of flying and selling everything with wings.

During the years that I was ferrying aircraft for Mac part-time, and going to graduate school full-time, he gave me tremendous encouragement in both areas. I gained experience flying with him that I could never have obtained elsewhere. One day, after my first attempt at trying to shepherd a Learjet to a distant city, I was so elated that I asked him how I could ever repay him for all the opportunities he had given me. "Professor," he said, "Someday you write a book teaching pilots about what makes airplanes work." This book is dedicated to Louis E. "Watcha" McCollum, aircraft salesman, mentor, and friend. Thanks for the ride, Mac, and I miss you.

Introduction

THE TIME IS RAPIDLY APPROACHING WHEN THOSE WHO ARE IN THE BUSINESS of preparing individuals for careers in aviation realize that it can no longer be business as usual. For 90 years, civilian flight training in the United States has focused on preparing pilots to fly in a single-pilot, single-engine, or light twin-engine environment. There is great resistance to change in some quarters, but the *ab initio* (from the beginning) movement is heartening. Clearly, the University Aviation Association, and many of its member universities, are beginning to meet the challenge of pilot training for the year 2000 and beyond. Individuals seriously interested in the education of aviation professionals should take heed and join forces with that outstanding organization.

The pilot has become less a practitioner of a skill and more a manager of systems. Whether that is an inherent good or evil caused by the evolution of aircraft matters little; it is simply a truth. We have evolved into an era where professional pilots must have a highly specialized college degree. They need analytical problem solving capability, outstanding written and oral communication ability, an almost visceral understanding of human factors, psychology, and interpersonal relations, a working knowledge of computer logic and operations, and an in-depth knowledge of the very complex systems that they must manage on board a complex aircraft. Few universities offer these subjects, let alone fixed-base operators and small flight schools.

Individuals interested in a professional aviation career are advised to apply to one of the handful of FAA Airway Science (AWS) institutions around the country. The program offers an innovative and unique opportunity for students to learn the critical subjects listed above, as well as many others, as they seriously prepare for a career in professional aviation. The AWS program is intellectually demanding and expensive, but it is one of the finest technical educations obtainable in this country. Students interested in locating an institution that offers AWS should contact the UAA, 3410 Skyway Drive, Opelika, AL 36801.

This book was written to support AWS and should help reciprocating-engine pilots prepare for the transition to turbine-powered aircraft. The book is intended to be a primer on turboprop and turbojet aircraft systems; it is not flight deck oriented. Avionics, flight deck instrumentation, specific operating procedures, and other operationally oriented details are intentionally omitted.

It is the intent of this book to give the reader a basic understanding of what individual systems are designed to do, and how they do it.

It is not the intent of this book to teach pilots how to fly turbine-powered aircraft. Pilots transitioning to turbine equipment need to receive type training in the specific aircraft to which they are transitioning.

Chapter 1 starts the book with a review of basic electricity. It is difficult to understate the importance of electricity on aircraft. Virtually every system is touched in some way by the aircraft's electrical system. A thorough understanding of the fundamentals of electricity is necessary to successfully understand how the aircraft operates. Each chapter after that looks at a different aircraft system and, as such, stands alone. The book may be read cover-to-cover or used as a reference manual. All basic system descriptions are essentially generic. They are intended to give the reader a basic understanding of how typical systems operate.

The Beech and Falcon aircraft systems at the end of the chapters are provided to give the reader specific examples of system operation by the pilot. The information provided was extracted from Beech Aircraft Corporation and Falcon Jet Corporation material and is intended to be used for educational purposes only. The information is not to be used for the operation or maintenance of any actual aircraft.

Hopefully this text will help pilots transition to more complex aircraft as well as help them gain a better understanding of that awesome mystery we call flight.

DAVID A. LOMBARDO
Shorewood, Illinois

1
Principles of electricity

TAKE A MOMENT TO LOOK AT YOUR SURROUNDINGS; ALMOST EVERYTHING you see can be connected to electricity in one way or another. The obvious things are appliances, lamps, the television and stereo, but what about the chair, table, carpet, and wallpaper? What if you are reading this book outside, in a park? It is almost impossible to escape the effect of electricity. The furniture and furnishings were almost certainly made with power tools and equipment in a well lit, air-conditioned factory. The grass was probably cut with a power lawnmower but even a push mower was built using power equipment. Quite simply, we would be back to the days of horse and buggy without electricity, yet it is one of the most misunderstood areas of our lives.

If asked which system of an airplane a pilot least understands, most would probably say the electrical system. Engines, propellers, flight controls, even hydraulics can be explained with some degree of confidence by almost any pilot. But an explanation of how electricity makes the lights go on, or how a battery "makes" electricity elicits a silent glance in a different direction.

While different aircraft employ electrical equipment differently, it is safe to say that virtually any aircraft system can utilize electricity in one way or another. In addition to the all-electric systems (avionics, lights, and ignition), hydraulic systems can use electric pumps, flight controls might use digital control inputs, fuel systems have backup electric pumps, and so on. Electrical conductors wend throughout aircraft not unlike the human nervous system. Electrical systems gather and display data for analysis, activate other systems, operate various motors, pumps and other equipment, and more.

Specific physics of electricity might be complex, but the basic principle of how electricity works is not. Given the important and far-reaching role it plays in aircraft, pilots should have a good understanding of elementary electrical circuits, components, and accessories.

MATTER

Everything is matter: to be a bit more specific, anything that has weight (mass) and occupies space. That includes any liquid, gas, or solid—the Pacific Ocean, air, a golden retriever, or even a molecule. A molecule is the very smallest particle of matter possible, regardless of its state or form, that is still identifiable as itself. If you take a molecule apart by dividing it into individual atoms, it loses its identity and ceases to be whatever it was before you took it apart. A substance that is made up of a single atom is called an element but most substances are compounds which means it takes two or more atoms to form the substance. Take, for instance, a molecule of water. Water is the compound H_2O: two atoms of hydrogen (H_2) and one atom of oxygen (O) (FIG. 1-1). If you were to remove one of the hydrogen atoms, the molecule would cease to be water.

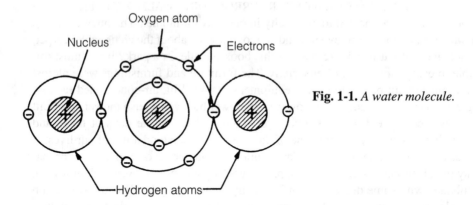

Fig. 1-1. *A water molecule.*

The atom

It is not difficult to understand that the atom is the smallest particle of matter. It is, in a very real sense, the basic building block of our existence because all matter, literally everything in existence, is made up of atoms. They are so small that they cannot be seen. In fact, if you lined up atoms side by side like a row of tennis balls, it would take about 200,000 atoms to make a row 1 inch long. An atom cannot be seen, but years of experiments have given scientists a pretty good description.

Actually, we know quite a bit about the atom. For instance, it is composed of a nucleus with one or more electrons orbiting around it in a manner somewhat reminiscent of planets around the sun. The nucleus has one or more protons and an equivalent number of neutrons. The notable exception to that rule is hydrogen, the simplest of all atoms, which has a nucleus containing only one proton and no neutrons, with a single electron orbiting around the nucleus (FIG. 1-2). In comparison, an oxygen molecule has eight electrons in two different orbits revolving around a nucleus with eight protons and eight neutrons (FIG. 1-3).

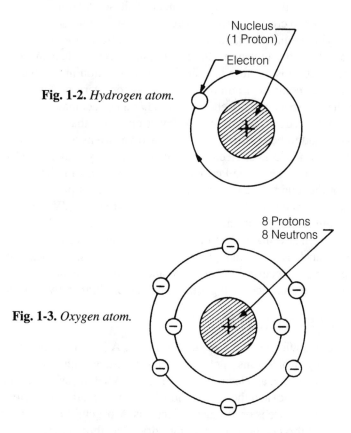

Fig. 1-2. *Hydrogen atom.*

Fig. 1-3. *Oxygen atom.*

Protons carry a positive charge and neutrons have no electrical charge. It seems as if the purpose of neutrons is to act as a sort of "glue" to keep the nucleus together. Even though they have no electrical charge, neutrons do happen to be equal in both size and weight to the protons with whom they share the nucleus. Oddly, electrons apparently have little or no weight. Scientists are uncertain if electrons are a particle of matter with a negative charge or simply a negative charge with no weight at all. Regardless, because individual neutrons and protons weigh about 1,845 times that of an electron, only the combined weight of the neutrons and protons is used to determine the weight of the atom.

Electrons. Electrons, which are the negative charge of electricity, revolve around the nucleus in orbits called *shells*. For some reason, certain electrons appear to be tightly bound to the nucleus and revolve in shells very close to it. Other electrons orbit farther away from the nucleus and are not so tightly bound; depending upon the atom, there might be several different shells. The farther away from the nucleus the shell of an electron is located, the easier it is to pull the electron free of the positive attraction of the protons. It is the controlled movement of those electrons, known as *free electrons*, that is used to create electrical current.

When trying to understand how electrons move from atom to atom it is important to understand how different atoms have different charges. When an atom has a neutral charge, one electron is orbiting in a shell for every proton in the nucleus; remember, while there is an equivalent number of neutrons for every proton, neutrons have no magnetic charge. So a neutrally charged atom has a perfect match between the number of protons and electrons. An atom with fewer electrons than protons—a shortage of electrons—is said to be *positively charged*, also known as a *positive ion*. Conversely, an atom that has an excess of electrons is *negatively charged* and called a *negative ion*.

If you were to refer to the Periodic Table of Elements in almost any college science book or dictionary, you would find that the number of protons for the various elements ranges from one for hydrogen to 103 for lawrencium. While that can add up to a lot of orbiting electrons, they are not randomly scattered around the nucleus. Well defined rules govern how many electrons can occupy a given shell. For instance, the shell nearest to the nucleus can hold no more than two electrons. The next closest shell can contain up to eight electrons, the next 18, and so on. More electrons equate to more shells to contain them, each shell getting successively farther away from the nucleus.

STATIC ELECTRICITY

Use of the term static electricity is very common. A less frequently used term is *dynamic electricity*, which runs appliances. Despite the terminology differences, electricity is still electricity; all electrons are the same. What the terms actually refer to are electrons at rest (static) and those in motion (dynamic). Static electricity is a condition of either too few electrons or too many electrons. Most children have seen the classic demonstration of this condition by running a dry comb through their hair, which causes electron transfer to the comb as a result of friction. The discharge is the result of the rapid movement of electrons in the opposite direction of the comb back to the hair, neutralizing each other. The result is little sparks.

While static electricity has no practical value, it can lead to serious problems because it occurs naturally as the result of friction and because it is uncontrollable. Two examples of potentially harmful static electricity are static buildup in an airframe as air molecules pass over it during flight and as a result of the friction of fuel flowing into a fuel tank during refueling. Two major problems are associated with the static buildup as a result of air molecules passing over the airframe. The first is its effect on the avionics, which can be heard as crackling and popping noises over the speaker. To minimize that problem, all electrically conductive sections of the airframe are bonded together to allow the static electricity to move easily along the length of the fuselage.

The distribution of a static electric charge on an irregularly shaped object is such that the charge moves toward areas that have the sharpest curvature or points. When the static electricity works its way to such a trailing edge, such as a wingtip or the fuselage tailcone, it is passed back into the air through a static wick that is attached to, and trails several inches behind, the airframe. While the static wicks manage to shed most of the static charge, some is always left in the airframe after landing.

The second problem associated with airframe static buildup occurs when a passenger steps off the airplane, which is not grounded because of the rubber tires, and contacts the ground. The entire force of the static buildup would suddenly transfer through the passenger to the ground. To eliminate that problem, tire manufacturers impregnate aircraft tires with carbon that causes the airframe to ground as soon as the tires touch the runway.

The other major aviation-related static electricity concern has to do with refueling. As fuel flows through the hose and into the tank, the molecules are in a constant state of friction, which causes static buildup. If proper precautions are not taken, a single static discharge spark could ignite the fumes pouring out of the fuel tank inlet causing a tremendous explosion. Fueling must always be conducted in accordance with safe operating procedures.

Electrostatic field

Every charged body has a force field around it called an electrostatic field. For purposes of illustration, this is depicted as lines extending outward in all directions from the body and ending at a position with an equal and opposite charge. To clarify the electrostatic field, the lines represent both the direction and the intensity of the force field. The number of lines emanating from the fields in FIG. 1-4 indicate field intensity while the arrowheads indicate the direction a small, positive test charge would move if it were exposed to the force field. For the positive force field, the direction would be outward because matching charges repel each other. For the negative field, the direction of motion is inward because opposite charges attract each other.

Carrying the previous illustration one step further, FIG. 1-5 shows the result of two, similar force fields acting upon each other when placed in close proximity. Again, note that similar fields repel one another and, even though the illustration is of positively charged fields, the same thing would happen between two negatively charged fields.

The way a charge is distributed on an object is also helpful in understanding how current is produced. Recall that a charge on an irregularly shaped object will collect at the points. Figure 1-6 illustrates how a negative charge would be distributed over a small, round metal disk. Because the disk offers uniform resistance across its entire

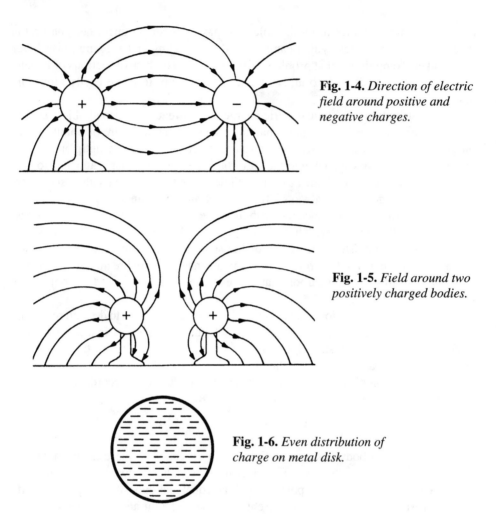

Fig. 1-4. *Direction of electric field around positive and negative charges.*

Fig. 1-5. *Field around two positively charged bodies.*

Fig. 1-6. *Even distribution of charge on metal disk.*

surface, all the electrons will mutually repel each other at the same intensity, causing them to assume a pattern of even distribution.

ELECTROMOTIVE FORCE

Now the practical part, electric current. As you might have already figured out, despite the common wisdom, electrons move from negative to positive, and they do so because of a pressure differential between those two points.

If there is a negative charge at one end of a conductor and a positive charge at the other end, there is an electrostatic field between the two positions. Electrons will be repelled from the negatively charged side and drawn to the positively charged side. Electron movement is called current flow through the conductor because of an electrical pressure differential. Again, the negatively charged side contains excess electrons and the positively charged side has a deficiency of electrons. The excess electrons will be repelled by the negatively charged side and drawn toward the positive side. It is not unlike the water analogy illustrated in FIG. 1-7. As long as there is an imbalance in the water pressure when the valve is opened, the higher pressure water will flow toward the lower pressure water until a differential is no longer present. It is not the pressure in tank A that causes the flow but the difference between the pressures in tank A and tank B.

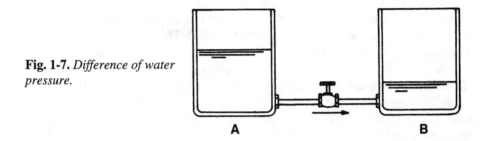

Fig. 1-7. *Difference of water pressure.*

A   B

The force in an electrical system that causes current flow is known as its *electromotive force* (emf). The symbol for emf is the letter E. The *volt* is the unit of measure of emf, which is simply the difference of electrical pressure between two points. The name volt was in honor of the Italian physicist Alessandro Volta (1745–1827) who made many contributions to the study of electricity. From a practical point of view, voltage is simply the measure of the potential difference between two points on the electrical system. For instance, a 24-volt battery has the potential difference from the negative post to the positive post of 24 volts.

CURRENT FLOW

Current flow, or more simply *current*, is the measure of the flow of electricity through a conductor. Current might flow in two different manners. When it flows in one direction only it is called *direct current* (dc). When current flows first in one direction, then reverses itself and flows in the opposite direction, it is called *alternating current* (ac).

Conductors. A conductor is nothing more than a collection of similar atoms (a substance) that can pass electrons from one atom's shell to the next. Virtually all metals, carbon, and water will conduct electrical current, however, the best conductors would logically be those substances that have the greatest number of free electrons. For that, and other reasons, silver, copper, gold and aluminum are the best choices, in that order. Because silver, and especially gold, are precious metals, most conductors are typically made of copper and aluminum. While aluminum has a slightly higher resistance than copper, aluminum is still more prevalent in aircraft because of lower weight. Even a cursory look behind any aircraft instrument panel will reveal yards of extruded aluminum conductor called *wire*.

Measuring current flow. One way to measure the flow of current would be to pick a point along the conductor and measure the number of atoms as they pass by, which is hardly a practical method. Instead, electrons are counted as a function of their combined electrical charge and measured in a manner similar to that used in measuring a liquid. For instance, the volume of water can be measured in gallons with the flow rate given in gallons per minute. Current flow, whose electrical symbol is I for *intensity of current flow*, is measured in *coulombs*. Its flow rate is coulombs per second, which is more commonly known as *amperes*. One coulomb is equal to 6.28 billion billion electrons. The coulomb was named for Charles A. Coulomb (1736–1806), a French physicist who conducted pioneering experiments with electricity, while the *ampere* (amp) was named for Andre M. Ampere (1775–1836), another French scientist. The flight deck instrument that displays amperage is called the *ammeter*.

Another very common misconception about electricity is that it travels at the speed of light. It would be more proper to say it reacts at approximately the speed of light. What does not happen is an electron that is released from the negative post of a battery races through the electrical system at 186,000 miles per hour toward the positive battery post where it arrives a millisecond later. Electrons actually travel at a comparatively leisurely pace, but when you activate a switch, the electrical system does respond very close to the speed of light.

Recall that current is the result of a series of free electrons moving from one atomic shell to another. As the electrons enter a shell, the atom suddenly has too many electrons and one is bumped out of the shell to the next atom. That electron then increases the number of electrons in the next shell, which causes one of those electrons to be bumped to the next atom, and so on down the conductor. Individually, these electrons do not move quickly but the reaction down the conductor is nearly instantaneous (FIG. 1-8).

 **Fig. 1-8.** *Electron movement.*

Think of it as a series of pool balls, touching side-by-side, in a pipe with a diameter just large enough to contain the balls. When a new pool ball is pushed in one end, the effect is instantly felt at the other end as one ball is pushed out. Even if the pipe were 50 feet long, the effect would be the same, but it might take minutes for that specific ball to work its way to the end of the pipe and be pushed out.

RESISTANCE

The process of moving electrons through a conductor, like moving pool balls through a pipe, is made more difficult because of friction, a natural resistance to movement. With electrons, the natural resistance is the result of the attractive force of the atom trying to hold the electrons in their orbits. The emf overcomes this resistance and initiates a current flow through the conductor.

The *ohm* is used to measure resistance and its symbol is R. It was named for a German physicist, George S. Ohm (1787–1854), who discovered the relationship between electrical quantities and formulated what is now known as Ohm's law. He discovered the number of amperes of current that flow through a conductor are determined by both the resistance of the conductor and the voltage. The formula for Ohm's law is:

$$I = \frac{E}{R}$$

Where:

I = current in amperes
E = potential energy in volts
R = resistance in ohms

If any two of these entities are known, it is a simple matter of solving for the unknown. For example, 24 volts applied to a conductor with 6 ohms of resistance will result in a current flow of 4 amps.

Four conductor components determine resistance:

- Type of material
- Total length
- Cross section
- Temperature

Type of material. The description of conductors explained that certain substances, notably those with numerous free electrons, make the best conductors. Conversely, it must be true that substances with a relatively low number of free electrons must be good at inhibiting electron travel. The best of those substances are called *insulators*; in order of resistance, they are: dry air, rubber, glass, ceramic, plastic, asbestos, and fiber compositions. Everything has the potential of being a conductor, but the current flow in some of these is so small that the flow is considered to be zero.

As emf increases excessively, the outer electron orbits of even the best conductors will begin to distort in an effort to keep their electrons in their shells. At some point, if sufficient emf is applied, the atomic structure will be strained beyond its elastic limit and a rupture will occur. When that happens, the insulator acts as a conductor until the excessive emf is removed.

A natural by-product of resistance is heat in the same manner as mechanical friction. In general, heat of resistance dissipates as quickly as it is created with little or no heat felt; however, excess current flow can exceed the ability of the conductor to dissipate heat into the surrounding air. The result is an overheating condition, burning insulation, and potentially starting a fire. To prevent the possibility of such a problem, it is important that a mechanic always selects the proper size of wire based on the intended current capacity.

Total length. The resistance of a conductor varies proportionally with its length. The longer the wire, the more the resistance. Figure 1-9 illustrates two pieces of the same wire but at different lengths. If you apply one volt to the 1-foot conductor, which

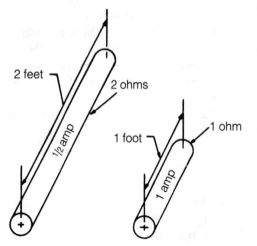

2 feet

2 ohms

½ amp

1 foot

1 ohm

1 amp

Fig. 1-9. *Resistance varies with length of conductor.*

has one ohm of resistance, the resulting current flow will be 1 amp. The same voltage applied to the 2-foot length would result in a current of ½-amp because the resistance to electron flow is twice as much as the 1-foot length.

Cross section. Wire resistance varies inversely with the area of cross section. *Cross section* is the diameter of a round wire, the most common form of conductor; however, a conductor could be square or triangular. If, for instance, the cross section of a conductor is doubled, the result will be a 50 percent reduction in the resistance to electron flow. The reason for this relationship is that the larger the cross section, the more area an electron has in which to move, and the less likely its progress will be impeded by a collision or by being captured by an atom.

Temperature. As a general rule, the resistance of most substances increases proportionally with an increase in the temperature surrounding the conductor. Select substances, such as carbon, have a resistance that is inversely proportional to the surrounding temperature, and yet a few others, such as constantan and maganin, have a resistance value that changes minimally with temperature. The conditions under which a conductor will be operating are carefully considered by the manufacturer when choosing which type of conductor to use. Wiring that is exposed to ambient temperatures in normal turbine aircraft operation are subjected to an extreme temperature variation from sea level to 36,000 feet.

POWER

One final unit of measurement is often used when discussing dc circuits. *Power* is equal to the value of the voltage multiplied by the current and the resulting unit of measurement is called the *watt* (P). The formula for determining wattage is: $P = IE$.

BASIC ELECTRICAL SYSTEM (dc)

All dc electrical circuits essentially consist of three basic components: A source of emf, resistance (such as a lamp), and conductors (commonly wires or a bus bar). A fundamental electrical system would look like that depicted in FIG. 1-10. A wire connects the negative side of the emf source (a battery), to one side of a resistive device (a light bulb), and then connects the other side of the light bulb to the positive side of the battery.

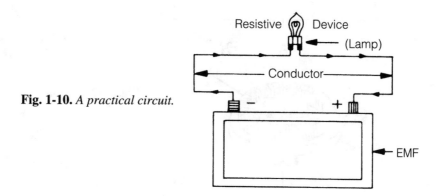

Fig. 1-10. *A practical circuit.*

The same circuit is displayed using standard electrical symbology in FIG. 1-11. From an operational perspective, the electrons depart the negative side of the battery (represented by the shorter vertical line), enter the lamp where the resistance causes the lamp to glow and give off heat, then back to the positive side of the battery through the wire. If no resistance were in the circuit, just a direct line between negative and positive, the flow of electrons would be so heavy and rapid that the conductor would rapidly overheat and burn. Another form of resistance, called a *resistor (fixed* or *variable)* is placed in a system simply to limit current flow; the most well known is a variable resistor. Also called *rheostat* and *potentiometer*, it allows variable control of circuit resistance.

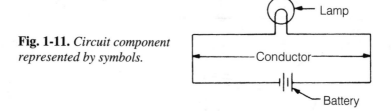

Fig. 1-11. *Circuit component represented by symbols.*

The rheostat, a variable, two-terminal resistor, controls the amount of current that flows throughout an entire circuit (FIG. 1-12). The potentiometer, a three-terminal, variable resistor, controls the amount of current flow through a load in a parallel circuit (FIG. 1-13). Rheostats are very commonly used as dimmer switches for lights and other electrical equipment; potentiometers are often used as volume controls and television brightness controls.

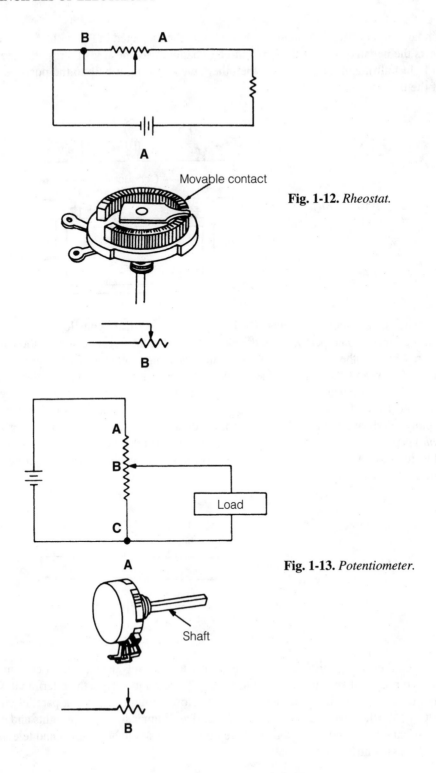

Fig. 1-12. *Rheostat.*

Fig. 1-13. *Potentiometer.*

Circuits (dc)

Three types of dc circuits are series, parallel, and series-parallel. From the pilot's perspective, the type of circuit by which a given accessory is powered will have a significant impact on the troubleshooting and operational implications.

Series circuit. The simplest of all electrical circuits, the series configuration is the basis for all other circuits. Figure 1-14 depicts a schematic for a single item series circuit that is commonly used in an aircraft. The switch permits simple, positive on/off control of the circuit by the crew. The load could be a cabin light, pitot heat, cooling fan, or even a galley coffeemaker. The fuse provides system protection in the event of a current overload that might jeopardize either the load or the wire. Fuses and circuit protection are discussed in more detail later in the chapter.

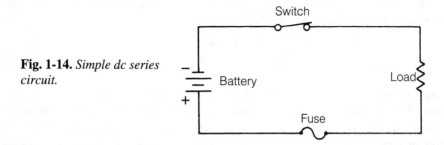

Fig. 1-14. *Simple dc series circuit.*

One characteristic of the series circuit is that regardless of the number of components, switches, or fuses, the current flow will be exactly the same throughout the entire circuit. If an ammeter were applied at any point along the circuit in FIG. 1-14, it would read exactly the same as at any other point. It is true that additional load on a circuit equates to additional circuit resistance, but whatever the resulting current flow, it will be evenly distributed throughout the circuit. Resistance in a series circuit is determined by the formula: $R_T = R_1 + R_2 + R_3$ where R_T is the total system resistance.

If the conductor is broken anywhere throughout the circuit, the entire circuit fails along with all the components on the circuit.

Parallel circuit. All elements of this circuit share a common voltage source while its multiple loads are divided among two or more parallel circuits (FIG. 1-15). Additional paths for current flow relate to lower the overall circuit resistance, which is distinctly different from the series circuit where any increase in resistance uniformly opposes current flow. In the illustration, points A, B, C, and D all have the same electrical potential. Points E, F, G, and H will also share a common potential. It is true that B to F might have a different current passing through its resistance than C to G, or D to H, but the voltage will remain the same throughout the system. The logic lies in the fact that the applied voltage between points A and E is the rated battery voltage. If, for instance, it is a 24-volt dc (Vdc) battery, that describes the potential between the negative terminal and the positive terminal, which includes the entire parallel circuit in between.

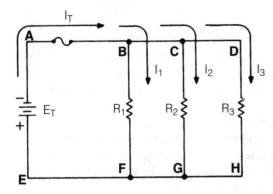

Fig. 1-15. *A parallel circuit.*

The total circuit resistance is the sum of each of the parallel circuits' resistances; therefore, in the illustration it would be

$$\frac{1}{R_T} = \frac{1}{R_1} + \frac{1}{R_2} + \frac{1}{R_3}$$

again, where R_T is the total system resistance. Similarly, system component current demand can be determined with the formula of $I_T = I_1 + I_2 + I_3$. The importance of understanding this relates to emergency electrical system management. If, for instance, a crew experiences an electrical generator failure, the proper response is to determine what electrical equipment creates the most severe electrical drain on the battery and shut down that equipment.

A review of the circuit amperage requirements for each component will give you an excellent idea of about how much current each piece of equipment draws. One of the highest users of current flow is communications transmitting. During the failure of an electrical system power source, the crew should try to keep radio transmissions to the absolute minimum dictated by safety. Also note that unlike a broken conductor in the series circuit, if there is a break in any one of the parallel circuits, only the circuit with the break will stop operating; the others will continue unaffected.

Series-parallel circuits. Discussing series and series-parallel circuits provides for an easy understanding of the basic principles of dc electrical systems. In reality, very few aircraft systems are either one. Instead, most circuits are a hybrid of the two, called *series-parallel*. This type of circuit consists of groups of parallel loads connected in series with other loads (FIG. 1-16). Note in the illustration that as the current leaves the battery it passes through a switch and goes directly to a load (R_1) then travels to parallel circuits R_2 and R_3. The latter two circuits are in series with R_1 while being parallel with each other.

PRINCIPLES OF MAGNETISM

Magnetism is the property of certain metals to attract certain other metals, typically ferrous materials made of iron or iron alloys. *Ferrous materials* include soft iron, steel, and alnico. In addition, *nonferrous materials* also have limited magnetic qualities:

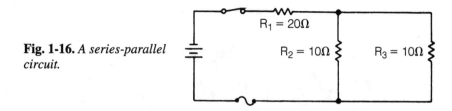

Fig. 1-16. *A series-parallel circuit.*

nickel, cobalt, and gadolinium. Everything else falls into the category of nonmagnetic with a few substances classified as *diamagnetic*, which means they are repelled by both magnetic poles.

Magnetism is invisible and is still not completely understood, but its existence has been known for many centuries. The discovery of magnetism is often attributed to the Chinese use of a rudimentary compass, circa 1100 A.D. It is true that the earliest known reference to the use of magnetism for navigation is made by a Chinese mathematician and instrument maker named Shen Kua (1030–1093). He discovered that a piece of this special ore, when hung by a thread or floated on a piece of wood in undisturbed water, would align in a north-south direction. Because of this characteristic the stone became known as a "leading stone," or *lodestone*. The recognition of magnetism, however, significantly predates its use as a navigation tool and dates back to Greece, circa 500 B.C. Thales of Miletus wrote about the stone that attracts iron and other pieces of the same material. Other than the planet earth, the lodestone, now known as *magnetite*, is the only natural magnet. All other magnets are artificially produced.

From a practical perspective, aside from its navigational value, the concept of magnetism was little more than a novelty until the eighteenth century. At that point magnetism and electricity became tightly interwoven. So much so, that it is now impossible to understand how certain electrical system components operate without an understanding of magnetism.

Magnetism has several important behavioral characteristics. For instance, simply rubbing an iron bar with a piece of magnetite will cause the bar to become magnetized. Magnets have opposite polarity at the ends and when the north-seeking ends of two different magnets are brought together, they repel each other. Conversely, when north-seeking and south-seeking ends are brought together, they attract each other. As with electricity, opposites attract. Because the earth is literally a giant magnet, the north-seeking end of a compass needle will point to earth's magnetic north pole while the south-seeking end will point to earth's magnetic south pole. In training, the pilot learns that neither of earth's magnetic poles is located anywhere near the geographic poles. The reason behind a compass' magnetic alignment with the poles lies in the magnetic field that surrounds all magnetic objects. Any such object will have a magnetic field between its poles that will arrange itself to accommodate the magnet's shape (FIG. 1-17).

The theory of magnetism holds that each molecule of a piece of iron is actually a small magnet with north- and south-seeking poles (FIG. 1-18A). While each is a miniature magnet, they are in a demagnetized condition arranged in a random order. By rub-

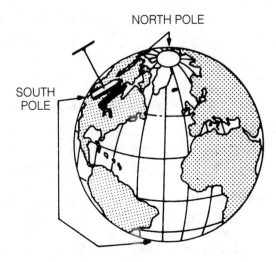

NORTH POLE

SOUTH
POLE

Fig. 1-17. *Magnetic field around magnets.*

bing the piece of unmagnetized iron with magnetite, the molecules rearrange themselves in a north/south-seeking order (FIG. 1-18B). At that point, with all the molecular magnetic fields aligned, they form a unified magnetic field for the object. Interestingly, it is possible to demagnetize the object by heating it or causing a sudden shock such as hitting it with a hammer. Apparently, both of these will cause misalignment of the molecules with a resultant marked decrease in the object's magnetism.

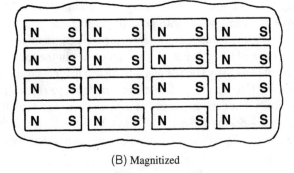

(A) Unmagnitized

Fig. 1-18. *Arrangement of molecules in a piece of magnetic material.*

(B) Magnitized

Magnets

For a number of reasons, natural magnets are not practical. Instead, electromagnetic technology requires the use of the more powerful artificial magnet. Two categories are *permanent* and *temporary* magnets. The permanent magnet retains its magnetism for very long time after the magnetizing force is removed. Temporary magnets rapidly lose all or most of their magnetism as soon as the magnetizing force is removed. One of the major concerns when choosing a substance to make a magnet is its retentivity. The amount of magnetism that remains after the external magnetizing force is removed is called *residual magnetism*. It is a major factor in electrical applications such as the generator. For that reason, substantial research has been done to find the best substance from which to produce a permanent magnet.

Hard steel was the original choice for permanent magnets but several alloys have proven to be much more effective. The best, alnico, is an alloy of aluminum, iron, nickel, and cobalt; remalloy and permendur are also considered excellent permanent magnets.

ELECTROMAGNETISM

The relationship between electricity and magnetism was discovered in 1819 by Hans Christian Oersted, a Danish physicist. He found that a compass needle would deflect when placed in proximity to a conductor through which current was flowing. If the current were suddenly shut off, the compass needle would swing back to north as if suddenly released. Because the conductor was made of copper, itself nonmagnetic, the phenomenon was attributed to a magnetic field around the conductor as a result of the electrons travelling through it. Figure 1-19 illustrates the phenomenon with a series of concentric circles around a conductor through which current is flowing. The illustration is simplified for clarity; in reality, the field would be continuous for the entire length of the conductor. As current flow increases, the magnetic field around the conductor expands outward; as it decreases, the field contracts toward the conductor. When the current stops, the field collapses and disappears.

When a conductor is looped many times, it is called a *coil*. The force field of a coil is such that there is a high concentration of flux lines through the center of the coil. To assist in concentrating the force field in the center, a soft iron bar is inserted. The result is called an *electromagnet* (FIG. 1-20). When direct current travels through the conduc-

Fig. 1-19. *Magnetic field around a conductor.*

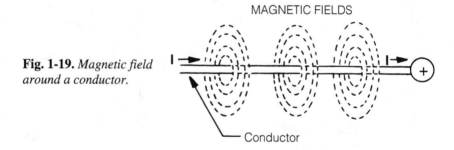

MAGNETIC FIELDS

Conductor

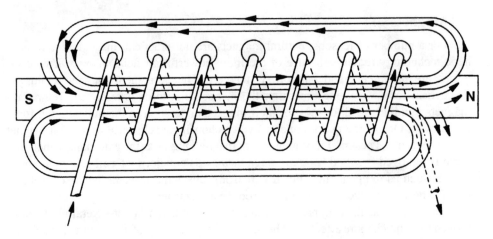

Fig. 1-20. *Electromagnet.*

tor, the coil becomes magnetized in the same manner as the iron core itself. The strength of an electromagnet is directly proportional to both the flow of current and the number of loops in the wire.

One of the most common practical applications of such a coil is the *solenoid* (FIG. 1-21). The T that appears in the center of the coil in the illustration would be a spring-loaded iron core that is held out of the center of the coil. When the coil is energized, the magnetic flux pulls the iron core (plunger) down into the coil. The movement of the plunger causes some action otherwise unrelated to the operation of the coil. For instance, in the retracted position the plunger might close the contacts on a starter motor. In that manner, the pilot can press a starter switch that operates a very low amperage solenoid that closes contacts and activates a very high amperage starter motor located outside of the flight deck.

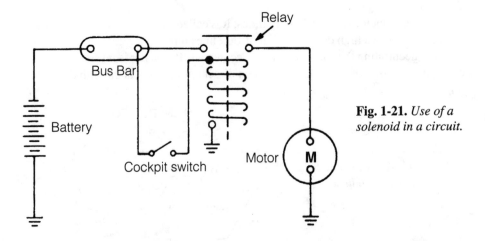

Fig. 1-21. *Use of a solenoid in a circuit.*

STORAGE BATTERIES

Aircraft have a wide range of electrical energy needs that span from the exterior pre-flight to the aircraft parking and securing checklists. While engine driven generators handle all the electrical requirements after engine start, there are many operational checklist and cabin comfort requirements prior to that time. In many cases, an auxiliary power unit (APU) or ground power unit (GPU) will be the best choice, but they are not always appropriate or available.

It is the job of the storage battery to reliably provide the necessary electrical power to fulfill system requirements when the engine-driven generator is not operational and APU or GPU power is not available. The battery is automatically recharged when the engine-driven generator is operational. Two types of storage batteries are used in aircraft: *lead-acid* and *nickel cadmium* (*nicad*). The lead-acid, which almost totally dominates the light aircraft industry, is overshadowed by the nicad in turboprop and turbojet aircraft; however, both batteries are common enough to warrant a brief discussion of each.

Lead-acid batteries

The lead-acid battery is very similar to its standard automotive counterpart. The cells are composed of positive plates made of lead peroxide and negative plates made of a spongy lead material. The cell is filled with a sulfuric acid and water electrolyte that totally engulfs all positive and negative plates, resulting in a chemical reaction. The battery discharges by changing stored up chemical energy into electrical energy. The generator, in turn, recharges the battery by supplying electrical energy, which is converted to chemical energy and stored in the cells. Charging and discharging does deteriorate the plates and consume the electrolyte, but this happens so slowly that a given battery can be recharged many times before it fails, provided it is properly maintained and serviced throughout its operating life.

Lead-acid cell construction. Figure 1-22 illustrates the parts of an individual lead-acid cell. Each plate has as its foundation a grid made of lead and antimony. A spongy coating of lead goes over some of the grids to form negative plates, and a lead peroxide compound is attached to different grids to form positive plates. The plates are arranged such that each positive plate is between two negative plates with porous separators in between every plate to prevent them from touching one another. The plates are then enclosed in a hard rubber cell which is filled with electrolyte through an opening at the top of the cell. Each cell has two terminal posts, one negative and one positive, and a nonspill vent cap. The cap is designed to permit venting of gases while preventing inadvertent chemical spill during unusual attitudes or inverted flight.

Operation of lead-acid cells. When a lead-acid battery is sitting dormant (no current flow), each cell has an open-circuit voltage of approximately 2.2 volts. Under normally anticipated conditions, the cell will hold that voltage for an extended period of time; however, the cell's natural tendency to slowly lose its charge will eventually lead to a condition of total discharge. As the cell approaches that point, the voltage will be-

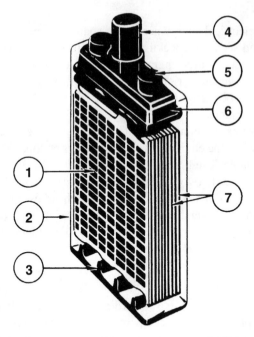

Fig. 1-22. *Lead-acid cell construction.*

1. Plates	3. Supporting ribs	6. Cell cover
2. Cell container	4. Vent cap	7. Separators
	5. Terminal post	

gin to rapidly drop. Simply said, the lead-acid battery does a good job of sitting and waiting to be used.

In a *closed-circuit* condition, where the battery is under load and supplying a current flow, the cell's voltage is gradually, but constantly, decreasing. The reason for the decrease is due to the *sulphation* of both the negative and positive plates. Sulphation is the result of the chemical process associated with discharge and, because it coats the plates, causes an increase in the cell's internal resistance. Eventually the coating on the plates becomes so thick that the electrolyte can no longer interact effectively with the plates and the battery is functionally discharged.

Lead-acid battery ratings. While each cell of a lead-acid battery has an approximate charge of 2.2 volts, they are simply rated as 2 volts. All the cells of the battery are connected together in series, which makes individual cell voltages cumulative. A 12-volt battery will have six cells; a 24-volt battery will have 12 cells.

Battery identification by voltage alone is insufficient from a practical perspective. A more important question is how long can you use the battery before it is completely discharged? Known as *battery capacity*, that indicator is rated in *ampere-hours* and it represents the amps supplied by the battery multiplied by the length of time the current will flow. For instance, a 50-ampere-hour battery will provide 50 amps for one hour, 25 amps for two hours, and so on. Unfortunately, while the rating system is convenient and easy to understand, it really does not work quite in the manner it suggests.

If, for instance, there is a very heavy discharge, the high current flow causes the battery to heat up. The hotter it becomes, the lower its efficiency, which causes the total ampere-hour output to decrease. A good example of that effect is trying to start your car in the winter. When the engine has been sitting out in −20°F temperatures and the oil is congealed to an almost solid mass, the strain the starter places on the battery can rapidly cause battery overheat resulting in discharge within a few minutes. To increase the ampere-hour capacity, it is possible to connect two or more batteries in parallel. The resulting system has the same voltage as any of the individual batteries but its ampere-hour rating is equal to the sum of all the batteries' ampere-hour ratings.

Two primary factors will shorten the life of a lead-acid battery: frequent, total battery discharge, and permitting the battery to remain in a discharged condition for an extended period of time.

Lead-acid battery testing methods. To determine the current state of charge, it is necessary to evaluate the condition of the active material on the plates. No convenient method makes the measurement directly; therefore, measuring the density of the electrolyte is necessary. The *hydrometer*, which looks similar to a large syringe, is used to draw up a sample of the electrolyte and measure its specific gravity. The more dense the electrolyte, the higher the state of charge. A specific gravity reading that ranges between 1.300 and 1.275 indicates a high state of charge. A reading of 1.275–1.240 indicates a medium charge, and 1.240–1.200 indicates a low charge. Readings should be taken with the electrolyte temperature between 90°F and 70°F, otherwise, correction factors must be applied.

Normal battery operation results in a slow loss of water over time. When the electrolyte level gets low, it is a simple matter of adding distilled water to bring the level back up; however, the state of the electrolyte should always be checked prior to adding water because the water does not instantly mix with the electrolyte. Using a hydrometer is not particularly difficult, but an individual inexperienced in its operation should get checked out by a mechanic before attempting to use it. Also, be very careful when working with the electrolyte because it is sulfuric acid. Splashed electrolyte will burn clothing (dissolving the material) and skin, and can cause blindness if it comes in direct contact with the eyes. Skin contact should be handled by thoroughly washing skin with cool water and then applying bicarbonate of soda to neutralize the acid. Wear old clothes that can be damaged if splattered by the battery acid.

Nickel-cadmium batteries

While both lead-acid and nicad batteries accomplish the same task, they are very different operationally and chemically. Despite the fact that nicad technology has been around for many years, it has only been the past 25 years or so that they have become very popular in aircraft. The primary reason has been expense. The initial cost of purchasing a nicad is significantly higher than its lead-acid counterpart; however, the benefits are equally significant. Nicads are the most common type of battery used in turboprop and turbojet aircraft, though they are not exclusive. Compared to lead-acid batteries, nicads have a much longer service life that helps keep total maintenance costs down and are able to discharge at a high rate without the voltage drop associated with its lead-acid counterpart. Nicads also accept high charge rates that result in a shorter recharge time, provide excellent engine starting capacity, and are very reliable.

Nickel-cadmium cell construction. The physical structure of the nicad is very similar to the lead-acid battery in that they both utilize individual cells containing positive and negative plates, plate separators, electrolyte, and a vent. That is where the similarity ends. Chemically and operationally there is significant difference between the two types of batteries.

In the nicad battery, the plates have a fine-mesh wire screen base upon which small granules of nickel powder are fused together to create a porous plaque surface. A coating of nickel-hydroxide is deposited on this surface to make a positive plate; cadmium-hydroxide makes a negative plate. Positive and negative plates are alternately placed side-by-side with a porous plastic strip separating them. The plate assembly is submerged in an electrolyte mixture of potassium hydroxide (KOH) and distilled water. The nicad's specific gravity range of 1.240 and 1.300 at room temperature is one major difference between lead-acid and nicad batteries. The specific gravity remains relatively constant regardless of the state of charge; therefore, it is impossible to determine the battery's state of charge from its specific gravity.

Operation of nickel-cadmium cells. Another major difference between the two types of batteries is the way they chemically react. When the nicad is in the process of being recharged, the negative plates lose oxygen and a metallic cadmium coat forms over them while the nickel-hydroxide on the positive plates becomes oxidized. This process will continue until either the charging current stops, or until all the oxygen has been removed from the negative plates leaving only cadmium. One of the by-products of the latter stages of the process is the formation of gas; a phenomenon which is increased if the cells are overcharged. The gas is the result of the water portion of the electrolyte decomposing into hydrogen on the negative plates and oxygen on the positive plates; therefore, totally recharging the battery results in gassing which means some water is used in the process.

The transformation of chemical energy into electrical energy happens during battery discharge as the chemical process reverses. Now, the positive plates give up oxygen to the negative plates in a process that causes some of the electrolyte to be absorbed by the plates. If you were to open the vent and look into a discharged nicad, it would appear as if it were very low on electrolyte. Never add water to a discharged battery. As the battery is recharged the electrolyte will come out of the plates and the level will rise.

In general, it is possible to interchange nickel-cadmium and lead-acid batteries with some advance preparation. It is critical to understand that chemically the two types of batteries are opposites. Installing a nicad to replace a lead-acid requires a thorough cleaning of the battery compartment with a neutralizing solution such as ammonia or boric acid to remove any trace of the sulfuric acid. The case must then be allowed to completely dry and finally be painted with an alkali-resisting varnish.

Servicing nicad batteries. Significant caution must be used when working with the electrolyte because it is very corrosive. Protective goggles, heavy rubber gloves and apron, and safety shoes must be worn. Skin exposure to the electrolyte should be given immediate attention by rinsing it with a solution of water, and either vinegar, lemon juice, or boric acid. **Warning:** When mixing electrolyte for a nickel-cadmium battery, always add the potassium hydroxide slowly to the water. Never pour water into potassium hydroxide because the resulting reaction can cause the chemical to explode. For this, and other reasons, nicad battery service is better left to the A&P mechanic who is properly trained and prepared to deal with it.

Thermal runaway. For all its virtues, the nicad has one potentially serious flaw: *overcharge runaway*, which is commonly referred to as *thermal runaway*. This is a condition in which the battery literally self-destructs. While it is a rare occurrence, every pilot who flies equipment with a nicad battery must be aware of this serious battery overheat condition.

A price is paid for the nicad's desirable traits of high discharge and recharge capability: low internal resistance. To make matters worse, there is an inverse relationship between temperature and both the cell's voltage and internal resistance. If for any reason a cell's temperature increases, its voltage and internal resistance will decrease. Three factors directly lead to increasing a cell's temperature:

- Excessively high discharge rate.
- Excessively hot ambient conditions, particularly with a poorly or improperly ventilated battery compartment.
- Deterioration of the plate separator material, which allows oxygen from a positive plate to interact with a negative plate where it will chemically interact with the cadmium and generate heat.

The beginning of the problem is an excessive discharge rate. Take, for instance, a night trip with several short legs flown in instrument meteorological conditions. Each stop requires battery-driven lights, radiant heat, and other accessories on the ground, and two engine starts. Each short leg heavily uses electronic navaids, electrothermal deice, lights, and accessories. In short, the battery is constantly being drained and recharged, which generates excessive battery temperature. Due to the way individual cells are installed in a battery case, the outer cells tend to dissipate heat through the sides of the battery case and, as a result, run slightly cooler than the middle cells. As battery temperature rises, it is the inner cells that get the hottest.

The generator attempts to recharge the battery by supplying sufficient current to the middle cells; their reduced resistance simply helps the generator. This excess current flow to the middle cells causes them to heat even more, further reducing their resistance. The reduced resistance, in turn, allows more current flow into the still hotter cells, and so on. The porous plastic strip between the individual plates breaks down fairly rapidly in such extreme conditions, further complicating the problem. At this point, shutting off the generator will stop the problem, otherwise the situation will begin to deteriorate rapidly.

When the middle cells get hot enough, their proximity to the adjacent cells will cause them to begin to overheat, too. Meanwhile, the inner cell's internal resistance and voltage becomes so low that the voltage of the good cells surrounding it will be higher by comparison and they will begin to feed the bad cells. As the inner cell's temperature increases, they begin to overheat the surrounding cells, which ultimately begin to be fed by the outer cells.

When any cell begins to receive voltage from a surrounding cell, isolating the battery from the generator will no longer have any effect; the battery is self-destructing. Pilot intervention cannot prevent this battery overheat condition; hence, the term thermal runaway. The only remaining solution is to land the aircraft as soon as possible. Consequences of this condition include fire in the battery box and literal meltdown with permanent airframe damage. Considering that some aircraft place the battery relatively close to a fuel storage tank, a precautionary landing is most appropriate.

Most aircraft with nicad battery systems have monitoring systems installed. These systems usually address overcharging one of two ways, either measuring the rate of battery charge (because excessively high rates indicate incipient runaway) or measuring and displaying battery temperature.

CIRCUIT PROTECTION

It is obvious from the explanation of thermal runaway that electricity can be a very dangerous item. Yet it is impossible to operate an aircraft without a significant electrical system. To satisfy the demand and to protect the aircraft and its occupants, two approaches are used: appropriate design and circuit protection.

When designing an aircraft system, engineers give considerable thought to its relationship and proximity to other systems. Electrical circuits are well insulated from other circuits, the airframe, and other systems. High current flow is, for the most part, kept off of the flight deck. Low voltage and amperage dc circuits connect cabin switches to remote solenoids that activate heavy amperage systems, such as starter motors. Even the best designed systems, however, cannot control what happens when the aircraft goes into regular operation.

Certainly the most serious problem that can arise in any electrical system or component is a *direct short*, or simply a *short*. The term refers to a situation where a break in a conductor, or other similar problem, is rerouting full system voltage directly to ground; the return side of the circuit. You will remember that Ohm's law states that as resistance decreases, current flow increases. Because the current is bypassing the intended load, a starter, the only resistance encountered is the conductor, which is practically nothing.

The result of this situation would be an inoperative starter because the current will take the path of least resistance, the battery will discharge very quickly and probably be destroyed in the process, and there would be a very real fire danger because the conductor would be carrying current far in excess of that for which it is rated. As the wire heated to a critical temperature, the insulation would melt, so too might the insulation of surrounding wires causing additional shorts, and if any of those wires were in proximity to a fuel or oil line, a major fire might result; therefore, all electrical systems incorporate various fuses, circuit breakers, and other protection devices that will minimize the danger associated with an electrical system that has had some form of breakdown.

Fuses. Fuses are placed in series with electrical circuits and consist of a thin piece of metal enclosed in a glass container. The purpose of the glass container is so the status of the fuse can be easily seen. Fuses are designed to allow current to pass through within certain specified limits. Aircraft fuses, which are very similar to automotive and household fuses, have an element made of an alloy of tin and bismuth. When the alloy reaches a predetermined temperature, the result of excessive current flow, the alloy melts, interrupting the circuit. This type of fuse is generally used to protect an individual component and is typically designed to handle between 1 and 10 amps though select fuses have greater capacity. When the fuse is "blown" it must be replaced with a new fuse. Unlike household fuses (usually found only in older homes) that screw into a fuse box, aircraft fuses are typically pushed into a fuse holder.

Another fuse made of copper is called a *current limiter* or *isolation limiter*; these super fuses protect entire electrical subsystems. Because of the high amperage rating, current limiters are often installed in such a manner as to make them inaccessible in flight.

Circuit breakers. This resettable fuse is designed to interrupt the circuit in the event that the current flow exceeds the breaker's limits. This protection device often replaces the more conventional fuse because a breaker "trips" and can be reset in flight simply by pushing it. When a circuit breaker trips, the pilot should wait a few minutes to allow for cooling before the reset. A nuisance trip might happen for no apparent reason; if the breaker trips again, shortly after being reset, the instance is not a nuisance; a problem in the circuit is causing the trip and appropriate action at the very least calls for not attempting another reset and continuing the flight with cautious monitoring of other electrical systems, or landing as soon as possible if the condition is serious, perhaps affecting a critical component or components.

Thermal protectors. This highly specialized fuse is designed to protect a motor. It automatically opens the circuit in the event that the motor temperature exceeds the protector's preset value. This is an *intermittent switch* in that it uses a *bimetallic strip*. When critical temperature is reached, the differential metals of the strip expand at different rates and open the circuit. When the temperature goes below critical, the metals strip returns to its normal condition and the circuit is again closed.

PRIMARY CAUSES OF ELECTRICAL CIRCUIT FAILURE

According to FAA Advisory Circular 65-9A, a few terms should be clearly understood with respect to electrical circuit failure:

- *Open circuit*   a circuit that is not complete or continuous. Figure 1-23 has examples of open circuits.
- *Short circuit*   a low resistance path that can be across the power source or between the sides of a circuit. Recall that the short usually creates high current

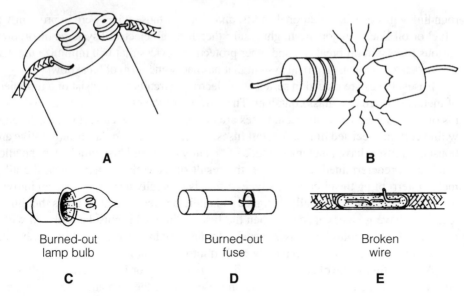

Burned-out
lamp bulb

C

Burned-out
fuse

D

Broken
wire

E

Fig. 1-23. *Common causes of open circuits.*

flow that will burn out or cause damage to the circuit conductor or components. Figure 1-24 has examples of short circuits.
- *Continuity* the state of being continuous, or connected together: not broken or does not have an opening.
- *Discontinuity* the opposite of continuity, indicating that a circuit is broken or not continuous.

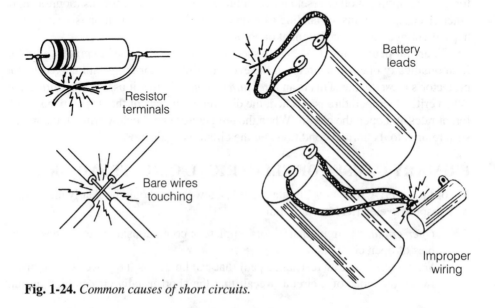

Resistor
terminals

Bare wires
touching

Battery
leads

Improper
wiring

Fig. 1-24. *Common causes of short circuits.*

ALTERNATING CURRENT

Alternating current (ac) has, for the most part, replaced direct current in the majority of electrical operations for several reasons. One of the most significant reasons is that ac can be transmitted for great distances far more efficiently because ac voltage can be increased and decreased with a *transformer*; therefore, it is possible to send very high voltage electricity over high tension wires and step down the voltage on location. From a systems perspective, ac equipment is generally smaller, lighter weight, and simpler in design than dc.

Comparison of ac/dc

Alternating current is very similar to direct current but does have some differences; dc flows only in a single direction with constant polarity, which means it only changes magnitude when the circuit is turned on and off; ac changes at specific intervals, going from zero to its maximum positive strength at a specific rate, then back to zero and down to its maximum negative strength (FIG. 1-25). The ac waveform is highly predictable, uniform and accurate. It is the waveform that is responsible for two effects that occur in ac but not dc: *inductive reactance* and *capacitive reactance*, which are subsequently detailed in this chapter.

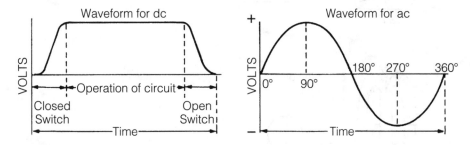

Fig. 1-25. *Voltage curves for dc and ac.*

GENERATING DIRECT CURRENT

Scientists knew for some time that an electric current that flows through a conductor will create a magnetic field; they did not know if the reverse would be true. Was a magnetic field capable of creating a current flow through a conductor? In 1831, Michael Faraday, an English scientist, proved the hypothesis and the electrical age was born.

Figure 1-26 illustrates how an electric current can be developed by a magnetic field. A cylinder has a conductor wrapped around it several times with the ends connected in a complete circuit. A *galvanometer* is hooked in series in the circuit. A galvanometer is an instrument that reacts to very small electromagnetic influences caused by the flow of a small amount of current. If a bar magnet is moved into the cylinder, the galvanometer needle will deflect in one direction showing current flow. When the magnet stops inside the cylinder and the relative motion between the magnet and cylin-

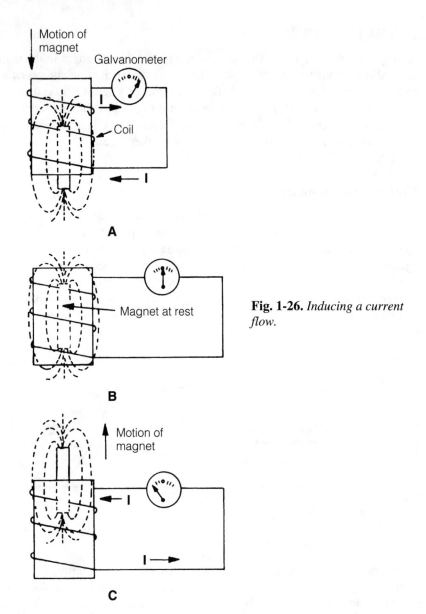

Fig. 1-26. *Inducing a current flow.*

der stops, the galvanometer needle goes back to the center showing no current flow.

When the magnet is drawn out of the cylinder, the needle moves in the opposite direction. It is the relationship between the conductor and the magnet that makes current flow. As such, the conductor and cylinder combination, called a *coil*, could be stationary with a magnet moving through it, or the coil could move inside a stationary magnet. In either case, it is the interaction of the coil moving through the magnet's lines of flux that creates current flow. The magnitude of the emf created is dependent upon three factors:

- The number of wires moving the magnetic field
- The strength of the magnetic field
- The speed of relative motion

The method of construction of a dc generator causes the emf in the circuit to travel in a single direction. A different technique is used in the ac generator, an *alternator*, to make the emf reverse direction.

GENERATING ALTERNATING CURRENT

Figure 1-27 shows a simple alternator with a loop conductor rotating in a magnetic field. As the loop rotates, the voltage varies as indicated. One full cycle is called a *sine wave* and is illustrated in position 5 of the graph. It indicates both the polarity and magnitude of the voltages generated. A cycle consists of two alternations: one positive and one negative. Frequency is a measure of the number of cycles that occur per unit of time.

Inductance. As ac moves through a wire coil, a magnetic field surrounds it. The field, created by the current flow, expands and collapses in unison with the alternating direction of flow. One of the undesirable but unavoidable by-products of that expanding and collapsing field is that it creates, properly termed *induces*, its own emf in the coil. The induced emf, called *counter-electromotive force* (cemf) tries to move in a direction opposite that of the applied emf.

Inductance is most pronounced in a coil, but even a straight wire has some minimal level of inductance. The factors affecting inductance are:

- Number of times the wire loops around the coil
- Cross-sectional area of the coil
- Type of material used to make up the core of the coil

Inductive reactance. Inductance is, in effect, induced emf "butting heads" with applied emf. How hard induced emf opposes the flow of current is called *inductive reactance*. Similar to circuit resistance, inductive reactance is measured in ohms. The greater the inductance in a coil, the stronger is the cemf, or opposition to the change in the value of the current.

Capacitance. The amount of electrical charge that can be stored in a capacitor is capacitance, which is measured in *farads*. A capacitor, originally called a condenser, is a device designed to store electricity in two conductors placed side-by-side but separated by a nonconductor called a *dielectric*. If a capacitor is in series with a dc source, as illustrated in FIG. 1-28, when the switch is closed, the B plate will become positively charged and the A plate will be negatively charged. The current flow in the circuit is at its maximum when the switch is first closed and diminishes proportionally to the rate that the capacitor charges. When the difference in voltage between plates A and B equals the dc source voltage, circuit current flow stops and the capacitor is fully charged. At this point

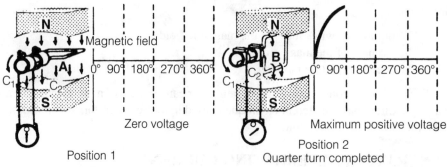

Position 1
Rotating conductors moving parallel to magnetic field, cutting minimum lines of force.

Position 2
Quarter turn completed
Conductors cutting directly across the magnetic field as conductor A passes across the North magnetic pole and B passes across the S pole.

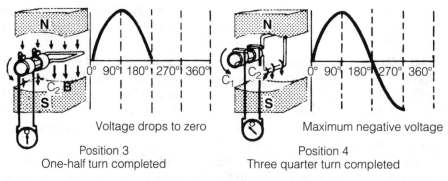

Position 3
One-half turn completed
Conductors again moving parallel to magnetic field, cutting minimum lines of force.

Position 4
Three quarter turn completed
Conductors again moving directly across magnetic field "A" passes across South magnetic pole and "B" across N magnetic pole.

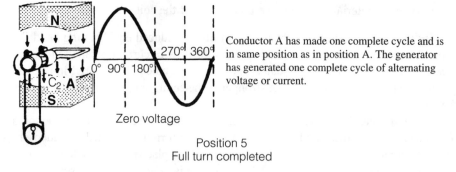

Position 5
Full turn completed

Conductor A has made one complete cycle and is in same position as in position A. The generator has generated one complete cycle of alternating voltage or current.

Fig. 1-27. *Generation of a sine wave.*

it is possible to open the switch with no effect on the capacitor's charge. The capacitor will hold that charge to the best of its ability until there is a demand for its voltage, however, given enough time the capacitor will eventually become discharged.

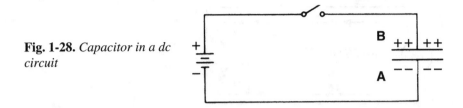

Fig. 1-28. *Capacitor in a dc circuit*

Three variables determine how much electricity can be stored by a capacitor: composition of the dielectric, plate size, and distance between plates. Plates are usually made of copper, tin, or aluminum often pressed into thin, rolled sheets. The storage capacity increases with the size of the plate area and decreases with the distance between the plates. Because the energy is actually stored in the space between the plates, the material of the dielectric has a major influence on the amount of electricity that can be stored. While a simple air-filled space will work, using a thin sheet of Bakelite (an early form of plastic) instead will increase capacitance five times. Listed from the greatest storage capability to the least, the common dielectric materials are:

- Crown glass
- Flint glass
- Porcelain
- Mica
- Quartz
- Common glass
- Isolantite
- Dry paper
- Hard rubber
- Asbestos paper
- Resin
- Air

Two primary categories of capacitors are fixed and variable. The fixed category is subdivided by type of dielectric: paper, oil, mica, and electrolytic.

The discussion to this point has been about a capacitor in a dc circuit; a capacitor's behavior is very different in an ac circuit. When ac is applied to the circuit, the electrical charge of the two plates continually changes (FIG. 1-29) with the alternating current. Similar to the dc application, current does not flow through the capacitor's insulator to complete the circuit but it does flow constantly throughout the remainder of the circuit. A by-product of capacitance, similar to inductance, is opposition to current flow. As with all forms of resistance, the opposition is measured in ohms and called *capacitance reactance*.

Phase of current and voltage in reactive circuits. When the sine waves of both the current and voltage pass through zero and reach maximum value simultaneously, they are *in phase* as depicted in FIG. 1-30A. When the current and voltage pass through

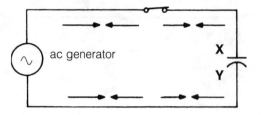

ac generator

X
Y

Fig. 1-29. *Capacitor in an ac circuit.*

zero at different times, they are *out of phase*. Circuits that contain only inductance result in a situation where the current reaches a maximum value after the voltage. In that case, the current wave lags voltage by 90°, or one-fourth a cycle (FIG. 1-30B). Circuits that contain only capacitance result in a situation where current reaches maximum value ahead of voltage by 90° (FIG. 1-30C); therefore, the amount that current leads or lags voltage is dependent upon the circuit's resistance, inductance, and capacitance.

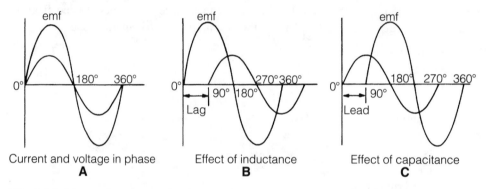

Current and voltage in phase
A

Effect of inductance
B

Effect of capacitance
C

Fig. 1-30. *Phase of current and voltage.*

MISCELLANEOUS ELECTRICAL COMPONENTS

A number of electrical components creep into every discussion with crew members and maintenance technicians. A thorough understanding of theory and operation is unnecessary; however, it is important to understand a few fundamentals.

Transformers. The purpose of a transformer is to change voltage. Depending upon design, a transformer will either *step up* (increase) or *step down* (decrease) voltage. A transformer consists of two coils that are independent of, but physically close to, each other (FIG. 1-31). Three basic pieces of a transformer are the iron core, a primary winding, and a secondary winding. The iron core provides a low reluctance circuit for the magnetic lines of force. The primary winding, which is wrapped around the iron core, is charged by the source voltage. The result is an expanding and collapsing magnetic field around the coil. As the primary field expands and collapses, it cuts through the secondary coil, also wrapped around an iron core, and induces a voltage in the core. Because it is the expanding and collapsing magnetic field that induces the voltage,

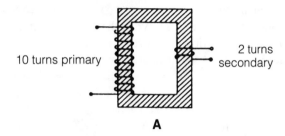

A

Fig. 1-31. *A step-down and a step-up transformer.*

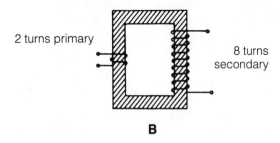

B

transformers only work with ac or pulsating dc. Standard dc current will not function in a transformer because its magnetic field does not expand and contract.

The determinant of how much voltage is induced in the secondary coil is based upon the *turns ratio*. The ratio compares the number of turns of the primary coil's winding to the number of turns in the secondary coil. The turn ratio in FIG. 1-31A is 5:1 because the primary coil has 10 turns and the secondary has 2; FIG. 1-31B has a ratio of 1:4. If the primary voltage in FIG. 1-31A is 20 volts, then the induced voltage to the secondary coil would be 4 volts. If, on the other hand, the primary voltage in FIG. 1-31B is also 20 volts, the induced voltage would be 80. For various reasons, no transformer is ever 100 percent efficient, so they are often rated as a decimal such as 0.9, meaning 90 percent, which is the transformer's actual efficiency percentage.

Transistors. This electronic device controls electron flow and converts ac to dc. It is very small, weighs next to nothing, is not subject to vibration associated problems, requires minimal voltage to operate, and will not pick up stray electronic signals. Transistors are *semiconductors* made up of two materials, each exhibiting different electrical properties. The term semiconductor comes from the fact that the quality of the materials is halfway between good conductors and good insulators.

Diodes. This is the electronic equivalent to the hydraulic system's check valve. A diode permits electron flow in one direction only. *Zener diodes*, also known as *breakdown diodes*, are used in voltage regulation. A somewhat more sophisticated version of the diode, it has a minimum threshold below which it will not allow current to pass.

When the circuit's potential equals the desired voltage the diode "breaks down" and allows the current to pass. When the circuit's voltage is below desired voltage, the zener blocks current flow. Transistors and diodes are basically rugged, but they can be easily destroyed by excess heat or by reversing polarity.

Rectifiers. This very specialized device converts alternating current into high-amperage, low-voltage direct current. In aircraft that employ alternators (ac producing generators) rather than a dc generator, some method is necessary to convert ac to dc to meet the requirements of the aircraft's dc systems. The rectifier is a low-cost, lightweight, low-maintenance, efficient means of meeting system requirements, accomplishing the task by simply limiting the direction of current flow—not allowing ac to alternate direction.

Inverters. This electrical component changes direct current into alternating current—the exact opposite of a rectifier. Inverters are employed when select components require ac to operate and the source of power is dc. Instruments, radios, radar, lighting, and certain accessories need ac. Select inverters provide two voltages, such as 26 and 115.

2
Electrical systems

THE TREMENDOUS DEMAND FOR ELECTRICAL ENERGY IN AN AIRCRAFT IS essentially handled by the generator. The battery, ground power unit, and auxiliary power unit have a limited place in fulfilling the requirement; it is clearly the task of the generator in the majority of operating conditions. Depending upon the operational requirements as determined by the manufacturer, either a dc generator or an ac generator (alternator) might be used.

From a functional perspective, a generator is a device that converts mechanical energy supplied by the powerplant into electrical energy, which is accomplished by electromagnetic induction. Chapter 1 explains that for electromagnetic induction to occur, it is necessary to charge the primary coil with alternating current so the rising and collapsing magnetic field cuts through the secondary coil. The result is to induce emf in the secondary coil, however it means that dc generators and alternators require an ac input to the primary coil. The two emf generating units are distinguished by the respective operation of and current produced by each type.

GENERATORS (dc)

Multiengine aircraft utilize a multiple generator system to meet the aircraft's electrical requirements. Select aircraft might operate two generators off a single engine, but it is more common that each engine will power a single generator. Twin-engine aircraft will have two generators, a three-engine aircraft will have three, and so on. Most commonly, all generators will be wired in parallel with a combined power output significantly higher than the electrical system's maximum demand including battery recharging. This redundant configuration allows for the loss of a generator with little or no significant impact on the electrical operating capability of the aircraft.

Recall from chapter 1 that when lines of magnetic force are cut by a conductor passing through them, the result is voltage being induced into the conductor. Two factors determine the strength of the induced voltage in the secondary coil:

- Speed at which the conductor rotates through the magnetic field
- Strength of the magnetic field

If the conductor forms a complete circuit, as depicted in FIG. 2-1, a current will be induced. The combination of the conductor and the magnetic field comprise a basic generator. The loop of wire rotating in the magnetic field cuts through the magnetic lines of force. At the point depicted in the figure, the plane of the loop is parallel to the lines of flux and the voltage induced in the loop causes the current to flow in the direction indicated by the small arrows. The induced voltage is at a maximum because the wires, which are 90° to the lines of force, cut through more lines of force per second than in any other position.

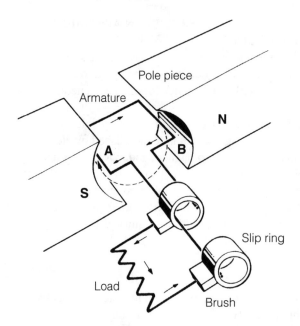

Fig. 2-1. *Inducing maximum voltage in a basic generator.*

This same situation occurs every 180° of rotation as first the A side of the loop and then the B side pass through the 90° point; therefore, as the loop approaches the vertical, or perpendicular, position, induced voltage begins to decrease to zero as the rotating conductor cuts through relatively fewer and fewer lines of flux. When the loop reaches the vertical position, there are no lines of force to cut because it is momentarily parallel to them, so there is no induced voltage. Then, as the loop continues to rotate, the induced voltage begins to increase, flowing in an opposite direction until it again peaks at the 90° position and again decreases to zero. This process produces an *alternating voltage*, so-called because of its continuous voltage reversal from positive to negative (FIG. 2-2). Given the nature of the process, it can be observed that all generators produce alternating current.

Fig. 2-2. *Output of an elementary generator.*

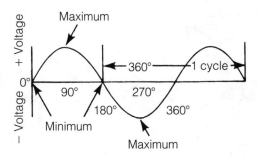

To use the voltage produced in an external circuit, it becomes necessary to connect the wire loop in series with the circuit. This can be accomplished by connecting the two ends of the loop to a pair of metal slip rings as illustrated in FIG. 2-1. A matching set of carbon brushes connected in series with the circuit ride on the slip rings transferring current to the circuit. Simpler yet, the slip rings can be replaced with a *commutator*, which is two half-slip rings matched together, but insulated from each other, to form one revolving, continuous cylinder with the brushes riding on the outside. This rotating part of the generator, the combination of the commutator and coil, is called an *armature* (FIG. 2-3).

The emf generated by the loop is ac, but the manner in which it is taken from the coil will determine if the generator provides ac or dc to the circuit. If slip rings are used

Fig. 2-3. *Basic dc generator.*

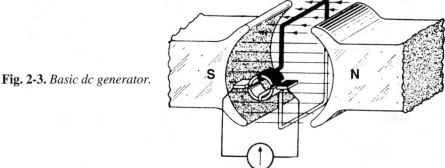

to take current from the loop, the result will be an ac generator (alternator). If a commutator is employed, it will be a dc generator because the commutator converts the ac to dc voltage. Figure 2-4 illustrates how it works. When the loop, which is rotating clockwise, reaches position A, no lines of force are being cut by the coil and, as a result, no emf is being generated. The black brush is shown just coming in contact with the black segment portion of the commutator while the white brush is just beginning to contact the white segment.

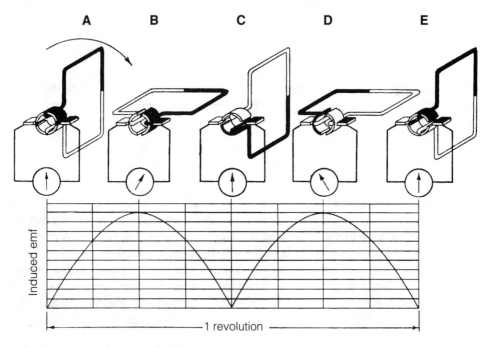

Fig. 2-4. *Operation of a basic dc generator.*

In position B, the maximum amount of flux is being cut, resulting in the greatest amount of induced emf. At that point, the black brush contacts the black segment while the white brush contacts the white segment. At that point, the deflection of the meter indicates a deflection to the right showing the polarity of the output voltage.

In position C, the loop has rotated a total of 180° and is once again not cutting lines of flux. The result is a drop-off of induced emf to zero. It is important to note the position of the commutator brushes with respect to the segments. The black brush at the 180° angle is contacting both the black and white segments on one side of the commutator while the white brush is contacting both segments on the other side. As the loop rotates just beyond the 180° point, the black brush contacts only the white segment and the white brush contacts only the black segment.

It is this switching of the commutator elements that assures the black brush always contacts the side of the coil moving downward while the white brush always contacts the side moving upward. As a result, the ac that flows through the loop is converted to dc, which always flows in the same direction through the external circuit. It is true that at the moment a given brush contacts both sides of the commutator there is, by definition, a short circuit, but the problem is alleviated by positioning the brushes so that the short occurs exactly at that point where there is no induced emf. That position is called the *neutral plane*.

One major problem associated with this system of converting ac to dc voltage is the voltage varies from zero to its maximum value twice during every revolution, resulting in a ripple effect. To greatly reduce that problem, more loops (coils) are added to the commutator (FIG. 2-5). As the number of coils and corresponding commutator segments increase, the effective voltage variation decreases to the point that the generator output voltage is a steady dc value. It is important to note that the actual output voltage of a given, single-turn loop is very small. Increasing the number of loops does not increase the output voltage, it only smooths out the ripple effect. Three things determine output voltage:

- The number of turns per loop. The more times a given loop is wrapped around the commutator, the more voltage that it will produce.
- Field current. The total flux generated by a given pair of magnetic poles. The more flux generated, the higher the induced voltage to the coil that cuts through it.
- The armature's speed of rotation. The faster the coils cut through the flux, the greater the induced voltage.

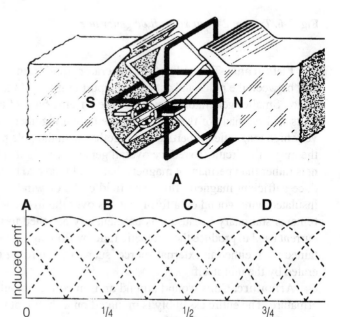

Fig. 2-5. *Increasing the number of coils reduces the Ripple effect.*

Parts of a dc generator

Three major components of the dc generator are the *field frame* (*yoke*), a *rotating armature*, and the commutator. Figure 2-6 depicts the various parts of a typical 24-volt aircraft generator.

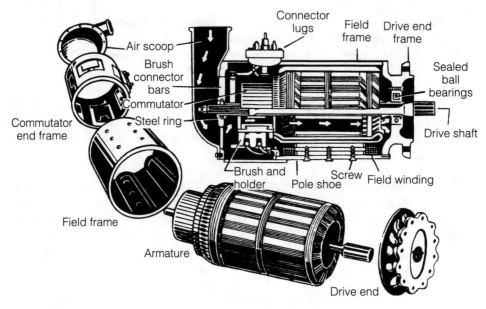

Fig. 2-6. *Typical 24-volt aircraft dc generator.*

Field frame. The body of the generator (field frame) completes the magnetic circuit between the poles and acts as a support unit for all the rest of the parts of the generator. Smaller generators typically have a frame built of a single piece of iron; larger units will be made of two parts that are bolted together. In either case, the frame has very high magnetic properties that together with the pole pieces form the major part of the magnetic circuit. To reduce overall generator weight, aircraft units use electromagnets rather than permanent magnets that would have to be very large and heavy to produce sufficient magnetic flux. The field coils, or winding, consists of many turns of insulated wire wound on a form that fits over the iron core of the pole. The field coil remains stationary because it is permanently attached to the field frame. The exciter current used to produce the magnetic field by flowing through the field coils can be obtained from either an external source, such as the aircraft battery, or from the dc generated by the unit itself.

Armature. Coils wound around an iron core commutator (FIG. 2-7) constitute an armature. The entire assembly is mounted on a shaft that rotates through the magnetic field of the stationary field coils.

Commutator. Wedge-shaped segments of hard-drawn copper, insulated from each other by thin pieces of mica, are located at the end of the armature (FIG. 2-8). The

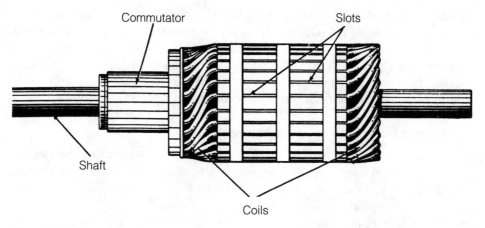

Commutator

Slots

Shaft

Coils

Fig. 2-7. *A drum-type armature.*

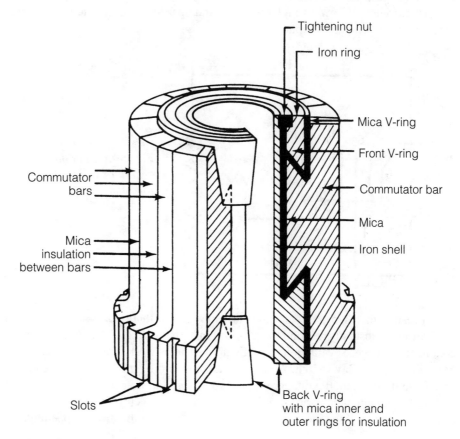

Tightening nut

Iron ring

Mica V-ring

Front V-ring

Commutator
bars

Commutator bar

Mica

Mica
insulation
between bars

Iron shell

Slots

Back V-ring
with mica inner and
outer rings for insulation

Fig. 2-8. *Commutator with portion removed to show construction.*

brushes ride over the surface of the commutator creating electrical continuity between the armature coils and the external circuit. A commutator has a limited life due to the wear resulting from the constant friction of the brushes rubbing across the surface. To increase their life as much as possible, the most common material used is high-grade carbon. It is a good cross between being soft enough to handle any irregularities in the surface of the commutator and being hard enough to afford maximum brush life.

Generator system support components

Voltage regulators. Aircraft electronic equipment demands that the electrical generating system provide a constant, well-regulated voltage. Of the three variables that determine generator output voltage, only the field current is easily controllable with a *voltage regulator*; a simplified schematic of such a system is depicted in FIG. 2-9. If the resistance in the field circuit is increased, less current flows through the field winding, which results in a decrease in the strength of the magnetic field. The rotating armature now turns in a weaker field with the result of a lower generator output voltage. On the other hand, if the resistance in the field circuit is decreased by the rheostat, a higher current will flow through the field windings. As the magnetic field increases, so too does the generator output voltage.

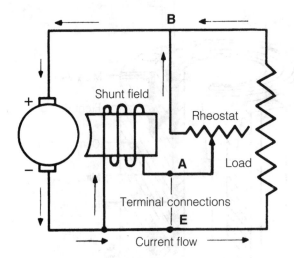

Fig. 2-9. *Regulation of generator voltage by field rheostat.*

In reality, the rheostat is a voltage sensing device that appropriately adjusts the field current to instantly respond to variations in output voltage. The result is a smooth, continuous voltage. Several voltage regulating systems work in the same manner.

Differential relay switch. A major concern regarding the health of a generator is protecting it from battery voltage. Without some form of protection, the brushes could weld to the commutator prior to engine start as the battery current flows through them onto the armature of the inoperative generator. To prevent this from happening, a *dif-*

ferential relay switch (*reverse current*) is incorporated. The reverse current relay is essentially an on/off switch that is controlled by the difference in voltage between the battery bus and the generator output.

The differential relay switch connects the generator to the electrical system's main bus whenever generator voltage exceeds bus voltage by at least 0.5 volt. If, on the other hand, bus voltage exceeds the generator output voltage, the reverse current relay opens and takes the generator *off-line*. Under normal operating conditions, the differential relays on all the generators do not close when the electrical load is reduced.

Take for instance a situation where a four-generator electrical system experiences a system load of only 50 amps. It is unlikely that all four differential relay switches will close; probably only two or three will go *on-line*. If a heavy load is then applied to the system, an equalizing circuit will lower the voltage of the generators already on-line while simultaneously raising the voltage of the generators that are off-line. This balancing act will allow all four generators to come on-line and equally share the heavy load. If all generators have been paralleled correctly, all the generators will stay on-line unless one or more generator control switches are turned off, or until engine speed falls below the minimum required to maintain generator output voltage.

Paralleling generators. Whenever two or more generators supply power to the bus, they must be operated in parallel. That means each generator supplies an equivalent proportion of the total ampere-load. For example, if three generators are meeting a 60-amp demand, each generator will provide 20 amps to the bus. The power supplied by a given generator to meet a load demand is called *ampere-load*. While it is true that power is measured in watts (the product of voltage and current), it is correct to use the term ampere-load because generator voltage output is considered to be a constant. If that is correct, then the power output is directly proportional to the ampere output of the generator.

Paralleling (equal load distribution) is accomplished among the generators with a special coil that is wound on the same core as the voltage coil of the voltage regulator. As minute differences occur between generators, the system senses the differences and inputs to the appropriate voltage regulator. A simplified schematic of a generator equalizer circuit is depicted in FIG. 2-10.

Generator maintenance

Six primary considerations of routine maintenance are:

- Security of the generator mounting
- Condition of electrical connections
- Dirt and oil in the generator
- Condition of generator brushes
- Proper generator operation and performance
- Voltage regulator operation

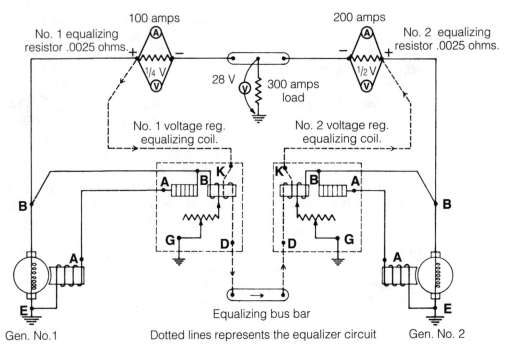

Fig. 2-10. *Generator equalizer circuit.*

ALTERNATORS

Recall that the generator converts mechanical energy into electrical energy, as a result of electromechanical induction. All generators require ac to operate and the type of current that a generator supplies to the circuit determines what the generator is called. An alternator provides ac. When comparing generators to alternators, it becomes readily apparent that the major difference between the two is the manner in which they are connected to the external circuit. Alternators are connected by slip rings and dc generators use a commutator.

Alternator types

Alternators can be classified several ways. One way is noting the component's excitation system. Aircraft alternators will use one of the following methods.

Direct-connected, direct-current generator. This system incorporates a dc generator on the same shaft as the alternator. The dc generator is used to produce the necessary voltage to excite the alternator's magnetic field. A variation of this system is to use the aircraft battery to excite the field. The obvious drawback to the latter method occurs if the battery is dead. In that situation, it is impossible to build up a magnetic field, which results in an inoperative alternator.

Transformation and rectification. This system uses the alternator's rectified ac output to excite its field windings. Because this system has no source of emf to excite the alternator field before the alternator produces its own emf, the unit must incorporate a residual magnet in the field to get the process started.

Integrated brushless. This system has a dc generator on the same shaft as the alternator, similar to the direct connected, direct current method. The difference between the two lies in the fact that the excitation circuit is completed by rectifiers rather than a commutator and brushes. The rectifiers are mounted on the generator shaft and their output is fed directly to the alternator's main field windings.

Another method of classifying alternators is by the number of phases. An alternator might be single-phase, two-phase, three-phase or more; an aircraft alternator is almost always three-phase.

The last common method of classifying alternators is by the type of stator and rotor: *revolving-armature* or *revolving-field*. The revolving-armature is built similar to the dc generator; the armature revolves within a stationary magnetic field; this alternator is relatively low power and not common in turboprop and turbojet aircraft. The revolving-field alternator is constructed with a stationary armature winding called a *stator* that engulfs a rotating-field winding called a *rotor*. The principal reason for this type of construction is the ease with which the stationary armature winding can be connected directly to the load. The elimination of slip rings or other sliding contacts results in advantages over the revolving-armature:

- Increased equipment life
- Reduced maintenance costs
- Elimination of arcing problem normally associated with high-voltage alternators

The rotating field must still incorporate slip rings to allow a source of ac to create the magnetic field; however, the voltage and current required by the rotating field are significantly less than the output of the alternator. As a result, the rotating field slip rings functionally work the same as those associated with any low-voltage generator.

Alternator-rectifier unit. This alternator is typically used in aircraft weighing fewer than 12,500 pounds. While used in dc electrical systems, and typically referred to as a generator, it is an alternator/rectifier. Recall from chapter 1 that a rectifier is an electrical circuit that changes ac into dc. This kind of unit is self-exciting without a permanent magnet. Initially, battery power is required to excite the magnetic field; however, excitation is derived upon operation from normal unit output. The alternator-rectifier is directly coupled to the engine through a flexible drive coupling and operates in a speed range of 2,100 to 9,000 rpm. Normal output voltage is 26–29 volts and 125 amperes.

Brushless alternator. This is perhaps the most efficient power generating device. Its design eliminates brushes by incorporating an integral exciter with a rotating armature that has its ac output rectified for the main ac field. The end result is no brushes. One such common unit is 3-phase, 4-wire and has a rating of 31.5 kilovolt amperes (KVA), 115/200 volts, and 400 cycles per second.

COMBINED ac/dc ELECTRICAL SYSTEMS

Most turboprop and turbojet aircraft use incorporated ac/dc electrical systems. Such aircraft typically use the dc system as the primary aircraft electrical system. It is common to employ two or more parallel dc generators: one per engine. Individual generators are commonly rated at 300 amperes.

The ac portion of the system often includes *fixed frequency* and *variable frequency* capability. Fixed frequency, single-phase ac powers frequency-sensitive equipment and is achieved through the use of inverters. The variable frequency portion of the system is supplied by two or more engine-driven alternators that provide three-phase ac power for special uses: ice protection of props, engine inlets, and windshields. Systems that are this extensive will also typically incorporate an auxiliary power unit to operate a generator when the engines are not operating.

Voltage regulation of alternators. Voltage regulation of multiple alternators is similar to that used to control dc generators. One major difference is that when two or more generators are operated in parallel, the governing factor is shared ampere-load; therefore, a synchronizing force is only required to equalize the voltage between all dc generators on-line. With the alternator, the equalizing force must synchronize voltage and frequency between alternators and assure that all alternators have the same phase sequence. Failure to properly balance the alternators prior to putting them into parallel operation on the main bus can result in significantly greater system damage than damage associated with the same failure in a dc generator operation.

Alternator constant-speed drive

Another major difference between alternator and generator operation is that alternators are frequently not directly driven by an engine. The problem is that many electronic devices that require alternating current also require that the voltage and frequency be very tightly controlled with little or no tolerance. The only practical solution to the problem is to assure that the rotational speed of the alternator remains absolutely constant, which is a difficult task because normal engine rpm might vary significantly depending upon phase of operation. The solution is a constant-speed drive system.

The *constant-speed drive* (CSD) unit (FIG. 2-11) allows the engine to power the alternator but assures that alternator rpm remains constant throughout the normal operating speed range of the engine. The CSD is functionally a hydraulic transmission, somewhat similar to that used in an automobile. Depending upon the manufacturer it might be controlled electrically or mechanically. It is designed to guarantee a constant alternator rpm within a specified engine rpm range; for instance, an alternator rotational speed of 6,000 rpm within an engine rpm range of 2,800 to 9,000 rpm. When the unit senses that alternator speed is beginning to slow, the CSD goes into overdrive and shifts gears appropriately to maintain the designated 6,000 rpm; if the unit senses an overspeed, it then goes into an underdrive condition to maintain the designated rpm.

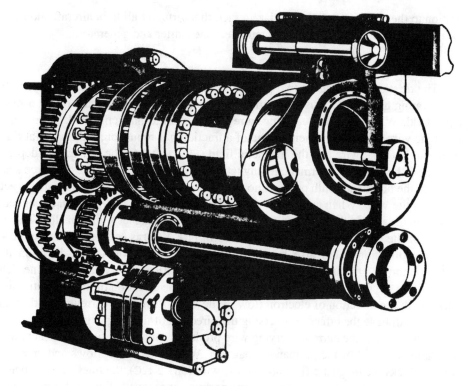

Fig. 2-11. *Constant-speed drive.*

ALTERNATOR MAINTENANCE

Alternators and generators are maintained in a very similar manner. The moving parts must be carefully inspected for wear because they are the weak link in both systems. Exciter brushes and slip rings wear out and must be replaced. From a practical point of view, a flight crew cannot perform maintenance on the power generating system. The crew's primary role is preventive maintenance by assuring strict compliance with all system operating requirements; it is particularly important to assure synchronization of alternators before putting them in parallel operation.

MOTORS (dc)

The beginning of chapter 1 stated that the electrical system spans the entire aircraft; few, if any, systems are untouched. Of all the uses for electricity, one of the most common is the dc motor. Its practical application is based upon its ability to convert dc energy into mechanical, rotary motion. An aircraft motor can pump fuel, hydraulic, and deicing fluids, extend and retract landing gear, crank engine starters, rotate anticollision lights, and serve as fluid control units, such as fuel shut-off valves. Studying the operation of the dc motor shall reveal that the motor is very similar in design and operation to the dc generator. So similar, in fact, that virtually all light aircraft and many twin-engine turboprops use a combination engine starter and generator.

Theory of the dc motor

The dc motor consists of two parts: a field assembly and an armature assembly. The rotating armature consists of current-carrying wires that are acted on by a magnetic field. When a current-carrying wire is placed in a magnetic field, a force acts upon the wire. The force is neither one of attraction nor repulsion, rather, it is at right angles to both the wire and the magnetic field (FIG. 2-12). As the illustration depicts, the wire is between two permanent magnets. The magnetic field's lines of force span between the north and south poles. If no current is flowing (FIG. 2-12A), no force is acting upon the wire. When current does flow through the wire (FIG. 2-12B), a magnetic field radiates around the wire and within the permanent magnet's lines of flux. The direction of the wire's magnetic field depicted in FIG. 2-12B is counterclockwise; however, the direction could also be clockwise, depending upon which direction the current is moving through the wire. An accepted electrical theory can determine field direction, in the meantime it is sufficient to say that the direction of the magnetic field varies with the direction of electron flow in the wire. Reversing current flow through the wire will have the effect of reversing the direction of the magnetic field.

In either case, the current-carrying wire produces a magnetic field that reacts with the stationary field of the permanent magnets. When the current flows and creates a counterclockwise magnetic field around the wire (FIG. 2-12C), the lines of force below the wire travel in the same direction as the permanent magnet's lines of force. The result is that the two magnetic fields at the bottom are additive; the lines of force at the top are in opposition and neutralize each other. The net result is the bottom field is strong while the top field is weak; therefore, the wire is pushed upward. If the current direction through the wire were reversed, the direction of the magnetic field around the wire would reverse, but the permanent magnet's field would remain unchanged. The net result in that case would be for the wire to be pushed down.

Fig. 2-12. *Force on a current-carrying wire.*

Force between parallel conductors. Whenever two force fields lie in close proximity to one another, they interact. Figure 2-13A depicts two side-by-side conductors with the resulting magnetic fields. The two black dots depict that the current flow is moving toward the reader; the cross in the center of the right wire in FIG. 2-13B shows the current flow is away from the reader. In FIG. 2-13A, because the current flow is the same direction for both conductors, both magnetic fields are clockwise. This causes the magnetic fields that lie between the two conductors to cancel each other with the result that the conductors are drawn together. In FIG. 2-13B, the left conductor's magnetic field is rotating clockwise while the right conductor's magnetic field is rotating counterclockwise. The result is that the field between the two conductors is one of reinforcement and the wires move in the direction of the weaker, outer field. In short, side-by-side conductors with current moving in the same direction are drawn together; side-by-side conductors with current moving in opposite directions are repelled.

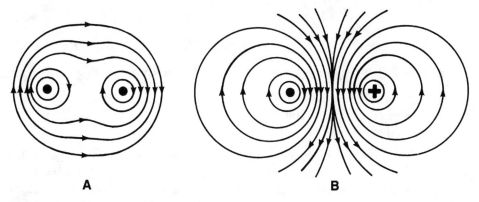

Fig. 2-13. *Fields surrounding parallel conductors.*

Developing torque. The fundamental principle behind the operation of a dc motor is simply this: When a coil in which current is flowing is put in a magnetic field, the resultant force makes the coil rotate. The direction of rotation is dependent upon the direction of current flow through the coil.

Refer to FIG. 2-14. Current flows away from the reader on side A, causing a counterclockwise magnetic field, and toward the reader on side B, causing a clockwise magnetic field. Again, determining the direction of the magnetic field is done through an accepted electrical theory and not of any particular importance. What is important is that the conductor's clockwise magnetic field on side B results in a tendency for side B to move down; the tendency for side A is to move up. This tendency will prevail, and the coil will rotate until it reaches the point where the coil's plane is perpendicular to that of the permanent magnet's lines of flux. At that point, the coil would stop rotating if it were not for the fact that the white coil, which is perpendicular to the black, would have simultaneously rotated to the position directly between the north and south poles of the permanent magnet. The white coil would now react exactly the same as the black coil did when it was in that position and the result would be another rotation bringing the black coil back between the poles of the magnet. This rotational tendency is *torque*.

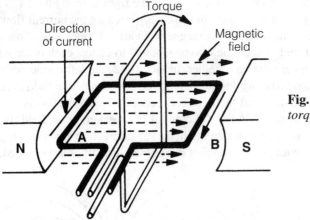

Fig. 2-14. *Developing a torque.*

The amount of torque that is developed by a dc motor varies with several factors:

- Strength of the magnetic field
- Number of turns in the coil
- Position of the coil in the field

The strength of the magnetic field depends upon the type of magnets used. Those used in dc motors are of the electromagnetic variety and are constructed of a very special type of steel. The result is an exceptionally strong magnetic field when an electrical current is induced.

The number of turns in the coil is also a very important factor in the amount of torque produced. Because torque acts on every turn, the more turns on the coil, the stronger the torque. The easiest way to increase the designed torque of a motor is to increase the number of turns of its coil.

Finally, it has been shown that as the plane of the coil parallels the lines of force, the torque is zero; as it cuts the lines of force at right angles, the torque reaches 100 percent. Any position in between results in a torque value somewhere between zero and 100 percent. The greater the number of coils on the armature, the greater the chance that at any given moment at least one coil will be perpendicular to the magnetic lines of force creating the maximum amount of torque; therefore, as the number of coils increases, so too does the torque.

Parts of a dc motor

The major parts of a dc motor include the armature assembly, the field assembly, the brush assembly, and the end frame (FIG. 2-15).

The armature contains the insulated copper wire coils, commutator, and soft-iron core on a rotatable steel shaft. The armature rotates within the stationary field assembly that contains the frame, pole pieces, and field coils. Select motors have as few as two poles; powerful motors might have as many as eight. The brush assembly is made

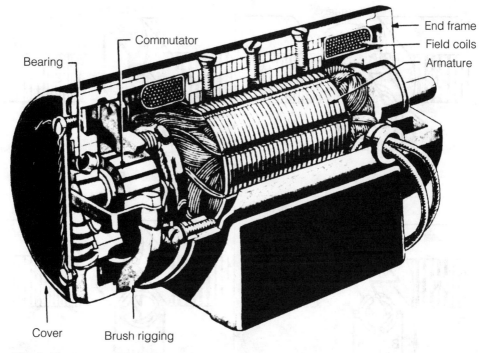

Bearing — Commutator — End frame / Field coils / Armature

Cover — Brush rigging

Fig. 2-15. *Cutaway view of a practical dc motor.*

up of small blocks of graphitic carbon that ride over the commutator of the armature. Finally, the end frame is the end of the motor, which is opposite the brush assembly. The end frame contains the bearing drive for the armature shaft and is often designed to mate to the unit that will be driven by the motor.

Recall that a coil of wire through which current flows will rotate if the coil is placed inside a magnetic field. At first glance, FIG. 2-14 would appear to be a practical dc motor, but one significant aspect is missing—the source of current for the coil. Until now it has just been assumed that current was flowing through the coil, but a careful analysis of the situation will reveal that if the coil were permanently attached to the terminals of a battery, for instance, the coil would simply rotate until it lined up with the magnetic field and then stop because the flow of current through the coil would travel in the same direction. To compensate for this problem it becomes necessary to devise a system that will reverse the current in the coil at exactly the same time as the coil parallels the lines of force. Doing so will create torque, again causing the coil to rotate; therefore, if the current can be made to reverse every time the coil is about to stop, the coil will continue to rotate as long as desired.

Figure 2-16 illustrates one method of solving the reversing current problem by using the same type of slip-ring assembly (commutator) employed in small generators. As the coil turns, each contact moves off one terminal and slides onto the other. This action causes the current flow to change direction every time the conductors parallel

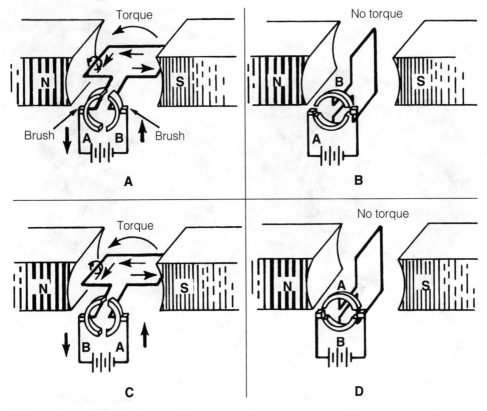

Fig. 2-16. *Basic dc motor operation.*

the magnetic field. The result is that the conductor next to the north pole of the magnet is always experiencing a current flow in the direction of the reader, while the current flow of the opposite conductor is always away from the reader. When the conductors are parallel to the force field (FIG. 2-16B and FIG. 2-16D), torque has reached its minimum value because the least number of lines of force are being cut. Two things prevent the coil from stopping at that position. The first is that the momentum of the coil keeps it turning until the coil reaches a position where the torque begins to increase again. The second is that many more than one coil are employed, so there is essentially no time throughout 360° of armature rotation when one or more coils are not cutting lines of flux and creating torque.

Types of dc motors

Three types of dc motors are series, shunt, and compound. They differ in the manner in which their field and armature coils are connected.

Series motors. A series motor consists of a field winding with relatively few turns of very heavy wire connected in series with the armature winding. That means that the current flowing through the field winding will also flow through the armature winding (FIG. 2-17). The result is that any increase in current causes the magnetic field of both the field and armature to increase simultaneously. The advantage of this configuration is that the reduced number of very large wires results in a relatively low resistance circuit. This enables the motor to draw heavy current during starting which results in the availability of a high starting torque.

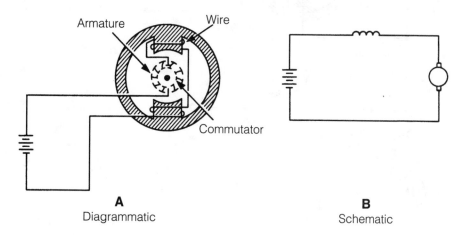

A
Diagrammatic

B
Schematic

Fig. 2-17. *Series motor.*

A phenomenon associated with the series motor is that its rotational speed is inversely proportional to its load. Such a motor will run at a very slow speed with a heavy load and very fast with a light load. In fact, it is possible, under conditions of very little or no load, to run so fast as to self-destruct; therefore, the series motor is best suited to applications that require high starting torque under heavy load conditions. The most common applications include engine starters, landing gear extension and retraction, cowl flap extension and retraction, and wing flap deployment.

Shunt motors. The field winding in this motor is connected in shunt (parallel) with the armature winding (FIG. 2-18). The resistance of the field winding is relatively high due to many windings. With the field winding connected directly to the power supply, the field current remains constant, not varying with motor speed. As a result, shunt motor torque varies proportionally with the current to the armature, and is not as strong as that produced by a series motor of equivalent size. On the other hand, the rotational speed of a shunt motor varies only slightly with changes in load and experiences only a modest speed increase in a no-load situation; therefore, the shunt motor is best suited for relatively low torque situations calling for a constant speed, such as fuel and hydraulic fluid pumps.

Compound motors. Finally, a compound motor is a cross between the series and shunt motors (FIG. 2-19). The shunt winding, which is connected in parallel with the armature winding, is composed of numerous turns of very fine wire. The series winding,

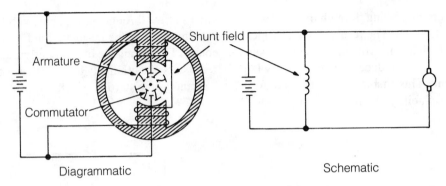

Diagrammatic   Schematic

Fig. 2-18. *Shunt motor.*

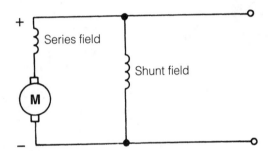

Fig. 2-19. *Compound motor.*

which is connected in series with the armature winding, is made up of a few turns of very large gauge wire. This combination provides a higher starting torque than the shunt motor, though lower than that obtained from a series motor of equivalent size. On the other hand, the relationship between rotational speed and load is not as sensitive as with a series motor. In short, the compound motor is a cross between the shunt and the series motors and is used in applications that are subject to sudden changes in load, or where high starting torque is required but for some reason a series motor cannot be employed.

Figure 2-20 is a graph comparing the load characteristics of dc motors against the motor's rotational speed.

Types of duty. It is very important for the flight crew to understand that different types of electric motors are designed to operate under different conditions. Some motors are specifically built for intermittent operation, while others are for continuous use. Intermittent motors, such as engine starters, have a very strict minimum cooling period between operations. It is common for the pilot's operating handbook to specify the minimum cooling time between starting attempts. For instance, the Beechcraft Super King Air B200 handbook specifies the following limitation with regards to the use of the engine starter: "STARTERS: Use of the starter is limited to 40 seconds ON, 60 seconds OFF, 40 seconds ON, 60 seconds OFF, 40 seconds ON, then 30 minutes OFF." Other motors are built for continuous duty and can be operated at the rated power for longer periods. A common example of such a motor is a fuel pump.

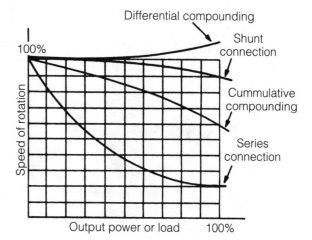

Fig. 2-20. *Load characteristics of dc motors.*

Reversing motor direction. To reverse the rotational direction of a motor, it is only necessary to reverse the direction of current flow to either the field windings or the armature. The result will reverse the corresponding magnetism, which causes the armature to rotate in the opposite direction. While it seems as if all that is necessary to reverse the direction of current flow is to simply switch the wires that connect the motor to the power source, it is not that simple.

Switching the lead wires to the motor will cause the current flow direction to reverse to both the field winding and the armature. Because the magnetic relationship will remain the same between the two, the motor will continue to rotate in the same direction. The current flow of one or the other must be reversed. The two most common methods of reversing current flow are using a split field motor or incorporating a double-pole, double-throw switch.

The *split-field* motor is a series motor with a split-field winding (FIG. 2-21). While there are variations in the exact methodology, essentially a single-pole, double-throw switch channels current flow to either of the two windings. If the switch is in position A, current flows through the lower field winding, which creates a north pole at the lower pole piece, and a south pole at the upper pole piece. If the switch is in the B position, current goes to the upper winding, which causes the magnetism of the field to reverse. The result is that the armature rotates in the opposite direction.

The double-pole, double-throw switch method can be designed to change the direction of current flow in either the armature or the field. The switch depicted in FIG. 2-22 is set up to reverse the current direction in the field windings, not the armature. The switch consists of two hinged contacts mechanically connected so that when one is moved, the other moves with it. When the switch is placed in the "up" position, the field winding is energized and a north pole is established at the right side of the motor. When the switch position is reversed, that is when both switches depicted rotate downward and contact the two "down" contacts, the current flow enters the field winding from the opposite side causing a reversal of polarity in the field only. Because the armature re-

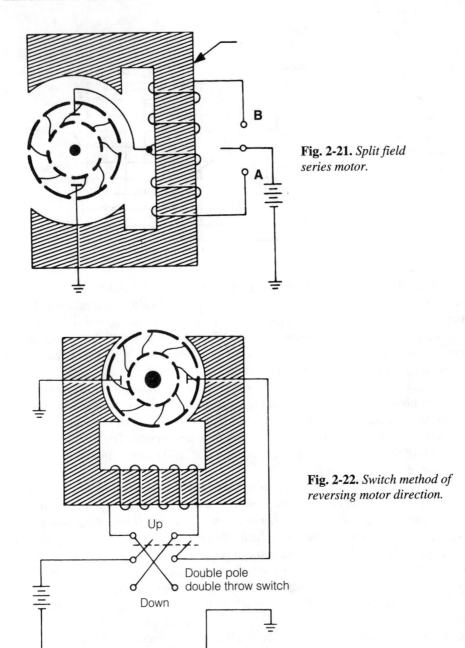

Fig. 2-21. *Split field series motor.*

Fig. 2-22. *Switch method of reversing motor direction.*

ceives the same current flow in either position, the relationship between the armature and field winding has changed and the armature will revolve in the opposite direction.

Motor speed. Variable-speed motors are easily adjustable by varying the amount of current that flows to the field windings. What is somewhat confusing, however, is that increasing current flow causes the motor to slow down, while decreasing current flow causes it to speed up. The explanation lies in a thorough understanding of *counter electromotive force* (cemf). Chapter 1 noted that when current is induced into a coil, the result is for a cemf to resist normal electron flow; therefore, the greater the induced current, the greater the resistance, and the slower the motor turns. Variable-speed motors might be either shunt or series-type.

The speed of shunt motors is controlled through a rheostat that is in series with the field winding (FIG. 2-23). If the crew wishes to increase the motor's rotational speed, the rheostat is adjusted to increase its internal resistance, which results in a lower current flow to the field. The reduced current flow causes a reduction in the magnetic field, which in turn causes a corresponding decrease in cemf. The lower cemf results in a momentary increase in the armature current, and torque, which causes the motor to speed up. As the motor speeds up, the cemf begins increasing and subsequently causes the armature current to decrease to its original value. When that happens, the motor will operate at a higher speed than previously. To decrease the motor's rotational speed, the resistance in the rheostat is decreased and the opposite occurs. The specific control mechanisms differ somewhat in the series motor, but from a functional point of view, the motor's rotational speed can also be controlled with a rheostat.

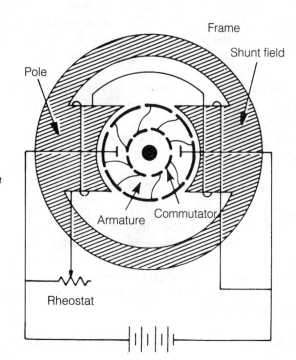

Fig. 2-23. *Shunt motor with variable speed control.*

Maintenance of dc motors

The primary preventive maintenance that can be performed by the flight crew is proper operation of all electric motors. Carefully adhere to the motor operating restrictions outlined in the limitations section of the pilot's operating handbook. If a motor does not operate, about the only thing a nonmaintenance technician can do is check the integrity and correctness of the motor's wiring, connections, terminals, fuses, and switches.

MOTORS (ac)

Despite the commonality of dc motors, there are several major advantages to ac motors. Primarily, ac motors cost less to buy, operate, and maintain. For one thing, most ac motors do not use brushes and commutators, which reduces cost, routine maintenance, and the potential for sparking. An ac motor is particularly well suited to a situation that requires a motor that will accurately hold a constant speed; less common are variable-speed ac motors. An ac motor might be rated in several ways: horsepower, voltage, current under full load conditions, rotational speed, number of phases, and frequency.

Types of ac motors

Two ac motors are used in aircraft: *induction* and *synchronous*. Either might be single-phase, two-phase, or three-phase.

The three-phase induction motor produces a large amount of power. As such, three-phase motors are typically used in situations requiring the movement of heavy loads such as starters, flaps, landing gear, and hydraulic pumps. Single-phase induction motors produce less power and are used in operating surface locks, intercooler shutters, and oil shutoff valves.

Three-phase synchronous motors operate at a constant synchronous speed and are used where accuracy is important, such as the operation of flux gate compasses and propeller synchronizer systems.

Single-phase synchronous motors are probably the most common of all because they are used in electric clocks and other precision, low-power-requirement equipment.

Maintenance of ac motors

From the maintenance technician's point of view the maintenance requirements of ac motors are even simpler than dc motors. From the flight crew's perspective, it is similar to dc motors; little can be done about a failure in-flight. Overall motor security and connectivity can be checked during preflight when the motor is readily accessible; however, caution must always be used when checking conductivity because some motors are hot-wired. Touching both terminals or leads simultaneously can result in severe electric shock.

MANUFACTURER DOCUMENTATION

(The following information is extracted from Beech Aircraft Corporation pilot operating handbooks and Falcon Jet Corporation aircraft and interior technical descriptions.

The information is to be used for educational purposes only. This information is not to be used for the operation or maintenance of any aircraft.)

Super King Air B200

Electrical system. The airplane electrical system is a 28-Vdc (nominal) system with the negative lead of each power source grounded to the main airplane structure. Dc electrical power is provided by one 34-ampere-hour, air cooled, 20-cell, nickel-cadmium battery, and two 250-ampere starter/generators connected in parallel. The system is capable of supplying power to all subsystems that are necessary for normal operation of the airplane. A hot battery bus is provided for emergency operation of certain essential equipment and the cabin entry threshold light circuit. Power to the main bus from the battery is routed through the battery relay which is controlled by a switch placarded BAT - ON - OFF, located on the pilot's subpanel. Power to the bus system from the generators is routed through reverse-current-protection circuitry. Reverse current protection prevents the generators from absorbing power from the bus when the generator voltage is less than the bus voltage. The generators are controlled by switches, placarded GEN 1 and GEN 2, located on the pilot's subpanel. **Note:** In order to turn the generator ON, the generator control switch must first be held upward in the spring-loaded RESET position for a minimum of one second, then released to the ON position.

Starter power to each individual starter/generator is provided from the main bus through a starter relay. The start cycle is controlled by a three-position switch for each engine, placarded IGNITION AND ENGINE START - ON - OFF - STARTER ONLY, on the pilot's subpanel. The starter/generator drives the compressor section of the engine through the accessory gearing. The starter/generator initially draws approximately 1100 amperes, then drops rapidly to about 300 amperes as the engine reaches 20 percent N_1.

Power is supplied from three sources: The battery, the right generator, and the left generator. The generator buses are interconnected by two 325-ampere current limiters. The entire bus system operates as a single bus, with power being supplied by the battery and both generators. There are four dual-fed sub-buses. Each sub-bus is supplied power from either generator main bus through a 60-amp limiter, a 70-amp diode, and a 50-amp circuit breaker. All electrical loads are divided among these buses except as noted on the accompanying Power Distribution Schematic (FIG. 2-24). The equipment on the buses is arranged so that all items with duplicate functions (such as right and left landing lights) are connected to different buses. Among the loads on the generator buses are the number 1 and number 2 inverters. Through relay circuitry, the INVERTER selector switch activates the selected inverter, which provides 400-hertz, 115-volt, alternating current to the avionics equipment, and 400-hertz, 26 Vac to the torque meters. The volt/frequency meter indicates the voltage and frequency of the alternating current being supplied to the avionics equipment.

The generators are controlled by individual voltage regulators which maintain a constant voltage during variations in engine speed and electrical load requirements.

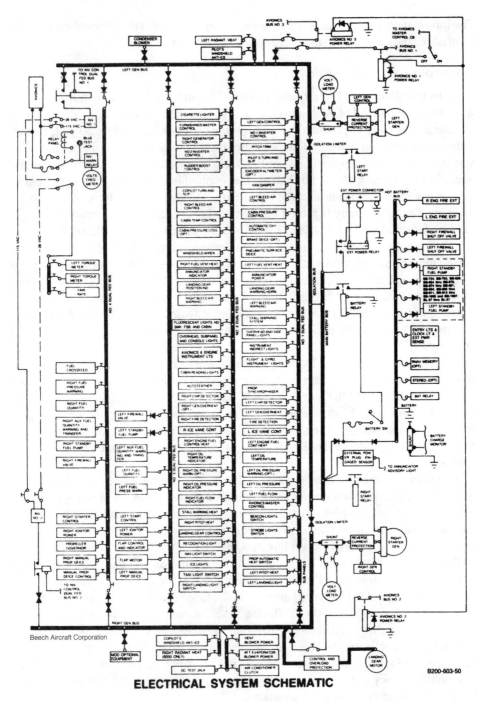

ELECTRICAL SYSTEM SCHEMATIC

Fig. 2-24. *Beech 200 electrical system schematic.*

The generators are connected to the voltage regulating circuits by means of control switches located on the pilot's left subpanel. The voltage regulating circuit will automatically disable or enable a generator's output to the bus. The load on each generator is indicated by the respective left and right volt/load meter location in the overhead panel.

Overheating of the nickel-cadmium battery will cause the battery charge current to increase. Therefore, a yellow BATTERY CHARGE caution annunciator light is provided in the caution/advisory annunciator panel to alert the pilot of the possibility of battery overheating. A Battery Charge Current Detector will cause illumination of the yellow BATTERY CHARGE annunciator whenever the battery charge current is above 7 amps. Thus, the BATTERY CHARGE annunciator may occasionally illuminate for short intervals when heavy loads switch off. Following a battery-powered engine start, the battery recharge current is very high and causes illumination of the BATTERY CHARGE annunciator, thus providing an automatic self-test of the detector and the battery. As the battery approaches a full charge and the charge current decreases below 7 amps, the annunciator will extinguish. This will normally occur within a few minutes after an engine start, but may require a longer time if the battery has a low state of charge, low charge voltage per cell (20-cell battery), or low battery temperature. This system is designed for continuous monitoring of the battery condition.

Illumination of the BATTERY CHARGE annunciator in flight cautions the pilot that conditions may exist that may eventually damage the battery. The operator should check the battery charge current with the load meter. This is accomplished by turning off one generator and noting the load on the remaining generator. Turn off the battery and note the load meter change. If the change is greater than .025, the battery should be left off the bus and should be inspected after landing. If the annunciator remains on after the battery switch is moved to the OFF position, a malfunction is indicated in either the battery system or charge current detector, in which case the airplane should be landed as soon as practicable. The battery switch should be turned ON for landing in order to avoid electrical transients caused by power fluctuations.

External power. For ground operation, an external power socket, located under the right wing outboard of the nacelle, is provided for connecting an auxiliary power unit. A relay in the external power circuit will close only if the external source polarity is correct. The BATTERY MASTER SWITCH must be on before the external power relay will close and allow external power to enter the airplane electrical system. The battery will also tend to absorb voltage transients when operating avionics equipment and during engine starts. Otherwise, the transients might damage the many solid state components in the airplane.

For starting, an external power source that is capable of supplying up to 1000 amperes (400 amperes maximum continuous) should be used. A caution light on the caution advisory annunciator panel, EXT PWR, is provided to alert the operator when an external dc power plug is connected to the airplane.

Falcon 50

Electrical system power supply. The electrical system is a 28-volt dc system supplied by three 9kW starter-generators and two 23 amp-hour batteries. The generators supply power to two dc buses through line contactors. Generator control switches are installed on the flight station overhead panel. Static voltage regulation and overvoltage protection are provided. In an emergency, the two buses can be connected by means of a switch.

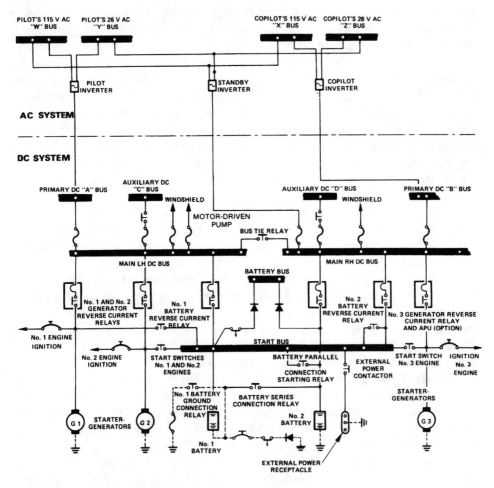

Fig. 2-25. *Falcon 50 electrical schematic.* Falcon Jet Corporation

The two batteries are connected directly to a battery bus. Each battery is connected to one main bus bar through a line connector controlled by battery switches. Engine start is by means of battery power and assisted by an optional APU. The other engine starts are then electrically assisted.

Under normal flight conditions ac power is supplied by two 750-VA inverters. A third standby 705-VA inverter is provided as a backup. A selector switch transfers the standby inverter to power either one of the two normal inverter systems.

In an emergency, load shedding switches disconnect on essential electrical equipment. All electrical systems are protected by circuit breakers on the overhead panel. Space provisions for additional circuit breakers is provided. Emergency batteries provide power to the standby horizon indicator and emergency lighting.

3
Turbine engines

IT IS DIFFICULT TO LINK A SPECIFIC INVENTION OR THEORY WITH THE BIRTH of practical aviation. The relative importance of the Montgolfier's balloons, Chanute's gliders, or the Wright's flying machine can be argued. While each of those machines had its impact on the air transportation industry, few would argue that the growth of the industry can be directly related to the development of the turbine engine.

Transport category aircraft rely solely on turbine engines for their power for many good reasons. The jet engine is far simpler than even the most up-to-date reciprocating engine and contains only about 25 percent of the moving parts. As a result, it can be designed and brought to production in far less time than a reciprocating engine. Operationally, the jet engine is also superior to the reciprocating engine. A turbine powerplant produces smooth, constant power rather than cyclical, intermittent power; the smoothness eliminates vibration, reduces structural attachment requirements, and allows significantly higher engine operating speeds. The implication of the latter being more power with less frontal area because a larger mass of air can be processed per minute due to the turbine's high rotational speed.

Given a jet engine and a reciprocating engine of approximately equal power, the jet engine will weigh about one-fourth, giving it a better power-to-weight ratio. The reduced powerplant weight will also reduce airframe structural requirements resulting in an overall airframe weight reduction. In addition, the jet engine operates much more efficiently at high altitude, further reducing airframe weight by eliminating the power-plant supercharging equipment necessary to allow reciprocating engines to operate in the less dense atmosphere.

TURBINE ENGINE HISTORY

Many people think turbine engine, or jet, theory is a recent idea; actually it dates back 2000 years. While countless individuals have thought about jet engine development since then, it is generally accepted that the father of the jet engine was an Egyptian scientist named Hero.

Around 100 B.C. Hero developed a working model of a machine that used the fundamental principles of a jet engine. Hero's Aeolipile (FIG. 3-1) had a large drum of water that was heated from below by fire. As the water boiled, the steam travelled through two pipes that were connected to, and served as pivot points for, a rotating sphere. The steam would fill the sphere, causing a slight increase in pressure, and would exit small pipes that acted as jets forcing the steam to travel in a direction parallel to the sphere's plane of rotation. The exiting steam would cause the sphere to turn about its axis as a result of the principle of equal and opposite reaction.

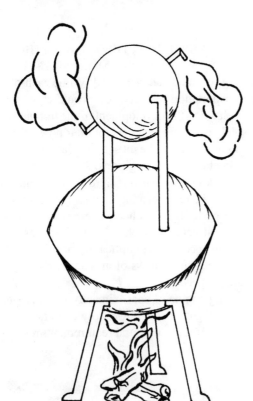

Fig. 3-1. *Aeolipile.*
Susan M. Lbardo

Practical uses for the reaction principle based upon the concept of using a stream of hot air, steam, or other fluids as a source of power would slowly be developed over the centuries. The invention of gunpowder allowed the Mongols to develop rockets and fireworks based upon the reaction principle as early as the thirteenth century. Scholars throughout the ages dabbled with the theory including Leonardo da Vinci, the Italian artist, scientist, and military engineer, in the sixteenth century. Regardless, it wasn't until the late seventeenth century, when the English physical scientist and mathematician Sir Isaac Newton developed the laws of motion, that the reaction principle could be practically applied.

PHYSICS REVIEW

To understand jet propulsion theory it is necessary to first understand basic principles of physics: force, acceleration, work, power, energy, speed, velocity, mass, and momentum. Many of these terms are often misunderstood and improperly used, yet they have very specific meanings. Pilots in general, and specifically turbojet aircraft pilots, should be intimately familiar with these terms and the laws of motion. Safe aircraft operation is predicated on an understanding of the implication of these terms.

Force

Force can best be described as either a push or pull acting on a body. The result will be either to make the body move, or to prevent it from moving. The measurement of force is expressed in units of weight such as pounds, grams, and the like. An example of a force is hitting a pool cue ball with the proper end of a cue stick, which causes the ball to move across the table.

In practicality, force is a vector quantity that means it has both magnitude and direction. In the case of the cue stick, which is propelled by the player's hand and arm, a force of perhaps 6 pounds hits the cue ball, causing it to move. How far it moves, called *work*, and in what direction, will be the combined effect of both the pounds of force imparted by the stick and the directional relationship between the stick's movement and the ball. Expert pool players who understand this relationship can make the cue ball roll perfectly straight, curve right or left, and even jump up off the table with a single hit by the cue stick.

Acceleration

Acceleration is the rate of change of velocity. A good example of acceleration is planetary gravity, a force that pulls bodies toward the earth in a very predictable manner. In fact, the effect of gravity is very predictable but different for every planet. On our planet, the gravitational constant, abbreviated g, attracts bodies toward the earth at a rate that causes the falling body to increase its velocity by 32.17 feet per second per second at sea level.

For instance, let's consider what would happen if we dropped an anvil out of an airplane that was flying at 36,000 feet. Regardless of the weight of the anvil, its falling speed would increase at the rate of approximately 32.2 feet per second per second so that one second after it was dropped it would be travelling at 32.2 feet per second, two seconds after drop it would be 64.4 feet per second, three seconds would be 96.6 feet per second, and so on.

Work

Work, a term used by virtually everybody on a daily basis, is typically used to describe any form of activity in which a person uses muscular or mental effort. From a scientific point of view, the term is much more specific. Work is done when a body moves over a distance as a result of a force acting upon it.

Contrary to popular opinion, when a pilot stands at the gate holding a 50-pound flight bag for 20 minutes, no work is actually being done. The pilot is merely exerting an upward force that acts contrary to the downward force of the weight of the bag. In fact, no work is performed when the pilot carries the bag at a constant velocity down the corridor. It is true that the bag is in motion, but the upward force exerted on it does not produce its horizontal motion. Only when a force acts upon the body and moves it has work been done. In this case, the pilot would actually be doing work when lifting the bag up off the floor or carried up stairs; therefore, only when a force acts upon a body and moves it has the force done work upon the body. The formula for work is as follows:

$$Work = force \times distance$$
$$W = Fd$$

To illustrate work, take the example where a pilot attaches a tow-bar to an airplane to get it out of a hangar. Due to the weight of the aircraft, the pilot is required to continuously pull on the tow-bar with 150 pounds of force to get the airplane to move 20 feet out of the hangar. Apply the equation to determine the amount of work produced:

$$W = Fd$$
$$W = 20 \text{ feet} \times 150 \text{ pounds (lbs) force}$$
$$W = 3,000 \text{ ft-lbs}$$

Remember, if the pilot forgets to release the brake before pulling on the tow-bar, no work will be performed even though energy is being expended.

$$W = 0 \text{ feet} \times 150 \text{ lbs force}$$
$$W = 0 \text{ ft-lbs}$$

Even when we determine how much work is being accomplished, we still do not have any idea how fast the work is being accomplished. It might have taken the pilot one minute to pull the airplane out of the hangar or perhaps one hour. All we know for sure is that it took 3000 foot/pounds of work to get the job done.

Power

The rate at which work is accomplished is called power. Power depends upon three factors: the distance the force moves, the force exerted, and the time required.

Let's say 500 pounds of cargo is supposed to be moved up a 20-foot ramp into the cargo hold of an aircraft. According to the formula for work, the ramp agent would exert the same amount of work loading the crates in 10 minutes as the agent would if the same crates were loaded in one hour; work does not take time into consideration. Anyone contemplating the task, however, realizes the difference between doing the job in 10 minutes rather than an hour; that difference is the amount of power it would take. The formula for power:

$$\text{Power} = \frac{\text{force} \times \text{distance}}{\text{tance}}$$

Using the ramp agent problem, the power required to move the cargo in 10 minutes is 1000 foot-pounds/minute; 60 minutes would only require 166.7 foot pounds per minute. Clearly, this type of task will be physically less demanding if it can be spread out over time.

$$\text{Power} = \frac{500 \text{ lbs} \times 20 \text{ ft}}{} = \frac{10000}{10} = 1000 \text{ ft-lb/min}$$

$$\text{Power} = \frac{500 \text{ lbs} \times 20 \text{ ft}}{} = \frac{10000}{60} = 166.7 \text{ ft-lb/min}$$

In practicality, when dealing with large power sources such as jet engines, the unit of foot pounds per minute is too small a unit to be useful. Instead, two major power measurement systems have been developed: English and metric systems.

English system. The English system equates power to the amount of work a horse can accomplish in a given period of time. The idea was originally conceived by James Watt, the inventor of the steam engine, who conducted experiments using a horse pulling a heavy wagon. Later, the wagon was replaced with a suspended 550-pound weight (FIG. 3-2). The end result is a calculation that tells us a 1-horsepower engine can raise 550 pounds, 1 foot in 1 second. By way of contrast, 1 horsepower is equal to 33,000-foot-pounds per minute, which means a 1 horsepower engine can raise 33,000 pounds, 1 foot in 1 minute (33,000 divided by 60 seconds = 550 foot pounds per second). Imagine trying to work with the power calculations of a 500- horsepower engine in foot pounds per minute; use the following horsepower formula.

$$\text{horsepower} = \frac{Fs}{550t}$$

Where:

horsepower = horsepower

F = force in pounds

s = distance
550 = ft-lb/sec/hp constant
t = time in seconds

The ramp agent problem can be computed to determine the horsepower necessary to complete the task:

$$\text{horsepower} = \frac{(500)(20)}{550 \text{ ¥ } 10} = \frac{10,000}{5,500} = 1.81 \text{ horsepower}$$

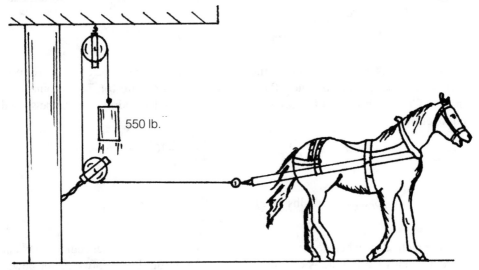

550 lb.

Fig. 3-2. *Horsepower.* Susan M. Lombardo

Metric system. Two units within the metric system are cgs and mks. The cgs units are related to ergs per second and are of no practical value from the pilot's point of view. Most of the metric calculations in aviation are done in meter-kilogram-second (mks) units. Work is measured in *joules*. Time is measured in seconds. The unit of power is *one joule per second* and is called a *watt*. The watt, named after the Scottish engineer James Watt, is still too small to work with under normal situations. As a result, power is normally measured in units of 1000 watts, called a *kilowatt* (kW).

Watt and kilowatt are commonly used terms but are usually associated with electricity. As a result, you might think of electrical power when you encounter them, but be careful because both are mks system units of power. Power can be measured electrically and mechanically. For instance, it is possible to measure the power consumption of an electric light bulb in horsepower or the power output of a jet engine in kilowatts. One horsepower equals 746 watts or 0.746 kilowatts; a good rule of thumb is 1 horsepower = ¾ kilowatts.

Energy

Energy is defined as the capacity for doing work and is divided into two categories: potential and kinetic.

Potential energy. Potential energy is the result of position, or *stored energy*. Position refers to the potential to convert the position of an object into energy. For instance, an airplane at altitude—by reducing the angle of attack of the aircraft and allowing it to descend, a pilot can increase the airspeed without having to change engine power. This is accomplished by converting the potential energy of altitude into airspeed.

Another example of the effect of altitude on potential energy can be seen with a bowling ball. Let's say a bowler is going to throw a 16-pound ball. The act of picking the ball up and raising it approximately 4 feet off the ground requires that the bowler expends 64 foot-pounds of work.

$$PE = WH$$
$$64 \text{ lbs} = 16 \text{ lbs} \yen 4 \text{ ft}$$

Where:
    PE = potential energy
    W = weight in pounds
    E = height in feet

If the ball is dropped by the bowler, it has the 64 foot-pounds of potential energy stored up, which will be released when it smashes down on his foot. A 10-pound sledgehammer is based upon the same principle.

Distortion of an elastic object is another form of potential energy. An example of this is a rubber band. Model airplanes that use a rubber band to power the propeller are converting the potential energy of the wound-up rubber into the rotational motion that causes the propeller to turn, which makes the airplane fly. A spring works in a similar manner. One last example of potential energy is chemical action. By releasing the energy stored in jet fuel in a controlled manner, the energy is converted into thrust, which causes the aircraft to move forward. Other forms of chemical energy are coal, wood, aviation gasoline (avgas), and propane gas.

Kinetic energy. Kinetic energy is the result of motion. A body has kinetic energy because it has velocity. Examples are a bullet being shot out of a rifle muzzle or water falling over a waterwheel. In both cases, the objects have energy that can be put to work. The rifle will impart an impact on its target far in excess of the simple weight of the bullet and the water will cause the waterwheel to turn, which will be used to power a machine.

A jet engine uses kinetic energy when exhausting engine gases strike the turbine wheel and cause the turbine to turn. The turning turbine wheel then powers the engine's compressor, various accessories, pressurization systems, and even a propeller in turboprop engines.

To determine kinetic energy, it is necessary to know two things: object mass and velocity.

$$KE = \frac{WV^2}{2g}$$

Where:

KE = Kinetic energy

W = Weight, lbs

V = Velocity, feet/second (fps)

g = acceleration due to gravity (32.2 fps^2)

To really understand the implication of kinetic energy to the pilot, take the example of a Boeing 727-200 landing in gusty conditions. With a gross weight of 165,000 pounds and a computed landing speed of 159 knots (269 feet per second), calculations show that the aircraft is generating 185,396,970 foot-pounds of kinetic energy on landing—a challenge for any brake system.

$$KE = \frac{(165,000 \text{ lbs}) (269 \text{ fps}^2)}{(2)(32.2)}$$

$$KE = \frac{11,939,565,000}{64.4}$$

$$KE = 185,396,970 \text{ ft-lbs of energy}$$

Speed

The rate of motion; the distance a body travels per unit of time: Speed is measured in units of distance per units of time. Common speed units are: miles per hour (mph), nautical miles per hour (knots), feet per second, kilometers per hour, and the like. In general, single-engine aircraft speed is measured in miles per hour, but advanced single-engine aircraft, all multiengine, turboprop, and turbojet aircraft speed is measured in knots. In addition, all FAA flight plans require airspeed computations in knots.

Velocity

Velocity is the rate of movement in a specified direction. Speed and velocity are often incorrectly considered interchangeable terms; there is a definite difference. The speed of a body tells you how fast it is moving and expresses the distance the object will cover in a specified time. Speed does not indicate in what direction the body is moving.

Uniform motion. A body is considered to be in uniform motion when the velocity is constant. That means the body moves the same distance in the same direction in each succeeding unit of time. An airplane that holds a constant heading, airspeed, and altitude would be an example of uniform motion.

Variable motion. If either the airspeed or direction vary, the motion is considered to be variable.

Mass

Mass is often incorrectly interchanged with weight. The mass of a body is the amount of matter that the mass contains; mass is a measurement of a body's inertia. Weight, on the other hand, is simply a measure of the effect of gravity on some amount of matter. The mass of an object remains constant anywhere in the universe, but the object's weight varies with the gravitational pull of the body nearby. For instance, on the moon an astronaut's video camera, which weighs 12 pounds on earth, would only weigh 2 pounds. The moon's gravity is ⅙ that of earth, but the mass of the camera remains unchanged.

Mass is computed by dividing the weight of a body by the gravitational acceleration. Recall that on earth the gravitational acceleration is 32.2 ft/sec². So, the formula would look like this:

$$M = \frac{W}{g}$$

Where:

    M = Mass
    W = Weight
    g = gravitational acceleration

Momentum

Momentum is mass multiplied by velocity (MV). The property of a body in motion determines how long is required to bring the body to rest under the action of a constant force. The formula is simple, but the implication is very important. No one doubts that a car traveling at 65 miles per hour will have a tremendous impact with anything it hits, but take for example a .45-caliber bullet that weighs only a few grams. When fired from a pistol, the mass of the bullet multiplied by the velocity yields several hundred foot-pounds of energy. Despite the small size of the bullet, more than enough momentum is present to knock the victim off their feet. This is an example of a small mass traveling at a high velocity.

On the other hand, a large mass traveling at a very slow speed can have a similarly devastating effect. For instance, when taxiing a Boeing 747, which weighs approximately 775,000 pounds, up to a jetway, the pilot must use extreme caution. Despite taxi speeds of only a few mph, if the pilot overshoots the mark, the momentum can cause the aircraft to destroy the jetway in slow motion.

NEWTON'S LAWS OF MOTION

First law of motion

A mass (an object) at rest will remain at rest and a mass in motion will remain in motion, in a straight line, unless it is acted upon by an unbalanced force. Countless examples in everyday life demonstrate that "a mass at rest will remain at rest." Buildings, for instance, do not suddenly move off of their foundations for no reason. When they do move, it is normally because of an outside force such as a tornado or earthquake. What is not so readily apparent in everyday life is that "a mass will remain in motion, in a straight line, unless it is acted upon by an unbalanced force." The most obvious example of a body remaining in motion is the Voyager spacecraft that—providing it does not collide with, or become influenced by, another body in space—will continue to travel through the universe unrestricted. On earth, the nemesis of perpetual motion is *friction*.

For centuries, perpetual motion has been the dream of physicists who have worked diligently to reduce friction to an absolute minimum wherever desirable. Engineers have designed airplanes and automobiles to approximate the soda-bottle shape (small in front tapering to large in the rear) that helps reduce the effect of air friction, which tries to inhibit movement. Another form of surface friction is when two surfaces rub against each other. Various types of bearings, such as ball bearings, have been created to fit between two surfaces that move in relation to each other, such as occurs with a wheel mounted on a car. Unfortunately, the ball bearings themselves move in relation to the surface creating their own friction, which attempts to slow down the relative motion and wear out the surfaces. To reduce the effects of friction, engineers have created stronger metals, smoother surfaces, and better lubricants, but it is unlikely that friction will ever disappear from an environment that includes gravity.

Second law of motion

If a body is acted on by an unbalanced force, the body will accelerate in the direction of the force and the acceleration will be directly proportional to the force and inversely proportional to the mass of the body. Simply put, Newton's second law of motion explains why a football flies through the air when kicked by a punter but a building does not budge when the punter kicks it with the same force.

The formula for Newton's second law of motion is:

$$a = \frac{F}{M}$$

Where:
a = acceleration
F = force
M = mass

By multiplying each side of the equation by M, the more common version of the formula, F = Ma, is derived. In an aircraft, F is the result of the expansion of burning gases in the engine; M is the mass of air that either is moved aft by the propeller or passes through the jet engine; the a represents how much the velocity of that mass of air changes as a result of being processed by the powerplant per unit of time.

The second law of motion dictates how much thrust must be produced by an aircraft engine to move the aircraft through the air. As the weight of an aircraft increases, for instance through the addition of fuel, bags and passengers, the required thrust to fly at a given airspeed must also increase. It is also necessary for the pilot to increase thrust in order to increase airspeed at any given weight.

Third law of motion

For every acting force, there is an equal and opposite reacting force. For example, when a sportsman fires a rifle at a target, the gunpowder in the bullet burns, causing the gases to expand within the rifle. Those gases cause a force acting rearward against the back of the rifle, which is perceived by the shooter as recoil against the shoulder. As a result, the bullet moves forward and the rifle moves backward with the same force.

The formula for Newton's third law of motion is:

$$M_1V_1 = M_2V_2$$

A balloon works in the same manner (FIG. 3-3). As you blow air into the balloon, the air stretches the rubber causing the balloon to get larger. At the same time, the pressure of the air inside the balloon becomes greater than the outside pressure because the inside air pushes with equal force against the entire inside surface of the balloon. If you let go of the stem and allow the air to escape, the air will exit with the same force as it has been pushing against the inside walls of the balloon. As a result, the balloon will move forward with the same force as that of the air flowing out of the stem. It is important, though confusing, to understand that it is not the escaping air pushing against outside air that makes the balloon move forward. It is the reaction of the escaping air pushing against the balloon that causes movement. The proof of this concept is demonstrated when a spacecraft moves through the vacuum of space as a result of the thrust produced by its rockets.

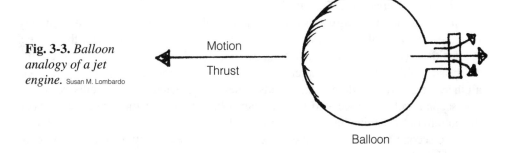

Fig. 3-3. *Balloon analogy of a jet engine.* Susan M. Lombardo

Motion

Thrust

Balloon

In the same manner that Newton's third law of motion causes the bullet, balloon, and spacecraft to move forward, so too does it allow an aircraft engine to move an airplane forward.

FACTORS AFFECTING THRUST

Because no machine achieves perfect efficiency, the amount of thrust a machine should be able to produce (gross thrust) is different from the amount of thrust actually produced (net thrust). This is true for aircraft powered by turbojet engines as well as a turbopropeller combination.

Numerous factors affect the thrust of a jet engine:

- Temperature of air entering compressor
- Pressure of air entering compressor
- Humidity of air entering compressor
- Aircraft airspeed (ram pressure or V_1)
- Designed engine operating speed (rpm)
- Compressor air bleed
- Temperature of air entering turbine inlet

A review of the formula for engine thrust gives insight into how it might be possible to increase the thrust output of a given engine. The formula for engine thrust is:

$$T_A = Q(V_2 - V_1)$$

Where:

T_A = thrust available

Q = mass airflow (density ¥ area ¥ velocity)

V_1 = inlet air velocity (fps)

V_2 = exit air velocity (fps)

It is obvious by the equation that the thrust output of an engine can be increased two ways: increase the relative velocity $(V_2 - V_1)$ and increase mass airflow (Q). When relative velocity is increased, the gases have a higher kinetic energy. As exhaust gas kinetic energy increases, wasted energy increases, which produces an opposite unintended effect: decreased engine efficiency.

It is possible to increase mass airflow in one of three ways. The first is to simply increase the size of the air intake to allow a greater amount of air to enter the engine. While this is a relatively simple solution, two problems are associated with this option. As the size of the engine inlet increases, so too does the parasite drag, causing an overall increase in airframe drag, which tends to negate the benefit of the increased thrust. The second problem associated with an increased inlet size is the clearance between the bottom of the engine intake and the ground.

The second method to increase Q is to increase the velocity of the incoming air mass. Unfortunately, for the same reason as stated when discussing the option of increasing relative velocity, this effectively causes a decrease in engine efficiency due to exhaust inefficiency.

The third method to increase Q is to increase the density of the air entering the engine. The fanjet engine employs this method by compressing intake air before it enters the compressor.

Therefore, the characteristics of the air entering a jet engine have a significant impact on that engine's operating efficiency. A given jet engine is designed to operate at maximum efficiency at a specific air density (temperature and pressure) and humidity. Deviations from the designed operating conditions have a major impact on engine operating efficiency, though humidity does not have as profound an effect on turbine engines as on reciprocating engines. As temperature decreases, engine operating efficiency increases. Conversely, as pressure decreases, engine operating efficiency decreases with it.

When an airplane climbs, the outside air temperature and pressure decrease, but the pressure decreases faster than the temperature. As a result, thrust decreases with altitude until approximately 36,000 feet (the beginning of the tropopause). It is at that point that the temperature stops decreasing and becomes constant; however, the pressure continues to decrease as altitude increases but with no offsetting temperature decrease, so the engine begins to dramatically lose thrust. As a result, 36,000 feet is typically the optimum altitude for turbojet engines.

Sea level—with the potential of maximum thrust output—is not desirable because the air density at sea level is very high and that literally tries to hold back the airplane. The net result is that an excessive amount of thrust is required to push the airframe through the thick density of the air.

Engine thrust is rated at sea level under standard conditions. If operated in conditions that result in an air density less than standard or a humidity higher than standard, the result can be significantly less thrust produced. On the other hand, in operations in conditions of higher than standard density or lower than normal humidity, the result would be more thrust produced.

Designed engine operating speed. Another way to increase thrust available is to increase engine revolutions per minute. The compressor of a turbine engine turns (*spools*) at a very high number of revolutions per minute compared to a reciprocating engine. The limiting factor of the rotation speed is preventing the rotor blades from travelling at the speed of sound. Think of yourself turning around 10 times per minute holding a string with one ball tied at 5 feet and a second ball tied at 10 feet. Though both balls are travelling at the same revolutions per minute the second ball must move faster through the air because it is farther away from the center of the turning arm; therefore, the smaller the engine, the higher the rotational speed possible with some turbojets operating as high as 50,000 revolutions per minute. With larger engines, the greater length of the rotor blade causes the blade tip to reach supersonic speed at a much slower engine rotation speed.

Turbine engine thrust performance differs from reciprocating engines in yet another way in that a change in thrust is not linear with revolutions per minute. For instance, reducing engine revolutions per minute from 100 percent to 99 percent can result in as much as a 5 percent loss of thrust in a turbine engine. Again, unlike reciprocating engine operations where in-flight, operational engine speed varies from idle to maximum, turbine engine speed seldom falls below 85 percent during any in-flight operation including approach to landing. This requires the turbine engine pilot to significantly plan ahead to prevent the need for large rpm changes. Turbine engine speed, which takes a significantly longer time to spool up from idle than a reciprocating engine, should not normally be slowed to idle speed while in flight.

Compressor air bleed. Another factor that affects engine thrust produced is the amount of air that is bled off of the engine compressor. The primary use of compressed air is mixing with fuel for combustion, another use is pressurizing the aircraft. Compressed air that is tapped off of the compressor prior to reaching the combustion chamber is known as *bleed air*. As the amount of bleed air increases, engine thrust decreases. As a result, takeoff restrictions are typically imposed on the pilot regarding the amount of bleed air allowed during takeoff. The reason is to assure maximum engine thrust on all operating engines in the event of an engine failure during takeoff and initial climb. Bleed air restrictions usually reduce the ability to air-condition or pressurize the cabin until a minimum safe altitude and airspeed are achieved.

Temperature of air entering turbine inlet. This is a practical limitation based upon the maximum temperature that the turbine blades can withstand before the heat begins to affect them. Pilots are critically aware of turbine temperature, especially during engine start. Exceeding the published maximum temperature limit requires that the engine be checked by a mechanic immediately and will often result in permanent damage to the engine. On rare occasion, upon initial engine start, the turbine temperature will rapidly rise and run away if the pilot does not act rapidly and correctly to prevent the runaway rise. This occurrence, known as a *hot start*, can cause catastrophic engine failure.

AIRCRAFT JET ENGINES

Turbine-powered engines can be classified various ways. Four engine classifications for aircraft are based upon the power produced:

- Turbojet
- Turboprop
- Turbofan
- Turboshaft

An airframe manufacturer chooses the type of powerplant based upon a number of reasons, not the least of which is the desired cruising speed of the aircraft. Fixed-wing aircraft with a cruising speed fewer than approximately 250 mph will typically use a reciprocating engine and propeller combination for reduced initial and direct operating cost benefits. As the design speed of the aircraft increases, so too does the cost effectiveness of using a turbine engine.

In the speed range of approximately 250–450 mph, the engine of choice is the turbine engine and propeller combination referred to as *turboprop*. For instance, small turboprop engines weigh approximately 400 pounds and produce about 1000 shaft horsepower for a power-to-weight ratio of about 2.5:1 compared to reciprocating engine ratios of less than 1:1. The major limiting factor of the turboprop is that it quickly loses power as drag increases, a significant problem at the lower, denser altitudes in which most turboprop aircraft fly.

As the design speed of the aircraft approaches 450 miles per hour the *turbojet* engine becomes the logical choice. Actually, with the advanced state of turbine engine technology, the turbine engine and fan combination, known as a *turbofan*, is the better choice and has all but replaced the older turbojet. The primary advantage of the turbofan is that it yields significantly more power at the higher subsonic airspeeds. A *turboshaft engine* is primarily found on helicopters with a different configuration, compared to other turbine engines, to transfer energy from the turbines to the rotors. Each engine classification is subsequently detailed in this chapter.

Direct comparisons between reciprocating engine/propeller combinations and turbine engines are difficult because power produced by a reciprocating engine is measured in horsepower and power produced by a turbine engine is measured in *pounds of thrust*; however, it is possible to calculate the pounds of thrust produced by the propeller, which can then be compared to the thrust produced by a turbine engine.

Another area in which turbine and reciprocating engines are often compared is fuel consumption. Pilots frequently cite the high fuel consumption associated with turbine engines as a serious drawback; however, a closer analysis of the situation will show that turbine engine aircraft typically have significantly greater payload carrying capability. It is usually the case that turbine engine aircraft consume fewer pounds of fuel per pound of useful load than reciprocating engine counterparts.

Turbojet engines

The turbojet engine was very simple in its original form as designed by Sir Frank Whittle. The basic purpose of the turbojet engine is to produce power by converting heat energy resulting from fuel combustion to rotating mechanical energy. Over the years, the original turbojet design has been altered to fulfill many different requirements including turbopropeller, turbofan, and turboshaft. A turbojet engine, which consists of very few moving parts, is still essentially simple. Free-flowing ram air enters the engine air inlet where it is efficiently channeled into the compressor. The compressor densely packs air molecules together and ships them to the combustion chamber.

Air under pressure flows from the compressor into the combustion chamber where fuel is added and ignition occurs. Unlike the reciprocating engine, the flame in the combustion chamber is self-propagating, which means once you light it the flame will continue as long as the fuel/air mixture is uninterrupted. As a result, jet engines do not use intermittently firing spark plugs. Instead, igniters fire only upon initial start-up and are deactivated when the engine is operating. The burning fuel/air mixture expands and is expelled out the rear of the combustion chamber and passes through the turbine.

The turbine is a fan-like structure through which the hot, expanding gases flow. The energy of the gas causes the turbine wheel to turn much like wind causes a pinwheel to turn. The turbine is fixed to a shaft that goes straight through the engine forward to the compressor. As the gases cause the turbine to turn, so too does the compressor rotate, which processes more air and continues the cycle. After the expanding gases pass through the turbine, they are exhausted through the engine tail pipe and jet nozzle which is shaped, also called *tuned*, to provide maximum thrust from a given amount of exhaust. The force of the expanding gases coming out of the exhaust pushes against the engine causing an equal and opposite reaction as described by Newton. This reaction pushes the engine and everything that the engine is attached to, such as an airframe, forward.

Turboprop engines

The turboprop engine is a derivation of the turbojet, designed to turn a propeller. The turbojet engine derives 100 percent of its thrust from its exhaust gases; the turboprop derives its thrust from a combination of propeller thrust and exhaust thrust. Depending upon the manufacturer, a turboprop engine will provide 5–25 percent of its thrust through exhaust and more than 75 percent of its thrust from the propeller.

Turbofan engines

The turbofan is a modification of the turbojet engine. It was designed to meet the ever increasing demand for more power and stricter noise standards by moving a significantly larger volume of dense, cold air. Equipped with larger air intakes, the turbofan processes several times more air than the normal turbojet engine providing partially compressed air to the compressor inlet. The result is more gas is expelled at slower velocity and reduced temperature, which results in higher thrust with less noise.

Figure 3-4 illustrates a turbofan consisting of three rows of compressor blades located ahead of the actual compressor. These fan blades extend several inches longer than the first set of compressor blades and direct air backward partly to the compressor and partly through ducts that bypass the compressor and flow between the nacelle and the engine. Also called a *forward-fan* engine, in reality, it is nothing more than a sophisticated propeller shrouded in the nacelle. The same principle applies to aft-mounted fans, in either case, the excess, dense, cold air that is moved significantly increases the overall engine thrust. The turbofan engine, which is an average of approximately 30 percent more fuel efficient, is rapidly replacing the turbojet in most air carrier and corporate aircraft applications.

Turboshaft engines

A turboshaft engine is closely related to the turboprop. In fact, it is defined as a turbine engine that powers something via a shaft other than a propeller. Turboshaft engines are commonly used in industry to operate electrical generating plants; the most common use in aviation is in helicopters. The two primary sections are the gas generator and the power turbine. The gas generator produces the necessary energy to keep the engine operating. Approximately ⅔ of the power produced goes to operating the compressor. The remaining ⅓ operates the turbine section, which powers a high-ratio reduction gearbox that causes the helicopter's rotor to turn.

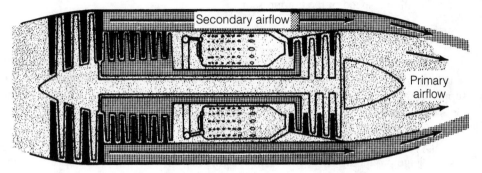

Fig. 3-4. *Forward turbofan engine.*

TURBINE ENGINE COMPONENTS

The development of power in a turbine or reciprocating engine requires intake, compression, combustion, and exhaust. The manner in which those events are accomplished in each engine is the primary reason behind the smooth operation of a turbojet engine as compared to its reciprocating counterpart. In the reciprocating engine, each event is a separate stroke in the cylinder meaning one power stroke for every four-stroke cycle. In the turbine engine all events occur simultaneously in separate sections of the engine. The difference between the reciprocating piston strokes and the smoothly rotating turbine engine is dramatic; a turbine is practically vibration free.

Turbine engines are made up of the following sections:

- Air inlet
- Compressor
- Combustion section
- Turbine
- Exhaust
- Accessory section
- Support systems (starter, lubrication, fuel, and auxiliary components, such as anti-icing, air-conditioning, and pressurization)

Air inlet duct

The purpose of the air inlet is to channel incoming air to the compressor with the least possible loss of energy. To accomplish that task, it is necessary for the engineer to design the intake in such a manner that it reduces airflow turbulence to a minimum. The result will be the maximum possible ratio of compressor discharge pressure to air inlet pressure. In addition to the design of the inlet, three other variables determine how much air will pass through the compressor:

- Ambient air density
- Airspeed of the aircraft
- Rotational speed of the compressor

Compressors

Two main functions of the compressor are to supply compressed air to the burners plus supply bleed air to fulfill various engine and airframe system requirements. To supply compressed air to the burners, the compressor increases the pressure of the mass of air that is channeled through the air inlet duct and routes it to the burners in the amount and pressure required.

Bleed air might be tapped off any of the various compressor stages with each successive stage producing air of increasing pressure and temperature. The specific stage for bleed air extraction is determined by the temperature and pressure required for the job. The bleed air ports are small openings in the case surrounding the compressor and located directly over the appropriate compressor stage. It is worth noting that as air is compressed, heat is a by-product. A turbine engine at sea level is operating with standard air temperature and pressure: 59°F and 14.9 pounds per square inch. The same air will have risen to more than 600°F and 167 pounds per square inch as it leaves the last compressor stage, which is before that parcel of air gets to the combustion section. Bleed air used to pressurize the cabin must first be cooled by an on-board refrigeration system. The primary purposes of bleed air are:

- Cabin pressurization
- Cabin heating
- Auxiliary drive units
- Deicing and anti-icing
- Control-booster servos
- Pneumatic engine starting
- Compressed air to turn instrument gyros

Compressors. Two major categories of compressors in turbojet engines are *centrifugal-flow* and *axial-flow*. Turbine engines are very often classified by the type of compressor so it is worth remembering that an axial-flow engine really refers to the manner in which the air moves through the compressor (FIG. 3-5).

The centrifugal-flow engine is named for its manner of air processing (FIG. 3-6). The engine channels incoming air into the compressor where the flow of air is turned 90° and accelerated outward by centrifugal action. The axial-flow engine, on the other hand, compresses air through a linear series of rotating and fixed blades. The important thing about the axial-flow compressor, which is also the reason for its name, is

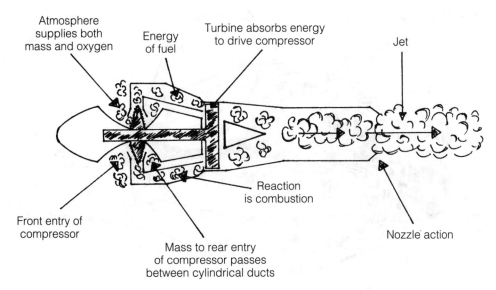

Fig. 3-5A. *Turbojet operating principles (Centrifugal Flow).*   Susan M. Lombardo

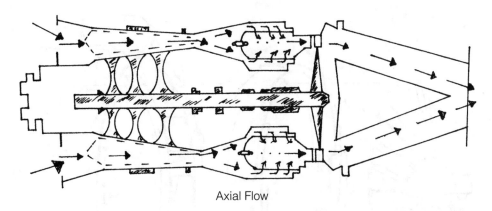

Axial Flow

Fig. 3-5B. *Turbojet operating principles (Axial Flow).*   Susan M. Lombardo

both airflow and compression are accomplished parallel to the plane of rotation, thereby avoiding energy loss as a result of changing direction.

Centrifugal-flow compressors. The primary components of the centrifugal-flow compressor are the *impeller* (rotor), a *diffuser* (stator), and a *compressor manifold* (FIG. 3-7). The impeller, which appears to be a larger version of the reciprocating engine supercharger impeller, is designed to catch the incoming air and accelerate it outward toward the diffuser. Depending upon the thrust requirements of the engine, impellers might be either single- or double-entry with the latter being two impellers back-to-back (FIG. 3-8).

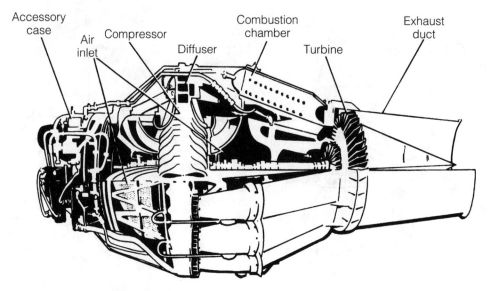

Fig. 3-6. *Centrifugal-flow engine.*

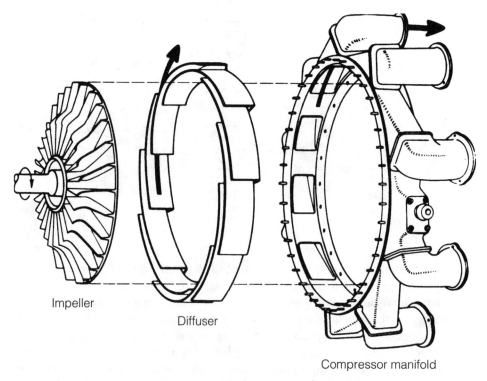

Fig. 3-7. *Centrifugal-flow compressor components.*

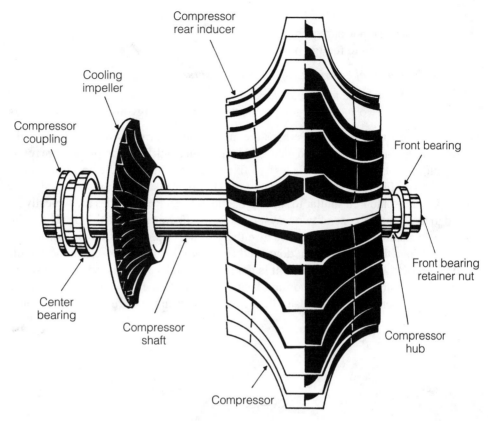

Fig. 3-8. *Double entry impeller.*

The diffuser is an annular chamber with a number of vanes that form a series of passages into the manifold. The purpose of the diffuser vanes is to direct the air from the impeller to the manifold at the gentlest angle possible to retain the maximum amount of energy. Another purpose of the diffuser is to provide the combustion chamber with air at the correct velocity and pressure for maximum efficiency. It does so by changing high-velocity compressor discharge airflow to a slower, but higher pressure airflow that will not blow out the flame in the combustion chambers. The manifold has one outlet port for every combustion chamber and assures that the airflow is evenly divided among all chambers. A compressor outlet elbow over each port changes the airflow from an outward to an axial direction, then channels it into the individual combustion chambers.

Advantages of centrifugal-flow compressors.
- High pressure rise per stage
- Good efficiency over a wide rotational speed range
- Simplicity of manufacture and relatively low cost

- Low weight
- Low starting power requirements
- More resistant to foreign object damage

Disadvantages of centrifugal-flow compressors.

- Large frontal area for a given airflow
- More than two stages are not practical because of energy losses between stages

Axial-flow compressors. Perhaps the most popular form of turbine engine is the axial-flow engine (FIG. 3-9). Two main elements of an axial-flow compressor are rotors and stators (FIG. 3-10). Rotor blades are fixed on a rotating spindle and force air to move rearward in the same manner as a fan. The rotor blades are preset at a specific angle and are contoured similar to small propeller blades. Rotor blades are typically affixed to the spindle in one of two ways: *bulb root* or *fir-tree root* (FIG. 3-11). The stainless steel vanes are then held in place by either screws, or peening, or locking wires, or the like. When the engine is shut down and cold, rotor blades fit loosely in the spindle and can easily be wiggled by hand. It is not uncommon to walk past an aircraft with a stiff wind turning the compressor and hear the sound of the rotor blades "clinking." The root's looseness accommodates heat expansion when the engine is in operation.

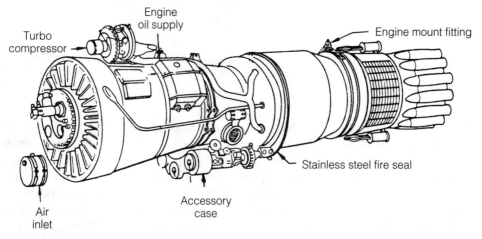

Fig. 3-9. *Axial-flow engine.*

The rotor, which revolves at a very high rate of speed during engine operation, catches the inlet air and moves it through a succession of rotor/stator stages. Each stage compresses the air and accelerates it to the next stage. As the velocity of the air increases, energy is transferred from the compressor to the air in the form of kinetic energy. The alternate set of blades next to the rotors is the stator blades. The stators act as a diffuser in each stage converting part of the air energy into pressure by slightly slowing the air. The

Front compressor
stator casing

Rear compressor
stator casing

Compressor
rotor

Compressor
rear frame

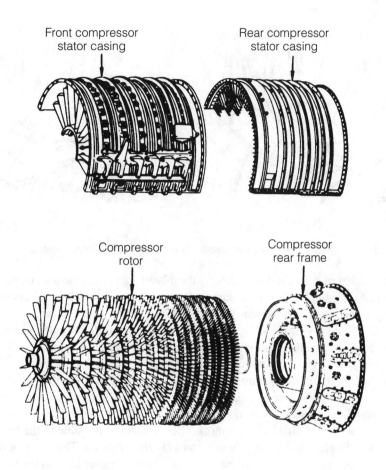

Rear compressor
stator casing

Front compressor
stator
casing

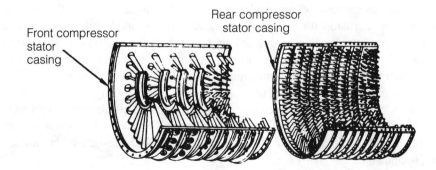

Fig. 3-10. *Rotor and stator components of an axial-flow compressor.*

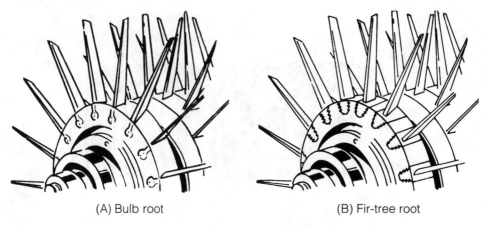

(A) Bulb root   (B) Fir-tree root

Fig. 3-11. *Common retention methods used on compressor rotor blades.*

total number of stages is determined by the mission of the engine with respect to required thrust. As the number of stages increases, so too does the overall compression ratio of the compressor. Axial-flow turbine engines generally have 10 to 16 stages.

The first set of stators, located before the first compressor stage, is referred to as *inlet guide vanes* (IGV). In some very sophisticated aircraft, such as military fighters, the inlet guide vanes might have computer-adjustable (variable) angles of attack to assure maximum efficiency throughout all phases of flight. A similar set of guide vanes, called *straightening vanes*, might be located at the compressor exit to stop the exiting air mass from rotating as it moves forward into the combustion chamber. The inlet guide vanes direct airflow in the most efficient manner possible to the first stage rotor blades by imparting the proper angle and a swirling motion. The swirling motion turns the air in the same direction as the engine rotation to reduce drag as the air contacts the first-stage rotor blades.

The stators are rows of nonmoving vanes attached to the inside of the compressor case. Stators alternate with, and fit snugly next to, the rotors. Similar to the rotors, the stators are preset at specific angles and are contoured. The stators catch incoming air and move it rearward. Because of their preset angle, stators also direct the flow of air in a manner that assures maximum compressor efficiency.

As the pressure ratio demand increases, it is possible to string two axial-flow compressors in succession, each driven by a separate turbine. This *dual spool* system is depicted in FIG. 3-12. Occasionally an engine will even feature a *triple spool* system. To differentiate spools, the first one in line after the air intake is designated N_1, the second one N_2, and so on.

Advantages of axial-flow compressors.

- High peak efficiencies
- Small frontal area for given airflow
- Straight-through flow, allowing high ram efficiency
- Increased pressure rise by increasing the number of stages

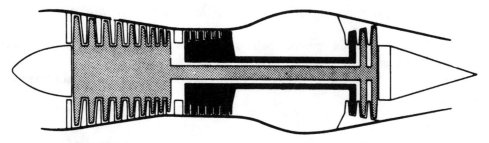

Fig. 3-12. *Dual spool system.*

Disadvantages of axial-flow compressors.

- Good efficiency only over a narrow rotational speed range
- Difficulty of manufacture and high cost
- Relatively high weight
- High starting power requirements
- Very susceptible to foreign object damage

Foreign object compressor damage. Concerns about foreign object damage are very valid with turbine engines. Centrifugal-flow compressors are significantly more solid and able to resist damage than the relatively small, separate blades of the axial-flow compressor. If, for instance, a bird enters the engine intake and gets to the compressor, a tremendous amount of damage can result as blades rotating as fast as 50,000 revolutions per minute try to chew up the object. Even large transport category turbofan engines turning at only a few thousand revolutions are potentially vulnerable to foreign object damage.

For example, a transport category, air carrier aircraft experienced an engine shutdown during takeoff roll. The pilot heard a loud bang accompanied by vibration, he pulled the engine to idle, shut down the engine, dumped fuel, and returned to the airport. The aircraft returned safely and the passengers deplaned normally, without injuries. Foreign object damage to the number 1 and 2 engines caused removal of the engines for further investigation and repair. The small, white arrow to the lower right in FIG. 3-13 indicates the foreign object's point of initial contact with a fan blade. The object broke the blade, causing the tip to fly freely throughout the case as if it were in a blender. This resulted in reactive subsequent damage to the entire fan.

The result of the incursion was as follows:

- Massive damage to the #1 engine fan section.
- Two fan blades in engine #1 were broken off inboard of the mid-span shrouds.
- All remaining blades exhibited heavy, hard-body impact damage.
- The number 1 engine core showed heavy evidence of metal splatter on the vanes.
- All of the blades in the compressor section displayed heavy impact damage.
- Number 1 engine's 16th stage bleed valve duct was torn open.

- Two of the combustor retaining pins had broken the safety wire and separated from the engine.
- Five of the accessory gearbox bolts were broken and the gearbox was loose but still attached.
- The first-stage guide vanes and high-pressure turbine module showed heavy evidence of metal splatter.
- Several high-pressure turbine stage-1 and -2 blades and stage-2 vanes were nicked.
- The low pressure turbine section revealed minimal stage-1 blade and vane damage with minor blade tip rub due to the initial unbalance cause by the fan blade separation.
- A number of engine fragments were retrieved from the runway.
- A hole was in the fan cowling at the 12:30 position.
- Minor fan blade fragment damage to the left side of the fuselage, and damage to seven passenger windows.
- Foreign object damage was also found in the number 2 engine as it ingested debris from the number 1 engine.

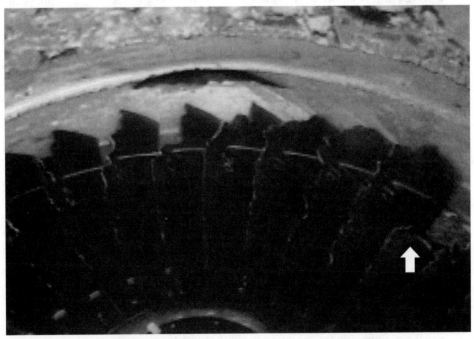

Fig. 3-13. *Fan foreign object damage.*

Investigators determined that the reason for damage to the engine was a bird strike. The sole trace of the origin of the incursion was a single bird feather that remained in the engine. The feather was sent to the Smithsonian Institution for verification of the species—the number 1 engine was struck by a Black American Crow, which has an average weight of 17 ounces.

Compressor stalls. It is important to remember that compressor blades are actually small airfoils that are functionally the same as a propeller or wing. The rotor blades are set at a given angle on the spindle and at a given angle on the stators in the engine case. The angle at which the incoming air meets those blades varies as a result of the air's velocity and the compressor's rotational speed. The blades' combined influence forms a vector that is called the *angle of attack*. An imbalance between those two vector quantities might stall the airflow.

By definition, a compressor stall occurs when the air mass travelling through the compressor slows down and stops. In extreme cases, the airflow might even reverse direction and begin to flow forward. On the flight deck, an incipient compressor stall might not be detectable—the first sign being a low pulsating sound. As the stall progresses, the sound intensity increases and if left uncorrected long enough will result in the sound of an explosion or loud backfire. Mild or transient compressor stalls do not necessarily do permanent damage to the engine. Severe stalls, also known as *hung stalls*, will dramatically degrade engine output at best and might result in permanent and even catastrophic engine failure. Compressor blades stall for numerous reasons:

- Anything that alters the proper operation of the compressor blades, such as blade failure, foreign object damage, and the like.
- A fuel mixture that is too lean.
- Abrupt aircraft movement, such as severe pitch up or down, resulting in a disruption to the normal engine intake airflow.
- Excess fuel flow.
- Engine operating speeds excessively above or below normal recommended operating speed.
- Damaged turbine blades that negatively impact the compressor's designed rotational speed.

The flight crew of one corporate jet experienced rapid cabin depressurization at altitude. In an effort to descend as rapidly as possible, the pilot pushed hard forward on the yoke. The result was to shear the airflow into the engines' intake and cut off normal intake air. Both engines immediately experienced compressor stall, which was verified by the crew as a deep, audible pulsing sound, fluctuating engine rotational speed as indicated by the N_1 and N_2 gauges, and an increase in exhaust gas temperature. Recognizing the problem, the pilot reduced the pitch angle and both engines recovered.

Combustion chambers

The combustion section raises the temperature of the compressed air that passes through and releases the potential energy contained in the fuel/air mixture. Most of the energy released turns the turbine wheel, which is connected by a long shaft to the compressor. (The turbine wheel is located downstream from the combustion chambers.) The energy that remains after turning the turbine becomes a high-velocity stream of hot gases that exit the exhaust of the engine providing forward thrust. You will recall that the temperature of the air as it leaves the compressor is approximately 600°F. After being processed by the combustion chamber, the air temperature is approximately 1600°F.

To accomplish its task of efficiently burning the fuel/air mixture, the combustion section must:

1. Mix the fuel and air in the proper manner for the ambient conditions to assure the best possible combustion.
2. Cool the hot gases to a temperature within the normal operating range of the turbine wheel.
3. Channel the exhaust gases in such a manner that it maximizes turbine wheel effectiveness.

Three combustion chambers are *can*, *annular*, and *can-annular*. All three have the same basic parts:

- Outer casing
- Perforated inner liner
- Fuel injection system
- Ignition system
- Fuel drainage system

Can combustion chambers. The can combustion chamber is used in axial-flow engines, but is particularly well suited for centrifugal-flow because the air is already divided up equally by the diffuser vanes. The number of burner cans varies among engine types but most engines use from 8 to 10. Each burner can is made up of an outer casing with an inner liner made of perforated stainless steel (FIG. 3-14). The individual cans are interconnected with a flame propagation tube that spreads the flame from one can to another during initial start-up because only two of the cans will have igniters.

A critical safety element in the design of the combustion chamber is a method for draining unburned fuel. When the engine is shut down, a portion of fuel invariably finds its way into the combustion chamber area after the flame has extinguished. If unburned fuel is not drained, it will eventually evaporate and leave gum deposits in the fuel manifold, fuel nozzles, and combustion chamber. Another potential problem is when a pool of unburned fuel collects in the bottom of a combustion chamber after shutdown. Such a situation might lead to an afterfire immediately following shutdown, or if a restart is attempted prior to the time the fuel evaporates, the possibility exists for a hot start.

Can burner liners have numerous perforations of different sizes and shapes. Each hole is designed to have a specific effect on the flame propagation within the liner. Air entering the combustion chamber flows through the various holes and divides into two separate airflows. The primary, or combustion, airflow enters the liner and mixes with the fuel, then is burned. The cooling or secondary airflow travels between the outer casing and the inner liner. It helps cool the inner liner and ultimately mixes with the combustion exhaust where it cools the exhaust gas from approximately 3,500°F to 1,500°F. This keeps the temperature low enough to prevent damaging the turbine wheel.

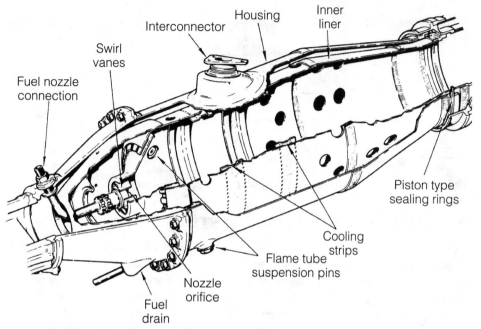

Fig. 3-14. *Can-type combustion chamber.*

Precision-drilled holes are also at the inlet end of the can where the fuel nozzle is located. The holes help atomize the fuel as it enters the combustion chamber; the finer the spray, the better the burn. In addition, louvers along the length of the inner liner direct some of the incoming air along the inside wall of the liner. This flow of air helps control the flame pattern, keeping it away from the liner to prevent liner overheating.

Annular combustion chambers. An annular combustion chamber consists of an undivided circular shroud extending around the outside of the turbine shaft (FIG. 3-15). The chamber might be constructed of one or more baskets. If two or more are used, they are located one outside the other in the same radial plane, which provides the name *double-annular* chamber. As with the can chamber system, two igniter plugs are usually mounted on the chamber housing.

Can-annular combustion chambers. This system was originally developed for the Pratt & Whitney JT3 axial-flow engine. The engine, which featured a dual-spool compressor, required a very heavy-duty and efficient combustion chamber. The can-annular combustion chamber is not as common as the other two types.

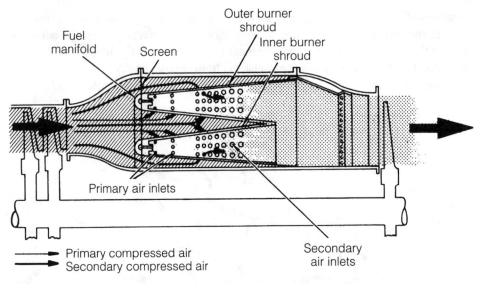

Fig. 3-15. *Components and airflow of a double-annular chamber.*

TURBINE ENGINE IGNITION SYSTEMS

Turbine engine ignition systems are significantly different than reciprocating engine counterparts. The system is designed to initially ignite the fuel/air mixture, which will then sustain combustion on its own. When ignition occurs, the system is deactivated. This very limited use results in reduced system maintenance and a longer life expectancy than a reciprocating engine system.

Another major difference relates to the ignition mechanism. The reciprocating engine's magneto is eliminated and the igniter can be powered with the normal aircraft electrical system; turbine system redundancy is maintained with two ignition systems (as if dual magnetos).

High-energy, capacitor-type ignition systems are used primarily on turbojet engines; electronic ignition systems, a derivation of the simpler capacitor system, are on turbojet and turboprop aircraft.

Turbojet engine ignition system. The capacitor discharge ignition system is composed of two identical and independent ignition units. An ignition unit is made up of two exciter units, two transformers, two intermediate ignition leads, and two high-tension leads. Both units are powered by the standard aircraft storage battery. Turbojet engines will easily ignite under standard, low-altitude conditions, but must also be capable of a restart in the very cold environment of high altitude; therefore, it is necessary for the ignition system to provide a high-heat-intensity spark. The system must be capable of arcing across a wide igniter spark gap, reliably, under a wide array of conditions. Redundancy is achieved because either ignition unit with its own independent igniter will provide the necessary spark to start an engine. Figure 3-16 illustrates one side of a typical ignition system.

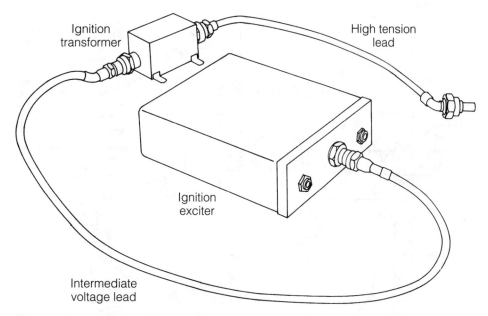

Fig. 3-16. *One side of a typical ignition system.*

Figure 3-17 depicts a functional schematic of a capacitor turbojet ignition system that operates on 24-volt dc supplied by the regular electrical system. The 24-volt power operates a dc motor that turns a multi-lobe cam and a single-lobe cam. Simultaneously, the 24-volt power goes to a set of breaker points that are operated by the multilobe cam. The breakers cause a high frequency interruption of the current that pulses through the primary winding of the auto transformer; the interruption causes an expanding and collapsing magnetic field around the winding, which is essentially the same as the process involved in a dc generator. As the magnetic field expands and collapses, a voltage in the secondary of the transformer is induced. The current then goes through a rectifier, which prevents it from backing up, and into a capacitor. As the pulses continue, the capacitor builds up a charge of approximately 4 joules.

When the capacitor reaches the proper charge, the contactor of a triggering transformer is closed by the single-lobe cam. Part of the charge goes through the primary of the triggering transformer and the capacitor that is connected in series with it. The current induces a high voltage in the secondary that ionizes the gap at the ignitor. When the ignitor is conductive, the storage capacitor discharges simultaneously the remainder of the accumulated charge with the charge from the capacitor in series with the primary of the triggering transformer.

All high-voltage portions of the triggering circuits are completely isolated from the primary circuits. The entire system is shielded against leakage of high-frequency voltage to assure no interference with navigation and communication equipment.

Electronic ignition system. This is a modification of the capacitor system and is used in turboprop and turbojet engines. Figure 3-18 shows the components of a typical

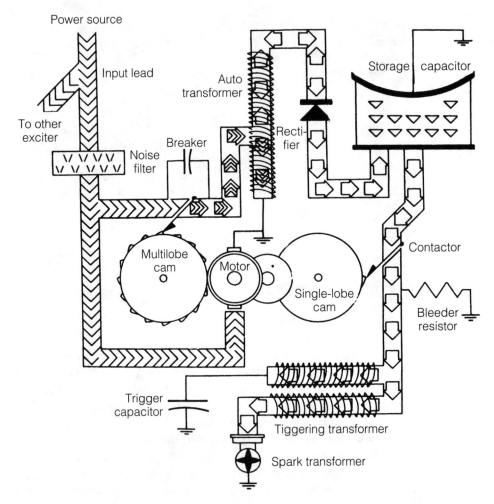

Fig. 3-17. *Capacitor-type ignition system schematic.*

electronic ignition system. The system contains a dynamotor/regulator/filter assembly, an exciter unit, two high-tension transformer units, two high-tension leads, and two igniter plugs. The dynamotor steps up the dc voltage supplied by either the standard aircraft battery or an external power supply. This voltage goes to the exciter unit that charges two internally located storage capacitors.

The voltage across the capacitors is stepped up by transformer units. The igniter sparks when the resistance of the gap is sufficiently low, in contrast to the voltage potential, to permit the larger capacitor to discharge across it. The second capacitor is low voltage, but very high energy, which results in a spark of great heat intensity capable of igniting the fuel/air mixture under any anticipated atmospheric conditions. A second advantage of this very hot spark is a cleansing action because the spark is hot enough

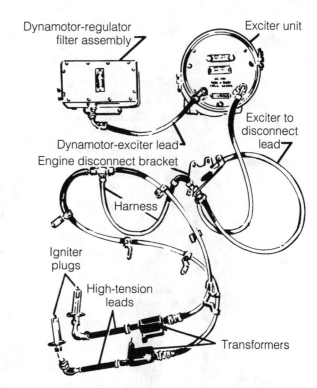

Dynamotor-regulator
filter assembly

Exciter unit

Exciter to
disconnect
lead

Dynamotor-exciter lead
Engine disconnect bracket

Harness

Fig. 3-18. *Typical electronic
ignition system.*

Igniter
plugs

High-tension
leads

Transformers

to burn away foreign deposits on the plug electrodes. The exciter is actually a dual unit that powers two otherwise independent igniter plugs. The igniters produce a continuous series of sparks until the engine starts. At that time, the flight crew deactivates the ignition system when the combustion flame is self-propagating.

Igniter plugs. A typical annular-gap igniter plug is illustrated in FIG. 3-19. This igniter is often referred to as a *long reach* because it sticks slightly deeper into the combustion-chamber liner to deliver the spark directly into the flow of the fuel/air mixture. Only a cursory look at an igniter plug will reveal that it is significantly different than the standard spark plug. The electrode is required to withstand a significantly higher energy. Despite the fact that it is designed to withstand the higher energy, it would still quickly erode if the system were left on for an extended period of time.

Another difference is that the electrode gap of the typical igniter is significantly larger than that of a spark plug because the operating pressures are lower; a narrower gap would cause the igniter to arc more easily. Finally, the problem of electrode fouling that is very common with spark plugs is all but eliminated in igniters because the very hot spark cleanses it.

A second igniter, the *constrained-gap plug* (FIG. 3-20) is used, though less frequently, in some engines. The constrained-gap has a lower operating temperature because it does not project as far into the combustion-chamber lining as the annular-gap

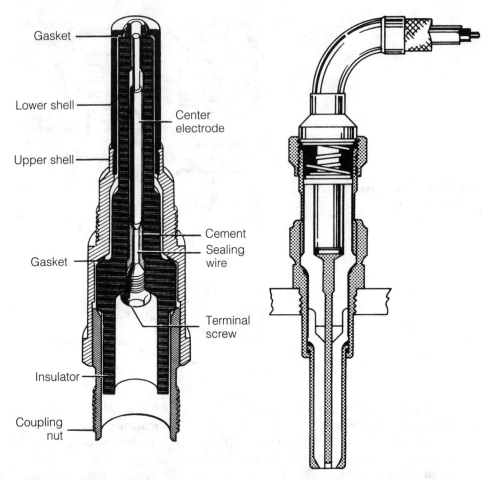

Fig. 3-19. *Typical annular-gap igniter plug.*   **Fig. 3-20.** *Constrained-gap igniter plug.*

igniter plug. For the spark to be able to effectively reach the fuel/air mixture, the igniter intentionally causes the spark to arc beyond the face of the combustion-chamber liner; essentially, the spark "jumps" into the flowing mixture.

TURBINES

The sole purpose of the turbine is to convert the kinetic energy of combustion chamber exhaust gases into mechanical energy to operate the compressor and other accessories. Approximately 60–80 percent of the total energy of the exhaust gases go to that purpose. The remaining 20–40 percent is used to propel the aircraft forward. The exact amount of the total energy that the turbine extracts depends upon the demand placed upon the turbine. As the load increases, the turbine extracts more kinetic energy from the exhaust gases, leaving less energy to use as thrust. From an operational point of view, that is a very important thing to understand. As demand on the compressor increases (pressurization, air-conditioning, and the like), less energy is left to push the airplane. Most operational checklists provide for compressor-load accessories, such as air-conditioning, to be shut off either during takeoff or in the event of an engine failure shortly after takeoff and during initial climb.

The two basic elements of the turbine assembly—stator and rotor—are similar to that of a compressor stage (FIG. 3-21 and FIG. 3-22). The stator, or turbine guide vanes, are placed directly aft of the combustion chambers and immediately forward of the turbine wheel. The turbine guide vanes serve two functions. The first function is to discharge the gas at as high a speed as possible. The vanes accomplish this because the individual blades are contoured and placed at such an angle so as to form what is effectively a series of small nozzles. As the air flows through this restricted area, a portion of the gases heat and pressure energy is converted to kinetic energy that is further converted to mechanical energy when the air passes over the rotor blades. At the same time, because the turbine guide vanes are set at a specific angle, the second function is accomplished, which is aiming the gas-flow at the rotors in the most effective manner.

A significant, ongoing concern on the flight deck is the temperature of the turbine, particularly during engine start. This is not an unfounded concern. All turbine-pow-

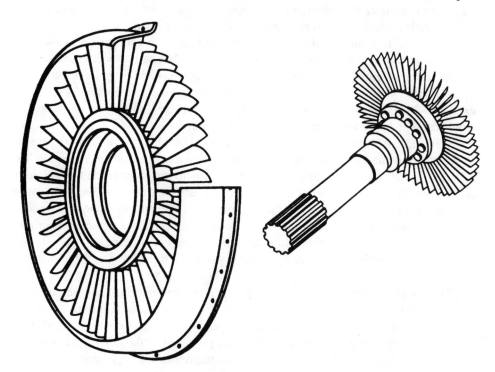

Fig. 3-21. *Stator element of the turbine assembly.*

Fig. 3-22. *Rotor element of the turbine assembly.*

ered aircraft have some sort of turbine temperature indicator on the instrument panel. Depending upon the manufacturer, the actual location of the temperature sensing probe varies, but the report is the same: turbine temperature. Again, depending upon the manufacturer, the flight deck instrument might be labeled *total turbine temperature* (TTT), *interstage turbine temperature* (ITT), *turbine inlet temperature* (TIT), or *turbine outlet temperature* (TOT).

The reason for concern is the vulnerability of the turbine wheel, which is dynamically balanced. The turbine is composed of blades affixed to a rotating disk that turns at a very high rate of speed. The faster it rotates, the greater the centrifugal loads on the turbine wheel, which attempt to make the wheel fly apart. Compounding the structural consideration is the very high temperature in which the turbine wheel must operate. High temperatures reduce the strength of the material.

Finally, due to the functional design relationship between the combustion section and the turbine section, the turbine disk rim tends to be more exposed to the gases passing through the blades. As a result, the turbine rim absorbs significantly more heat from the gases than the blades. In addition, the rim absorbs heat from the turbine blades through normal conduction, adding temperature gradient stresses to the other inherent problems. Methods are available to relieve some of these stresses, but it is important to remember that adherence to temperature limitations as outlined in the aircraft operating handbook is an absolute operational requirement.

Fixed shaft turbine configuration. Two basic turbine configurations are *fixed shaft* and *free turbine*. In the case of the fixed turbine, the shaft goes directly from the turbine to the compressor and then to a reduction gearbox and prop shaft (FIG. 3-23). The reduction gearbox is necessary because turbine engines can operate at speeds in excess of 20,000 revolutions per minute that, if connected directly to the propeller, would cause the prop tips to exceed the speed of sound.

A fixed shaft engine, such as the Garrett TPE-331 is able to more closely respond to the power demands of the pilot by providing instantaneous power response. The TPE-331 draws in ambient air and compresses it with a two-stage centrifugal compressor. Exiting the second-stage diffuser, air is directed into the annular combustion chamber and mixed with fuel. The fuel/air mixture is ignited and a continuous combustion is maintained. The expanding gases enter the turbine nozzle area, experiencing further flow acceleration due to the convergent turbine nozzle design. Nozzle-directed airflow impinges upon the first-stage turbine rotor, causing the rotor to rotate.

The hot gases continue to flow through the remaining nozzles and turbine rotors, and exit the powerplant to the atmosphere as exhaust. Rotational turbine motion is transmitted to the compressor section and the gearbox through a common fixed shaft. Approximately ⅔ of the mechanical shaft power produced by the gas generator is used to drive the compressor. The gearbox converts the remaining high speed/low torque energy into low speed/high torque energy needed to drive the propeller and engine accessories. Advantages of a fixed shaft system include controlled descent capability and rapid reverse thrust availability.

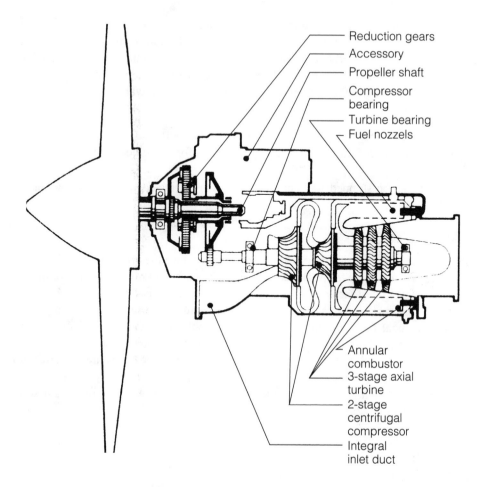

Reduction gears
Accessory
Propeller shaft
Compressor
bearing
Turbine bearing
Fuel nozzles

Annular
combustor
3-stage axial
turbine
2-stage
centrifugal
compressor
Integral
inlet duct

Fig. 3-23. *The Garrett TPE331 turboprop engine.* <small>Garrett Engine Division, Allied-Signal Aerospace Company</small>

Controlled descents. Positive, direct shaft control prevents windmilling overspeed and provides superior propeller braking effects at or near flight idle. Consequently, as a precautionary or emergency action, the pilot can maintain a fast rate of descent at a slow airspeed.

Rapid reverse thrust. Fast and positive propeller reversing are a consequential bonus of the fixed shaft engine design. On the ground, with the prop lever in the reverse range, the pilot has substantial control over the prop especially during landing roll or aborted takeoff.

Free turbine configuration. In the case of a free turbine system, a second turbine (T_2) is located behind the turbine (T_1) that is connected to the compressor. Because considerable energy is extracted from the exhaust gas by turning the T_1 turbine, it is often referred to as the *high pressure turbine*; T_2, which is downstream from T_1, is called the *low pressure turbine*. T_2 is turned by the exhaust air after passing T_1, and is connected to a second shaft that runs through the hollow T_1 shaft. This T_2 shaft connects the second turbine to a reduction gearbox, typically located in the air inlet area, which allows the connection of a propeller. Because a free turbine system allows the propeller to operate independent of the compressor, there are several advantages:

- Better control of prop speed.
- Easier engine starting due to elimination of the friction of the propeller gearbox.
- Low prop speed during taxi reduces noise and foreign object damage by surface debris.
- Prop can be stopped without shutting down engine for passenger loading/unloading.

MAIN BEARINGS

The main bearings serve a very important function because they support the entire span of the main engine rotor. Depending upon the manufacturer, the length of the engine and the weight of the engine rotor, the exact number of bearings varies. Additionally, the number of bearings will be greater for a multispool, axial-flow compressor than a centrifugal-flow compressor engine. The smallest engines might utilize as few as three bearings while larger, multispool engines will have more than six.

Generally speaking, ball bearing or roller antifriction bearings are preferred for use as main engine bearings. Two major drawbacks are a susceptibility to foreign matter damage and a tendency to fail without warning, which are offset by a number of significant benefits:

- Relatively low cost
- Low friction
- Precision alignment
- Ease of replacement
- Functionally well suited to the operation
- Work well in the normally high operating temperatures
- Ease of maintenance
- Able to simultaneously handle the loads imposed by the rotation of the rotor and the loads created by producing thrust.

EXHAUST SECTION

The exhaust section, which is located directly behind the turbine section, has two primary purposes: produce exhaust gas high exit velocity and minimize exhaust gas turbulence. The section consists of two major components: *exhaust cone* and *exhaust nozzle*. Select aircraft, such as a center-mounted, single-engine fighter, might also have an exhaust duct or tailpipe to channel the exhaust gas through the fuselage and out the aft portion of the aircraft.

The purpose of the exhaust cone, or *exhaust collector*, is to channel all exhaust gases as they are discharged from the turbine blades and combine them into a single, cohesive gas stream (FIG. 3-24). To accomplish this task, the cone decreases gas velocity slightly while increasing gas pressure. The gases, still swirling as a result of their encounter with the rotating turbine, pass the fixed struts that support the inner cone. In addition to holding the inner cone in place, the fixed struts also serve as a stabilizing influence on the gas and minimize the swirling motion. If this were not done, the exhaust gases would exit the engine at an approximate 45° angle. The inner cone helps to stabilize the outflowing gas and reduces airflow turbulence that would otherwise result in unwanted drag.

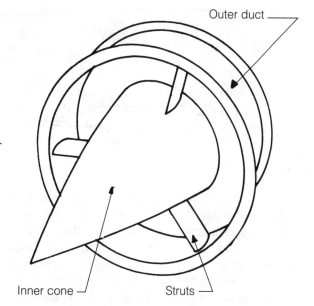

Fig. 3-24. *Exhaust collector with welded support struts.*

The exhaust nozzle is the aft opening of the turbine engine exhaust duct. The purpose of the nozzle is to fine-tune the density and velocity of the exhaust gases as they leave the engine. But as the diameter of the nozzle varies, so too does engine performance and gas temperature. For more conventional aircraft with a fairly well defined mission, such as corporate operations, fixed diameter systems are employed. In that case, the actual diameter of the nozzle reflects the engineer's calculation of maximum performance under average anticipated operating conditions. For high performance aircraft with a wide operating envelope, such as military fighter and attack aircraft, this is often done with variable nozzles controlled by an on-board computer system. Such a system varies the diameter of the opening to maximize exhaust efficiency for all conditions of flight.

THRUST REVERSERS

Significant increases in operating weights of air transport and corporate aircraft have made stopping on a runway an important consideration for aircraft designers. It is increasingly difficult for conventional disk brakes to stop high-speed, high-gross-weight aircraft in relatively short landing distances at some airports. The turboprop aircraft uses a highly effective propeller-reversing system discussed in chapter 2, but the turbojet and turbofan aircraft do not have the luxury of large, powerful propellers to create a reverse thrust. As a result, an airflow reversing system is incorporated to reverse the exhaust gases as they leave the nozzle (FIG. 3-25).

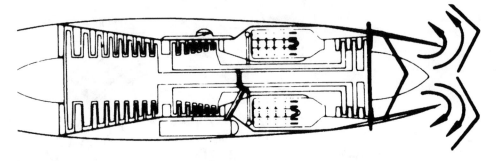

Fig. 3-25. *Operation of the thrust reverser.*

Most thrust reverser systems are capable of producing approximately 45 percent of the rated engine thrust in a reverse direction, which adds approximately 20 percent to the friction braking capability on dry, flat runways. On runways that have limited or nil braking action, reversers provide up to 50 percent of the braking effectiveness; however, all aircraft operating manuals will specify a ground speed below which thrust reversers should not be used. The limitation deals with potential foreign object damage rather than aerodynamic efficiency. When slower than the published speed, the forward airflow from the reverser is moving faster than the aircraft and the airflow will push surface sand, dust, and other debris ahead of the aircraft only to be ingested as the engine intake passes through it.

The majority of thrust reverse systems simply provide a braking force on the ground, but some are capable of being activated in flight; the effect can be dramatic. Extreme caution must be exercised when activating thrust reversers in flight and the aircraft operating handbook must be followed precisely. In-flight thrust reverse can cause the aircraft to descend at a tremendous rate and many aircraft do not allow their use in flight. The most common method of preventing accidental in-flight activation of the thrust reverser system is to use an electrical override system connected to a squat switch on the landing gear. In such aircraft, it is impossible to activate thrust reverse without weight on the landing gear; however, for higher performance aircraft, some form of slowing the aircraft down in-flight is necessary. In addition to thrust reverse systems, other options include *dive brakes*, *wing spoilers*, and *speed brakes*. Two types of thrust reversers are *mechanical-blockage* and *aerodynamic-blockage*.

Mechanical-blockage systems. Reverse thrust is accomplished by deploying a removable obstruction, often referred to as *clamshells*, in the exhaust gas airflow at the rear of the exhaust nozzle. The gases exhaust normally, encounter a physical barrier and are deflected in a forward direction thereby effectively reversing the thrust of the engine. Now, as the engine thrust is increased by the pilot, it is flowing in the opposite direction and slows the aircraft.

Aerodynamic-blockage systems. Thrust reversing is accomplished by blocker doors and cascade-turning vanes either in the fan exhaust or the turbine exhaust areas. The blocker doors deploy, causing the exhaust thrust airflow to reverse. The angle of the blocker vanes deflects the airflow through the cascade vanes that further direct the airflow forward. This produces a reverse flow of the gases similar to the mechanical-blockage system. It is important to note that while both systems deflect the gases forward, they do not direct them parallel to the longitudinal axis of the engine. Doing so would cause the hot exhaust gases to be reingested by the engine, which could easily lead to a compressor stall. Instead, the gases are deflected at an optimum angle that provides maximum reverse thrust while preventing reingestion through the engine intake.

ENGINE NOISE

Turbine engine noise has become a major issue of airport operations. Certain airports have imposed night landing restrictions; other airports have attempted to ban certain types of aircraft altogether because of elevated noise profiles. In response to the growing concern, the federal government has enacted legislation requiring ever-increasing restrictions with respect to noise produced by aircraft engines. Clearly, a noise problem is associated with turbine engines. Several sources of noise are associated with a turbine engine; the most objectionable is noise produced by the engine's exhaust.

The noise is the audio byproduct of the turbulent exhaust air mixing with the relatively calm surrounding, ambient air. Exiting the exhaust nozzle is a high-velocity, relatively low-turbulence gas that produces a high-frequency noise, but the gas does not travel very far. As the gases dissipate farther aft of the nozzle, the turbulence increases and the interaction with the still air causes a low-frequency noise. Low-frequency sound travels farther than high-frequency sound and, as a result, this is the noise that reaches the human ear. Whales, for instance, emit a very low frequency sound and can communicate with other whales halfway around the earth.

Noise levels will vary from engine to engine and are directly proportional to engine thrust and the amount of work the engine performs on the air while passing through; therefore, a larger engine can move the same amount of air as a smaller engine but do it with less work; hence, it will be quieter at an equivalent thrust output.

Turbofans are quieter than turbojets at equivalent thrust outputs because of the way a turbofan processes air. Turbofans must have larger turbines to supply the necessary power to drive the fan. These larger turbines reduce the velocity of the exhaust gas. Because exhaust gas noise is proportional to exhaust velocity, this yields the desirable result of reducing the overall noise level.

ACCESSORY SECTION

The primary purpose of the turbine engine accessory section is to power equipment required for the operation and control of the engine. A secondary purpose is to support other aircraft systems. The accessory section consists of an accessory case with mounting pads for individual, engine-driven equipment, and a gear train driven by the engine rotor. Because the rotor turns at such high speeds, an accessory drive shaft with a reduction gear coupling interfaces the rotor with the individual accessories. Many engines also use the accessory section as an oil sump and distribution point to lubricate the engine bearings and other moving parts.

Typical equipment mounted on the accessory section includes:

- Electrical generators
- Hydraulic pumps
- Fuel pumps
- Fuel control unit
- Auxiliary fuel pump
- Oil pump and sump
- Tachometer
- Engine starter

The engine starter is not operated by the accessory section; the starter is mounted there for better proximity to the engine.

ENGINE STARTING SYSTEMS

Many types of starter systems have been developed for turbine-powered engines; three basic types are electric motor, air turbine, and combustion. The purpose of a starter is to get the air cycle started by turning the compressor at a high enough rate of speed to supply adequate airflow to support combustion in the burners. With a multiple-spool engine, only the high-pressure compressor needs to turn. Once fuel is added and ignition occurs, the starter must continue to assist the engine in spooling up to a sufficient rotational speed to assure that the engine is self-sufficient. Deactivating the starter system during the start-up procedure is referenced in the operating handbook and is usually related to a specific compressor speed as indicated by the N_1 tachometer.

Electric motor starter systems. Two types of electric starters are *direct-cranking* and *starter-generator*. Both systems are very similar except the starter-generator is not disconnected after engine start; the starter-generator produces electricity when the engine is operating. The direct-cranking system, which resembles a reciprocating engine system, serves only as a starter and is disconnected when the engine has reached a self-sustaining speed. The starter-generator system is very popular on corporate turboprop aircraft because one unit fulfills two requirements, weighs less than two separate pieces of equipment, and reduces the number of spare parts in inventory.

Air turbine starter systems. This type of system is commonly used on three- and four-engine aircraft and with larger engines where the necessary torque to turn the engine's compressor would require a very large electrical starter motor. Air turbine starters are capable of providing the necessary torque with a unit that is as little as one-quarter the weight of an equivalent electrical starter. The system operates by directing high pressure air onto an axial flow turbine contained within the starter system. The turbine drives a reduction gear that is connected to a starter clutch (FIG. 3-26). Most aircraft are designed to utilize one of three possible air sources to operate the system: *auxiliary power unit*, *bleed air* from another engine, and a *ground power unit*.

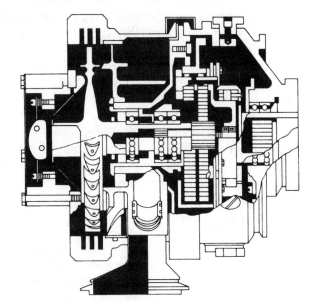

Fig. 3-26. *Cutaway of an air turbine starter.*

An auxiliary power unit (APU) is a small, on-board turbine engine usually located in the aft fuselage section of the aircraft. This lightweight engine, which can be easily started with an electrical starter, is used to power the aircraft systems when the main engines are shut down. The APU typically provides heat, air-conditioning, electricity, and compressed air, but is not capable of providing any in-flight thrust. By tapping off air from the APU's compressor and routing it to one of the air turbine starter systems, it is possible to start a main engine.

The bleed air source method simply allows the bleed air of any operating engine to be channeled to the air turbine starter system of any inoperative engine. It is the same system that permits using the APU as an air source.

The ground power unit (GPU) is often confused with an APU. As the name implies, the ground power unit is always a ground-based engine and the APU is located on-board the aircraft. The GPU also provides bleed air to one of the engine air turbine starters. After the first engine is started, subsequent engines can be started using the bleed air of the operating engine.

From a practical point of view, when making an intermediate stop to pick up or drop off passengers, standard procedure is usually to shut down only the engine(s) on the side of the cabin door. This is done to protect the passengers from flying debris and potential hearing damage. Normally, the engine(s) on the opposite side of the aircraft would be operating at idle. On aircraft that do not use electrical starting systems, when the door is closed, the inoperative engines would then be started using the bleed air source method.

If, on the other hand, the stop was going to be longer than a quick turnaround, keeping any main engine running would be expensive. If there are no passenger comfort considerations, such as maintaining air-conditioning in the summer or heat in the winter, it is best to shut down the aircraft entirely. Subsequent starting would be with a GPU if available. If no GPU is available, or if heating or air conditioning is a requirement, the aircraft can be maintained on APU power.

Combustion starter systems. The fuel/air combustion starter was primarily developed for use in regional air carrier aircraft operating over short stage lengths and into smaller airports without GPU availability. The system, which can be used in turboprop and turbojet engines, contains a turbine-driven power unit with auxiliary fuel, ignition, and air systems. Normally, the fully automatic system provides a fast and efficient start by using the combustion energy of conventional jet fuel combined with compressed air stored in a rechargeable bottle.

WATER INJECTION

High ambient air temperature is a significant performance problem associated with turbine engines. When the compressor inlet temperature is high, the result is a serious loss of thrust in the turbojet/turbofan and power in the turboprop. Normally, this does not present a major problem if the airport has adequate runway length. At airports with shorter runways, the pilot can manage and configure passengers, baggage, and fuel to reduce the load and increase the aircraft performance; however, a limited number of aircraft that operate routinely at airports with short runways and/or high ambient temperatures might have a water injection system installed to increase engine thrust during takeoff and initial climb. Essentially, water is pumped directly into the compressor inlet, or diffuser case as appropriate, where the water increases the airflow density, reduces hot section temperatures, and permits an increased fuel flow to develop more thrust.

REQUIRED MAINTENANCE

Unlike reciprocating engines, turbine engine maintenance varies widely among engine types. Manufacturer-recommended procedures should be followed. Turbine engine maintenance falls into two categories: cold section and hot section.

Cold section

The primary concern of the cold section is the compressor. By far, the most significant problem associated with compressor maintenance is foreign object damage, which ranges from blade erosion due to frequent ingestion of sand or other particulate matter to ingestion of a large, solid object such as a bird. While ingesting a bird, loose bolt or rock will almost certainly cause catastrophic engine failure, continuous blade erosion from small particulate matter should not be discounted. Taxiway and runway environments are littered with dirt, grit, sand, and the like. A turbine engine ingests enormous quantities of ambient air, especially on takeoff. Some erosion and minor damage to individual axial compressor blades is tolerable and repairable if caught early enough.

Particulate matter that goes through the compressor is hurled outward by centrifugal force, causing a slow buildup on the casing, rotor blades, and stator vanes. Compressor blades are meticulously engineered to provide maximum aerodynamic efficiency. As foreign matter builds up on them, the compressor becomes less efficient and engine performance begins to degrade. This is analogous to the buildup of ice on a wing and eventually, if allowed to continue, the turbine engine will no longer accelerate adequately and the exhaust gas temperature will begin to increase. Eventually, the engine will fail.

Hot section

A hot section inspection includes the combustion section, turbine section, and exhaust section. The major determinant of the life expectancy of a turbine engine is based upon the status of its hot section. Manufacturers recommend specific, periodic hot section inspections based upon the total number of hours the engine has been operated and the total number of times the engine has been started; starts are also called *cycles*. The importance of strict compliance with manufacturer's inspection and cleaning recommendations cannot be overstated, primarily because cracking is the most often occurring problem.

It is hard to imagine a more conducive environment for crack propagation. Hot section components are subject to thermal shock during starting, elevated operating temperatures, elevated rotational speeds, and elevated linear loads resulting from thrust.

MANUFACTURER DOCUMENTATION

(The following information is extracted from Beech Aircraft Corporation pilot operating handbooks and Falcon Jet Corporation aircraft and interior technical descriptions. The information is to be used for educational purposes only. This information is not to be used for the operation or maintenance of any aircraft.)

Super King Air B200

Engines. The Beechcraft Super King Air B200/B200C is powered by two Pratt & Whitney Canada PT6A-42 turbopropeller engines, each rated at 850 SHP (FIG. 3-27).

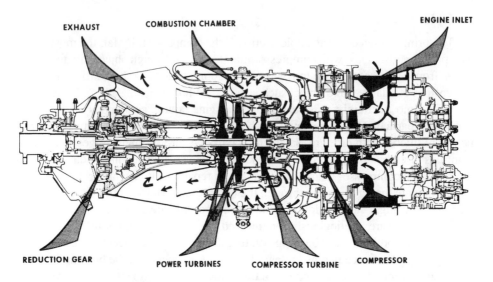

EXHAUST COMBUSTION CHAMBER ENGINE INLET

REDUCTION GEAR POWER TURBINES COMPRESSOR TURBINE COMPRESSOR

Fig. 3-27. *Pratt & Whitney PT6A-42 turbopropeller engine.* Beech Aircraft Corporation

Each engine has a three-stage axial flow, single-stage centrifugal-flow compressor, which is driven by a single-stage reaction turbine. The power turbine—a two-state re-action turbine counter-rotating with the compressor turbine—drives the output shaft. Both the compressor turbine and the power turbine are located in the approximate center of the engine, with their shafts extending in opposite directions.

Being a reverse flow engine, the ram air supply enters the lower portion of the nacelle and is drawn in through the aft protective screens. The air is then routed into the compressor. After it is compressed, it is forced into the annular combustion chamber, and mixed with fuel that is sprayed in through 14 nozzles mounted around the gas generator case. A capacitance discharge ignition unit and two spark igniter plugs are used to start combustion. After combustion, the exhaust passes through the compressor turbine and two stages of power turbine and is routed through two exhaust ports near the front of the engine.

A pneumatic fuel control system schedules fuel flow to maintain the power set by the gas generator power lever. Propeller speed within the governing range remains constant at any selected propeller control lever position through the action of a propeller governor, except in the beta range where the maximum propeller speed is controlled by the pneumatic section of the propeller governor.

The accessory drive at the aft end of the engine provides power to drive the fuel pumps, fuel control, the oil pumps, the refrigerant compressor (right engine), the starter/generator, and the tachometer transmitter. At this point, the speed of the drive (N_1) is the true speed of the compressor side of the engine, 37,500 rpm, which equals 102.6 percent N_1.

The reduction gearbox forward of the power turbine provides gearing for the propeller and drives the propeller tachometer transmitter, the propeller overspeed gover-

110

nor, and the propeller governor. The turbine speed on the power side of the engine is 30,000 rpm. After reduction gearing, the propeller rpm is 2000.

Engine torque at the propeller shaft is indicated by a torquemeter located inside the first stage reduction gear housing. The torquemeter is a hydromechanical torque measuring device. It consists of: a ring gear and case (helical splines between ring gear and case), torquemeter cylinder, torquemeter piston, valve plunger and spring, differential pressure sensor and servo transmitter combination, and servo indicator calibrated to indicate ft-lbs.

Torque at the power turbine shaft and the resisting torque at the propeller shaft gears is converted from rotary motion by the helical splines to a translating motion at the piston face. A change in torquemeter oil pressure results from the piston translation. The valve plunger and spring maintains oil pressure proportional to engine torque. The differential pressure sensor uses a bellows system to sense differences between torquemeter oil pressure and a reference pressure. Bellows movement drives the transmitter servo. The electric signal from the transmitter drives the servo motor in the torquemeter indicator. Torque is indicated by indicator needle position on a calibrated dial.

Propulsion system controls. The propulsion system is operated by three sets of controls: the power levers, propeller levers, and condition levers. The power levers serve to control engine power. The condition levers control the flow of fuel at the fuel control outlet and select fuel cutoff, low idle, and high idle functions. The propeller levers are operated conventionally and control the constant speed propellers through the primary governor.

Power levers. The power levers provide control of engine power from idle through takeoff power by operation of the gas generator (N_1) governor in the fuel control unit. Increasing (N_1) rpm results in increased engine power.

Propeller levers. Each propeller lever operates a speeder spring inside the primary governor to reposition the pilot valve, which results in an increase or decrease of propeller rpm. For propeller feathering, each propeller lever lifts the pilot valve to a position that causes complete dumping of high pressure oil, allowing the counterweights and feathering spring to change the pitch. Detents at the rear of lever travel prevent inadvertent movement into the feathering range. Operating range is 1600 to 2000 rpm.

Condition levers. The condition levers have three positions: FUEL CUTOFF, LOW IDLE and HIGH IDLE. Each lever controls the fuel cutoff function of the fuel control unit and limits idle speed at 56 percent N_1 for low idle, and 70 percent N_1 for high idle.

Propeller reversing. When the power levers are lifted over the IDLE detent, they control engine power through the Beta and reverse ranges. **Caution:** Propeller reversing on unimproved surfaces should be accomplished carefully to prevent propeller erosion from reversed airflow and, in dusty conditions, to prevent obscuring the operator's vision.

Condition levers, when set at HIGH IDLE, keep the engines operating at 70 percent N_1 high idle speed for maximum reversing performance. **Caution:** Power levers should not be moved into the reversing position when the engines are not running because the reversing system will be damaged.

Beechjet 400A

Engines. The Beechjet 400A is powered by two aft pod-mounted JT15D-5 turbofan engines manufactured by Pratt & Whitney Aircraft of Canada, Limited (FIG. 3-28). The engines are lightweight, twin spool, front turbofan jets having a full-length annular bypass duct.

The low compressor consists of a low compressor fan followed by a primary gas path booster stage. A concentric shaft system supports the high and low rotors. The inner shaft supports the low compressor fan and booster stage and is driven by a two-stage turbine supported at the rear. The speed of this assembly is designated as N_1 rpm. The outer shaft supports the high-pressure centrifugal compressor impeller and is driven by a single-stage high-pressure turbine. This speed is designated as N_2 rpm.

The JT15D-5 will produce 2,900 pounds of static thrust on a standard day at sea level and a maximum continuous thrust of 2,900 pounds. All intake air passes through the low compressor fan. Immediately aft of the fan, the airflow is divided by concentric ducts. Most of the total airflow is bypassed around the engine through the outer annular bypass duct and is exhausted at the rear. Air entering the inner duct passes through a booster stage and is compressed by the impeller. The high pressure air then passes through a diffuser assembly and moves back to the combustion section.

The combustion chamber is a reverse flow design to save space and reduce engine size. Most of the air entering the chamber is mixed with fuel and ignited while the remainder streams down the chamber liner for cooling. Fuel is introduced by 12 dual-orifice nozzles supplied by a dual manifold. The mixture is ignited initially by spark igniters that extend into the combustion chamber at the 5 and 7 o'clock positions.

After start, combustion becomes self-sustaining. The hot gases expand in the reverse direction and pass through a set of turbine guide vanes to the high-pressure turbine. As the expanding gases move rearward, they pass through another set of guide vanes and enter the two-stage low-pressure turbine. The greater portion of the remaining energy is extracted there and transmitted by the inner shaft to the forward mounted fan. The hot gases are then exhausted into the atmosphere.

An accessory gearbox is mounted on the lower side of the engine's intermediate case and is driven by a tower shaft from the bevel gear to the N_2 shaft. Its function is to turn the engine during starting, and to drive the accessories for the engine and airplane systems. The following components are driven by the accessory gearbox: dc starter/generator, N_2 speed sensors, oil pump, fuel pump, hydromechanical metering unit (HMU), and hydraulic pump.

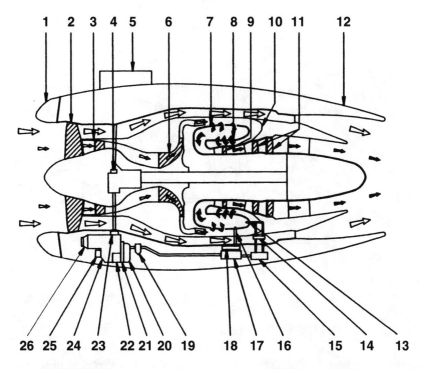

1. INTAKE DUCT
2. LOW COMPRESSOR
3. BOOSTER STAGE
4. N_1 SPEED SENSOR
5. ENGINE ELECTRICAL CONTROL
6. CENTRIFUGAL COMPRESSOR IMPELLER
7. DIFFUSER
8. COMBUSTION CHAMBER
9. HIGH COMPRESSOR TURBINE
10. ANNULAR BYPASS DUCT
11. LOW COMPRESSOR TURBINES
12. EXHAUST DUCT
13. FUEL INJECTOR
14. EMERGENCY FUEL SHUTOFF VALVE
15. FUEL FLOW DIVIDER VALVE
16. IGNITER
17. OIL-TO-FUEL HEAT EXCHANGER
18. EXCITER
19. FUEL FLOW TRANSMITTER
20. HYDROMECHANICAL METERING UNIT
21. FUEL PUMP
22. HYDRAULIC OIL PUMP
23. N_2 SPEED SENSOR
24. ACCESSORY GEARBOX
25. ENGINE OIL PUMP
26. STARTER-GENERATOR

Fig. 3-28. *Pratt & Whitney JT15D-5 turbofan engine.* Beech Aircraft Corporation

Falcon 900B

Engines. General. The powerplant installation consists of three Garrett TFE731-5BR-1C turbofan engines with 4,700-lb thrust (uninstalled, sea level, ISA +10°C conditions) fitted with core mixer exhaust nozzles (Figure 3-29).

113

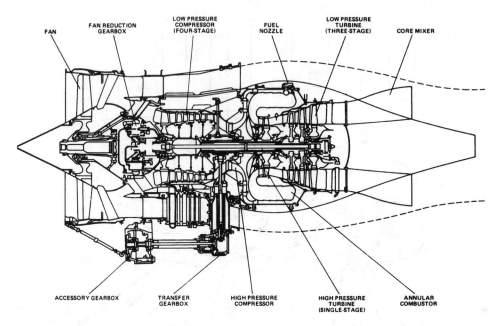

Fig. 3-29. *Garrett TFE731-5BR-1C turbofan engine.* Falcon Jet Corporation

Side engine installation. The nacelles are located on either side of the rear fuselage, with the engine main mounts located near the baggage compartment area. Inlet leading edges are anti-iced and are made of plated aluminum to permit polishing. Movable cowls provide easy access to engine components.

Center engine installation. The center engine is held in place by three top mounts. The inlet leading edge and S-duct are anti-iced. Plated aluminum is used for the inlet leading edge in order to facilitate polishing. The S-duct inspection door is permanently hinged to facilitate inspection and to protect against inadvertent opening while the center engine is operating. An integral hoist system is installed to facilitate engine removal for maintenance. Movable cowls provide easy access to engine components.

The afterbody includes a hydraulically operated, fixed-pivot, target thrust reverser. The thrust reverser deploys or stows in less than three seconds and can be used down to zero knots airspeed. A light is illuminated on the flight deck master warning panel whenever the thrust reverser is not stowed.

114

4
Lubrication and cooling systems

TWO LUBRICATION SYSTEMS ARE USED IN TURBINE ENGINES: *WET-SUMP* AND *dry-sump.* Most systems are dry-sump in conjunction with axial-flow engines; select wet-sump systems are used with centrifugal-flow engines. One of the primary differences between wet-sump and dry-sump systems is the wet-sump stores oil in the engine; the dry-sump uses an external oil tank mounted on, or near, the engine. In the dry-sump system, the hot, return oil is cooled by an oil cooler that uses either an air/oil or fuel/oil heat exchanger system.

Of all the points on a turbine engine that use oil as a coolant, the most critical is the exhaust turbine bearing. With exhaust gases leaving the combustion chamber well in excess of 1000°F, this fast-turning bearing requires a high level of cooling efficiency. In addition to oil cooling, certain engines also add the cooling effect of airflow. The air can be supplied in a number of ways. One method is by a cooling air impeller that is mounted on the compressor shaft just aft of the main compressor. Other engines add air-pumping vanes to the front of the turbine wheels to channel cooling airflow over the turbine disk, which reduces heat radiation to the bearing surface. Some axial flow engines will use the cooler, pressurized fourth or fifth stage bleed air.

Whatever method is used, there are several advantages to using supplemental air cooling, for instance, it reduces or eliminates the need for an oil cooler in a wet-sump system. A significant amount of heat can be dissipated with cooling air, thus, reducing the heat that must be absorbed by the oil. Additionally, using cooling air allows a reduction in the quantity of oil necessary to meet the total cooling requirement. Engines that use only oil cooling invariably require an oil heat exchanger, though it is smaller and lighter than such a system would be on a reciprocating engine of equivalent power.

WET-SUMP SYSTEM

Few wet-sump lubrication systems are associated with centrifugal-flow engines, but a typical system is illustrated in FIG. 4-1. Many of the components of the wet-sump system are similar to those of the dry-sump, but the major difference is the location of the oil reservoir. The reservoir of a wet-sump system is located either in the accessory gear case and front compressor bearing support casing, or it is mounted on the bottom of the accessory case. In either instance, wet-sump reservoirs are an integral part of the engine and will hold the majority of the engine oil supply.

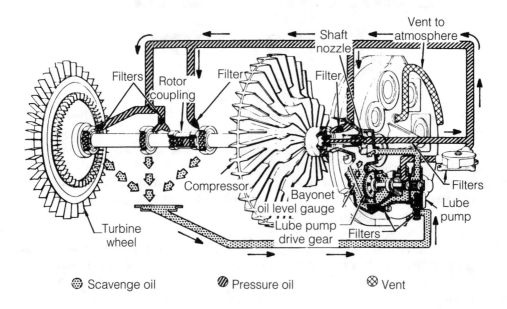

Fig. 4-1. *Turbojet engine wet-sump lubrication system.*

The system illustrated is typical of engines using wet-sump systems in that the bearing and drive gears in the accessory drive are all lubricated by the splash method. The remaining lubrication points are fed by oil that is sprayed directly onto the appropriate bearing or coupling in a manner similar to that used in the dry-sump system. One other difference between the two types of systems is that the wet-sump pressure system is of variable pressure in which the pump outlet pressure depends on engine rpm.

Scavenged oil goes back to the reservoir (sump) through a combination of pump suction and simple gravity as the oil flows to the lowest point of the engine which is the sump.

DRY-SUMP SYSTEM

Dry-sump systems incorporate an oil tank and cooler arrangement that is mounted somewhere on the engine (FIG. 4-2). It has the advantages of being able to hold a large quantity of oil, easily control oil temperature, and allow the engine design to remain streamlined and efficient. Engines that store oil internally lose some aerodynamic efficiency in their shape because a sump area must be included at the bottom of the engine for oil storage and cooling.

This information includes those components that are typical of dry-sump lubrication systems. Not all components will be found with every system. While it is correct to say that dry-sump systems utilize an oil reservoir or tank that is external to the engine, there must be a small sump located in the engine to which the oil drains, is collected, and pumped to the main tank. This sump typically contains the oil pump, scavenge and pressure inlet strainers, scavenge return connection, pressure outlet ports, oil filter, and various mounting bosses for pressure and temperature indicating units.

Figure 4-3 depicts a typical external oil tank. The swivel assembly connected to the oil outlet is designed to provide a constant supply of oil to the engine in any aircraft attitude. The horizontal baffle mounted in the center of the tank helps contain a constant quantity of oil in the lower half of the tank regardless of the aircraft's attitude. The total tank capacity is greater than the space under the baffle; therefore, as oil enters the upper half of the tank, the oil flows through the flapper door to the lower half of the tank keeping the lower half constantly full with the excess oil stored above. The door does not permit oil to travel back into the upper half in the event of turbulence or an unusual attitude, though it does provide for heat expansion; therefore, the oil in the lower half of the tank is essentially trapped and can only find its way out through the swivel assembly intake port. Some tanks also include a de-aerator tray for separating air from return oil.

Oil pump. The purpose of the oil pump is to move oil from the tank to the specific parts of the engine that need lubrication. Select pumps include pressure supply and scavenge elements in one pump; others are either a pressure supply pump or a scavenge pump. Which type of pump, or how many pumps, are used with any given engine depends upon the specific requirements of the engine. For instance, axial-flow engines have long rotor shafts that require more bearings than their centrifugal-flow counterparts. As a result, axial-flow engines require a greater number of supply and scavenger elements or pumps with a greater capacity.

Scavenge elements typically have a greater pumping capacity than pressure elements; that is, the system is able to take used oil out of the bearing sump faster than it is able to pump fresh oil to the bearing. The pressure pump supplies the site with a constant flow of fresh oil to assure the part is constantly bathed, but the scavenger will

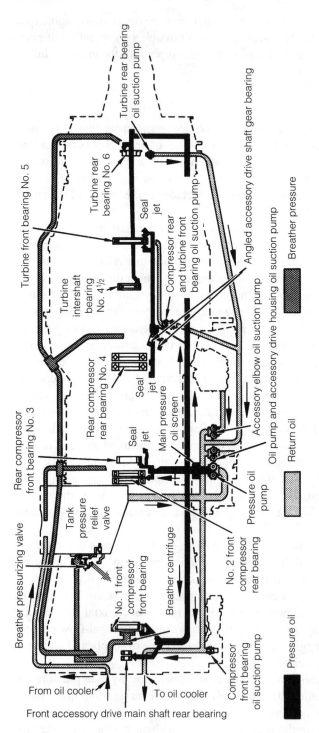

Fig. 4-2. *A turbojet dry-sump lubrication system.*

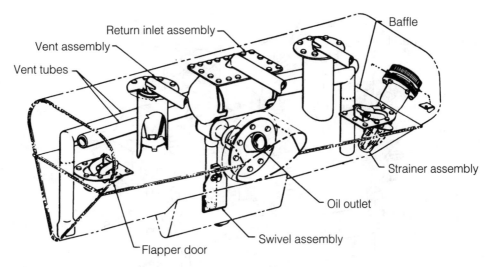

Fig. 4-3. *Oil tank*

draw away the used oil so quickly that it prevents any oil accumulation in the bearing sump. This prevents oil accumulation that might inhibit flow to the bearing.

Most oil pumps can be separated into three types depending upon design: *gear*, *gerotor*, and *piston*. A gear pump is the most common on aircraft powerplants and a piston pump is the least common; the piston pump is not discussed.

Figure 4-4 illustrates a common, two-element, gear oil pump. The illustrated pump has a single pressure element and a single scavenger element, but an actual gear pump might include several pressure and scavenger elements.

A relief valve can be seen on the discharge side of the pump. The valve limits the pump's output pressure by sending bypass oil back to the inlet side of the pump when output pressure exceeds a predetermined limit. Directly below the relief valve, in the diagram, is the shear section. The gear to the far left is driven by the engine. If the pump's gears seize, the shear section would break and automatically disconnect the pump from the engine's accessory drive-gear.

Gerotor pump. This pump is somewhat similar to the gear-type pump. A gerotor typically contains a single pressure element and two or more scavenging elements all driven by the same shaft. The pressure is determined by the engine's rpm; therefore, when the engine is at idle, there is minimum pressure, and when the engine is at maximum rpm, there is maximum pressure. All elements are essentially the same shape, but the capacity of the elements can vary by changing the size of the gerotor elements. A pressure element, for instance, might have the pumping capacity of 3.1 gallons per minute (gpm); a scavenge element has a 4.25 gpm capacity.

Figure 4-5 illustrates a typical set of gerotor pumping elements. Each set of gerotors is separated by a steel plate that essentially makes each set its own pump consisting of inner and outer elements. The small star-shaped inner element has external lobes

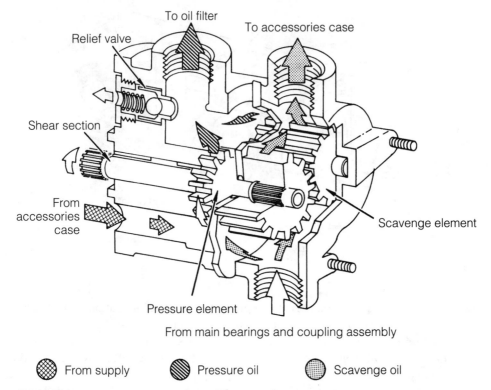

Fig. 4-4. *Cutaway view of gear oil pump.*

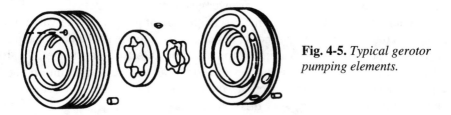

Fig. 4-5. *Typical gerotor pumping elements.*

that fit within the internal lobes of the outer element. The small element fits on the pump shaft and drives the outer, free-turning element. The outer element then fits inside a steel plate that has an eccentric bore. The possible combinations allow for many arrangements of pump and scavenge capability.

Filters. One of the most important facets of engine lubrication is filtration. Turbine engines rotate at very high speeds on antifriction ball and roller bearings. Even small amounts of particulate matter could rapidly lead to deterioration of the bearings and potential engine destruction. In addition, oil flows through small drilled holes or core passages; small particulate matter could clog these passages, preventing oil flow, resulting in lubrication failure of one or more bearings.

The main oil filter strains oil leaving the pump to lubricate the bearings. Typical main filter components include a housing, an integral relief (bypass) valve, and filter element. The bypass valve assures that oil will be available for lubrication even if the filter element becomes clogged. Similar to a hydraulic bypass valve, it will open at a predetermined pressure that is slightly above normal system operating pressure. The pump continues to put out oil under pressure regardless of the condition of the filter.

If the filter becomes clogged, two things happen: First, oil flow will stop downstream from the filter, and second, the pressure between the pump and the filter will begin to increase because the oil has no place to go. When the pressure increases to the predetermined value the bypass valve opens and allows oil to flow around the filter. The obvious drawback is that the bearings are now being lubricated with unfiltered oil.

Probably the most common type of filter is the replaceable laminated paper element that is also used in hydraulic systems. Another type is made up of a series of spaces and screens that forces the oil through screen filters to an outlet port. Another very common system is the filtering assembly depicted in FIG. 4-6. Oil enters through an inlet port, is forced through a removable, internal, stainless steel screen, and then exits through an outlet port.

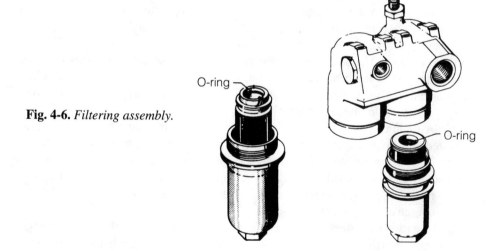

Fig. 4-6. *Filtering assembly.*

Secondary oil filters are located throughout the system and are used for various special applications. Unlike the main oil filters, these typically only filter oil at isolated positions. The finger screen filter, for instance, is used to strain scavenged oil. "Last-chance" filters, on the other hand, are fine mesh screens that strain oil immediately prior to its being sprayed out a nozzle onto a bearing.

Oil pressure relief valve. The purpose of the oil pressure relief valve is to set a maximum limit to the system pressure. The relief valve is set to a predetermined value and will automatically bypass oil back to the inlet side of the pump if system pressure exceeds the maximum limit. This is particularly necessary in systems that employ an oil cooler because coolers have very thin walls to maximize the heat exchange. Excessive oil pressure could rupture the heat exchanger walls causing a massive and catastrophic oil leak.

Oil jets. Also called *nozzles*, the oil jets are located directly next to the bearings and parts to be lubricated. The jets are connected to the pump via various pressure lines. The nozzle is designed to atomize and spray the oil directly on the appropriate spot. Some systems include an air/oil mist that is obtained by tapping bleed air from the compressor to the oil nozzle outlet. Such a system is most likely to be used with ball and roller bearings, though the solid oil spray is still considered to be the more effective of the two methods.

One concern regarding oil jets is that the tiny tip orifice is easily clogged. This is the most likely place for system contamination to cause problems; hence, the "last-chance" filter. These filters are as inaccessible as they are effective. Last-chance filters can only be cleaned when the engine is disassembled for overhaul. The problem is if the filter clogs, the tip will not spray oil on the bearing, which will rapidly lead to bearing failure. Prevention of the problem is accomplished by the maintenance technician faithfully and regularly cleaning the main lubrication system filter to get rid of contamination before it reaches the nozzles.

Lubrication system gauge connections. Oil temperature and pressure sensing hardware is connected to the oil system to provide flight deck indications. The oil pressure gauge indicates the pressure at the output port of the pump as the oil moves toward the nozzles. The sensor is located between the pump and the nozzles. The oil temperature gauge indicates the temperature of the oil at the engine's pressure inlet. Oil temperature can be obtained two ways: a thermocouple fitting in the oil line or an oil temperature bulb inserted in the oil line.

Lubrication system vents. Oil tank air pressure and outside air pressure are equalized with vents that are on the tank or accessory case, depending upon the type of engine. The vents are also called *breathers*. High altitude aircraft systems usually route the vent through a check valve that is preset to maintain approximately 4 psi within the tank to provide a positive pressure on the oil to assure flow to the oil pump inlet regardless of the ambient pressure.

Lubrication system check valve. The purpose of the check valve is to prevent oil from seeping into the engine after shutdown. Check valves are normally spring-loaded ball-and-socket and generally require approximately 5 psi to unseat and allow oil to flow through. The problem is that when the oil pump stops operating, the entire system is essentially in a static condition. Gravity, acting upon the oil lines, might cause the oil in some lines to reverse direction and flow backward through the system. If left unchecked, this would allow oil to flow back into the various sumps underneath the bearings, the engine's accessory gearbox, the compressor rear housing, and the combustion chamber. Unwanted accumulated oil can lead to excessive loading of the accessory drive gears during engine start, contamination of the cabin pressurization air, or internal oil fires.

Lubrication system thermostatic bypass valves. This valve is integral to an oil cooler system. The valve maintains the proper oil temperature by varying the proportion of the total oil flow passing through the oil cooler. Figure 4-7 shows the cutaway view of a common valve with a spring-loaded thermostatic element, two inlet ports, and one outlet port. The valve is spring-loaded because the pressure drop through the oil cooler could become too great because of denting or clogging of the cooler tubing. In such a case, the valve will open, bypassing the oil around the cooler.

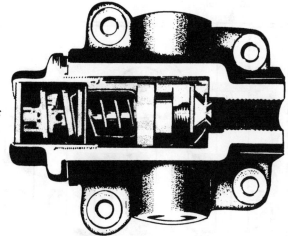

Fig. 4-7. *A typical thermostatic bypass valve.*

Oil coolers. Wet-sump systems typically do not include oil coolers because they use cooling air supplied by an auxiliary impeller on the rotor shaft, or by bleed air. In any case, the centrifugal-flow engine has few bearings to lubricate, which means the oil does not get excessively hot. In addition, intake air that enters the engine flows around the accessory case that cools the oil reservoir.

The axial-flow engine's dry-sump does require an oil cooler for several reasons. First, the design of the engine is such that air cooling must be done with bleed-air, but that is not as effective as the forced air cooling that is available from a centrifugal-flow engine's auxiliary impeller because bleed air is already heated by compression. Second, axial-flow engines have more bearings to be cooled than centrifugal-flow engines, which means the oil gets hotter. Third, because of the design of the axial-flow engine and associated oil tank, the engine intake air does not flow around the oil reservoir. As a result, the only method of cooling the oil is through an oil cooler system.

Two basic oil coolers are *air-cooled heat exchangers* and *fuel-cooled heat exchangers*. Air-cooled heat exchangers are normally installed at the forward end of the engine and lie directly in the intake air stream. Air-cooled heat exchangers are similar to but slightly smaller than those found on reciprocating engines.

Figure 4-8 is a functional schematic of a fuel-cooled heat exchanger cooler. This system serves two functions when heat is exchanged from the oil to the fuel. First, the system cools the hot lubricating oil; if the oil does not require cooling, a thermostatic valve causes the oil to bypass the cooler. Second, the system warms the fuel and thereby preheats it for combustion; fuel moving toward the engine passes through the heat exchanger.

An aerodynamic advantage of the heat exchanger is a smaller frontal area for the engine and nacelle, which reduces drag, because the cooler is mounted on the aft portion of the engine.

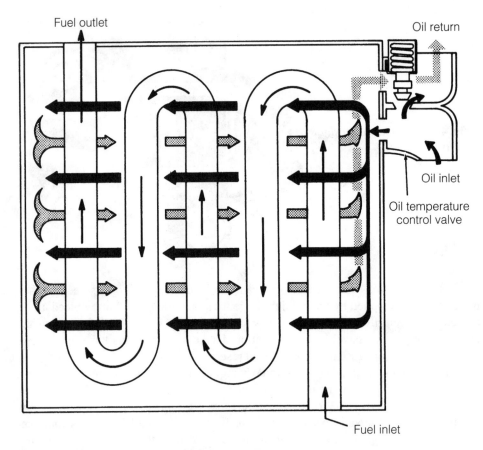

Fig. 4-8. *Fuel/oil heat exchanger cooler.*

TYPICAL DRY-SUMP SYSTEM

A typical turbojet, self-contained, high-pressure, dry-sump lubrication system is illustrated in FIG. 4-9, depicting the pressure, scavenge, and breather subsystems. The pressure system supplies oil to the main engine bearings and accessory drives. The

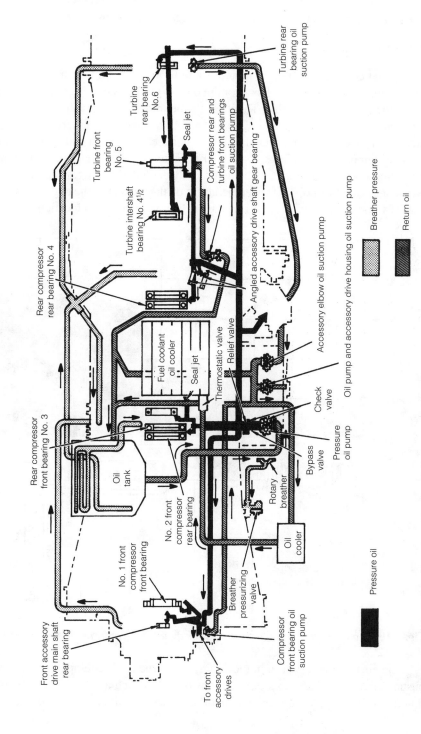

Fig. 4-9. *Typical turbine engine dry-sump lubricating system.*

scavenger system collects the oil in the bearing sump areas and routes it back to the oil tank that is connected to the inlet side of the gear pressure pump.

COOLING SYSTEMS

All internal combustion engines require some form of cooling. Unlike the reciprocating engine, the turbine engine has a continuous burning process that takes place entirely within the casing of the engine. The ideal air/fuel ratio for combustion is 15:1. Internal temperatures would soar to over 4000°F if additional cooling air were not added. In reality, a significantly greater amount of air is channeled through and around the engine to reduce temperatures to fewer than 1500°F. Figure 4-10 illustrates the temperatures at the engine's outer skin when it is being properly cooled. (The engine depicted is a dual axial-compressor, turbojet.) The temperatures on the outside of the engine case are substantially cooler than temperatures inside the engine at the same location.

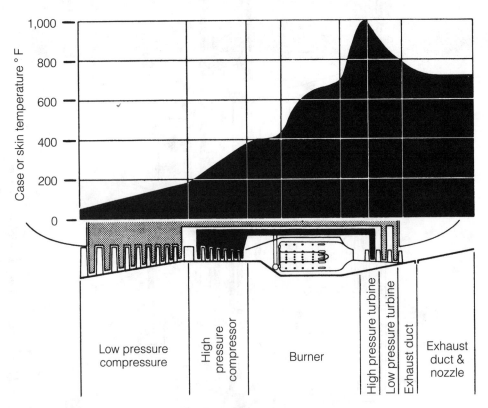

Fig. 4-10. *Typical outer-case temperatures for dual axial-compressor turbojet engine.*

The air that moves through the engine helps reduce the temperature of the combustion-chamber burner cans that are constructed in such a manner as to encourage a thin film of cooling air to flow over both the inner and outer surfaces of the can. Different combustion chamber configurations manage the airflow differently to accomplish the same effect. The net effect in all cases is for a large amount of relatively cool air to mix with the exhaust gases aft of the burners to reduce gas temperature prior to entering the turbine area. In addition, some engines also provide cooling air inlets in proximity to the turbine case, which allows ambient air to also cool the turbine case.

The engine exterior and engine nacelle are also cooled with an airflow that goes between the engine case and the nacelle (FIG. 4-11). Engine compartments are commonly divided into two sections. The forward, engine air inlet section, is built around the inlet duct; the aft section is constructed around the engine itself. A fume-proof seal divides the two sections, which prevents the fumes from potential fuel or oil leaks in the forward section from being ignited by the heat of the aft section. Ram air provides sufficient airflow to cool both sections.

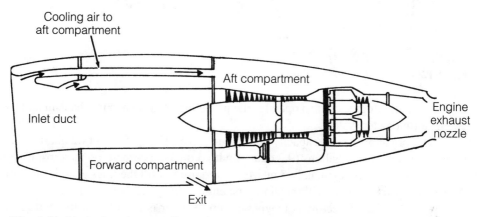

Fig. 4-11. *Typical engine nacelle cooling arrangement.*

MANUFACTURER DOCUMENTATION

(The following information is extracted from Beech Aircraft Corporation pilot operating handbooks and Falcon Jet Corporation aircraft and interior technical descriptions. The information is to be used for educational purposes only. This information is not to be used for the operation or maintenance of any aircraft.)

King Air C90

Engine lubrication system. Engine oil, contained in an integral tank between the engine air intake and the accessory case, cools as well as lubricates the engine. An oil radiator, located in an air duct that is below the pitot air duct for the engine and also a part of the lower pitot cowling, keeps the engine oil temperature within operating limits (FIG. 4-12). Cooling air for the oil radiator enters a flush NACA duct below the pitot air duct for the engine and passes through the radiator in a path roughly parallel to the

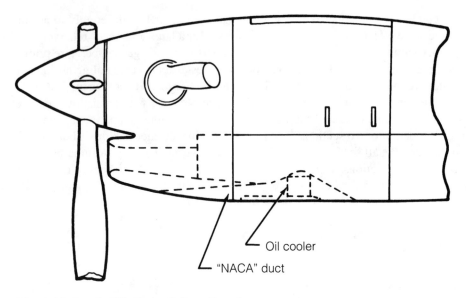

Oil cooler

"NACA" duct

Fig. 4-12. *Beech C90 Pitot cowling oil cooler.* Beech Aircraft Corporation

engine air path. Engine oil also operates the propeller pitch change mechanism and the engine torquemeter system.

The lubrication system capacity per engine is 3.55 U.S. gallons (14.2 quarts). The oil tank capacity is 2.3 gallons (9.2 quarts) with 5 quarts measured on the dipstick for adding purposes. Approximately 5 quarts are required to fill the lines and oil radiator. Approximately 1.5 quarts will remain in the engine oil system when drained.

Magnetic chip detector. A magnetic chip detector is installed in the bottom of each engine nose gearbox. This detector will activate a yellow light on the annunciator panel, L CHIP DETECT or R CHIP DETECT, to alert the pilot of possible metal contamination in the oil supply.

Beechjet 400A

Oil system. The system supplies cooled, pressurized oil for lubrication and cooling of engine bearing and accessory drive gears and bearings. An integral oil tank on each engine has a capacity of 2.03 U.S. gallons, of which 1.34 U.S. gallons are drainable. Recommended oils are listed in Pratt & Whitney Service Bulletin 7001.

Oil drawn from the oil tank by the pressure oil pump is ducted through a check valve to the pressure relief valve for oil pressure regulation (FIG. 4-13). The oil is also passed through the oil cooler and oil filter. Excess oil pressure at the oil filter outlet opens the pressure regulating valve and some of the oil is bypassed and ducted externally through a second check valve to the oil pressure pump inlet. From the filter, oil is routed to the engine bearings and accessory gearbox. If the filter becomes clogged, a bypass valve opens allowing lubrication to continue and the L or R O FLTR BYPASS an-

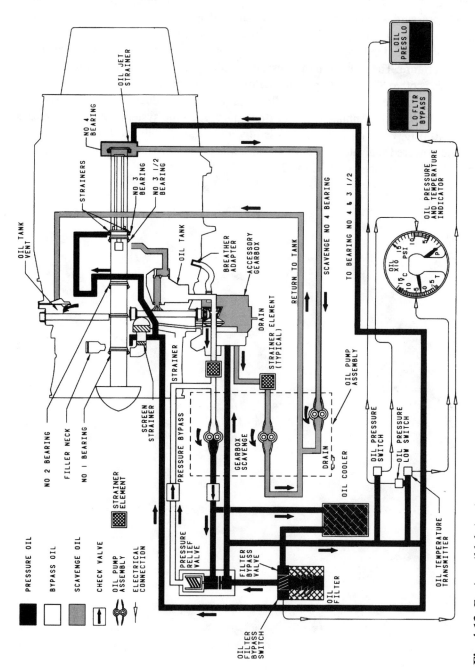

Fig. 4-13. *Engine oil lubrication system.* Beech Aircraft Corporation

nunciator will illuminate. Circulated oil from the #4 bearing area and accessory gearbox is returned to the tank by two scavenge pump elements in the oil pump assembly.

Oil pressure is sensed by a transmitter and is displayed on two ac-powered indicators with dual pointers (pressure and temperature) on the instrument panel. The scale ranges on the dual indicators are 0 to 150 psig for oil pressure and –50°C to 150°C for oil temperature. The minimum oil pressure at idle rpm is 40 psig. The normal operating range is 60 to 83 psig. Oil pressure below 60 psig is undesirable and should be tolerated only for the completion of the flight, preferably at reduced power settings. The L or R OIL PRESS LO annunciator will illuminate when the system pressure decreases below 40 psig. Oil temperature is sensed by a resistance bulb and transmitted to the dual dc powered indicators. Minimum oil temperature for engine start is –40°C and normal operating range is 10°C to 121°C.

5
Aircraft propellers

EARLY ATTEMPTS AT POWERED FLIGHT FACED MANY PROBLEMS, NOT THE least of which was how to convert engine power to thrust for forward movement. Some of the earliest propeller designs were nothing more than flat boards designed to push air rearward, much like a fan. Eventually it was discovered that the same principle of lift that causes a wing to fly could work effectively with the propeller. Over the years, aircraft became heavier, began to fly higher, travelled farther distances, and flew at faster speeds. Propeller technology regularly pushed the limit of the envelope to accomplish those goals by increasing the number of blades, using ever-thinner and lighter metal instead of wood, and by allowing the pilot to control blade pitch to maximize efficiency for each condition of flight.

Engineers soon discovered, however, that there were limitations to how fast a propeller could turn. As the rotational speed increases, so too does the centrifugal force acting upon the propeller blades trying to pull them out of the hub. But the answer was not necessarily making the blades longer for increased thrust while turning them at a slower speed—that answer ran into trouble with the speed of sound. Even though blade revolutions per minute (rpm) could be reduced, the increase in distance between the prop tip and the hub made the arc of the prop tip significantly longer. This is the same phenomenon you experience when swinging a rubber ball at the end of a string; the revolutions per minute might remain the same, but the longer the line, the faster the ball (at the end of the string) must move to cover the increased distance. With a propeller, as the tips approach the speed of sound, the blade efficiency begins to decrease and fluttering and vibration begin to increase. If they turn fast enough, structural damage is the result.

With the advent of the turbine engine, there was a brief time when it appeared that propeller-driven aircraft might vanish. But the smooth running, highly reliable turbine powerplant, efficient in the cold temperatures of high altitudes, did not fare as well at the lower altitudes frequented by the reciprocating engine and propeller combination; however, an increasing number of operators wanted an aircraft with the reliability and efficiency of a turbine engine that could be used at altitudes conducive to trip stage lengths of a several hundred miles or fewer; thus, the turbine engine and propeller powerplant combination was born that has come to be known as the *turboprop*. This hybrid could routinely operate in the 15,000–25,000-foot altitude range.

Depending upon the objectives of the specific propeller, two or more metal blades—typically aluminum alloy or steel—might be attached to a central *hub*. The hub is the end of a shaft that goes to a reduction gearbox that reduces the very fast rotational turbine speed to a manageable propeller speed. From a functional point of view, a propeller blade—or a helicopter rotor—is simply a revolving wing. Through careful aerodynamic engineering, each rotating blade produces maximum thrust throughout its full length. The combined thrust of all blades, depending upon the application, will pull or push the airframe. Props that pull, known as *tractors*, are far more common; however, pusher prop applications are becoming more numerous. One of the earliest pusher applications was the in-line, multiengine Cessna 337 Skymaster with one tractor and one pusher prop, both powered by reciprocating engines. Pilots that did not hold a multiengine rating and flew the Skymaster had to obtain a special endorsement on their certificate: center line thrust. Unfortunately, the 337 was the only center line thrust light twin ever built. Turboprops rule the pusher prop market with luxurious aircraft such as the Beech Starship and Piaggio Avanti.

PROPELLER THEORY

An aircraft moving through the air creates *drag*, a force that resists the forward motion of the aircraft. For an aircraft to be capable of flying straight and level, it is necessary to oppose that force with *thrust*: an equal but opposite force. The amount of work accomplished by the thrust is equal to the thrust multiplied by the distance the aircraft moves. Work is not a unit of measurement that is readily usable in an aircraft. Instead, we use power, which is equal to thrust times velocity, and in turboprop aircraft the power expended by the thrust is called *thrust horsepower*.

Terminology

Propellers are not 100 percent efficient due to *slippage*. Propeller slip is the difference between geometric pitch and effective pitch as displayed in FIG. 5-1. *Geomet-*

ric pitch is defined as the ideal distance a propeller should advance in one revolution. *Effective pitch* is defined as the distance actually advanced.

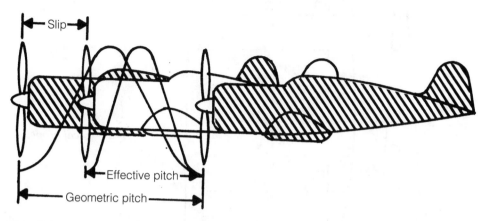

Fig. 5-1. *Effective and geometric pitch*

Propeller blade thickness decreases as distance from the hub increases (FIG. 5-2). The angle of the blade also changes along the same dimension as depicted by the 6-inch-*station* cross sections in the illustration. To assist in accurately defining a location on a propeller blade, a system was developed using blade station numbers that are measured in inches from the center of the blade hub. Several terms are also displayed in FIG. 5-2. The thickest part of the blade is called the *shank*, the lowest portion of which is known as the *blade butt* or *base*. The shank provides stiffness to the blade and is fitted into the propeller hub to securely hold the blade in place. The blade tip is the last 6 inches of the opposite, thin end, and is usually rounded at the very end.

The cross section of a propeller blade is illustrated in FIG. 5-3. The cross section easily reveals a strong resemblance to a wing section. The imaginary *chord line*, used as an engineering reference, goes from the *leading edge* to the *trailing edge* of the section. The leading edge, which is the thickest part of a blade section, is the front of the blade and that part that first meets the air as the blade rotates. The curved side, called *cambered*, is the blade back and equates to the top of a wing. The back of the blade faces forward, when the propeller is mounted on the engine, creating lift out in front of the propeller in the direction of aircraft travel. The other and flatter side of the blade is called the *face* because it faces the pilot.

Pilots will often incorrectly allude to "pitch" when referring to the angle between the chord line of the blade and the propeller's plane of rotation. The proper term for that angle is *blade angle* (FIG. 5-4). The typical constant-speed, reversible, full-feathering turboprop will have a blade angle range of approximately +90° (feather) to –15° (reverse). (Feathering is fully explained underneath the **Flight operation mode** heading in this chapter.)

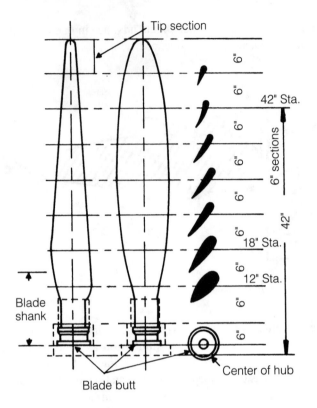

Fig. 5-2. *Typical propeller blade elements.*

Tip section

42" Sta.

6" sections

42"

18" Sta.

12" Sta.

Blade shank

Center of hub

Blade butt

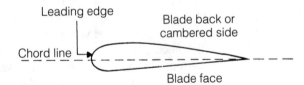

Leading edge

Blade back or cambered side

Chord line

Blade face

Fig. 5-3. *Cross section of a propeller blade.*

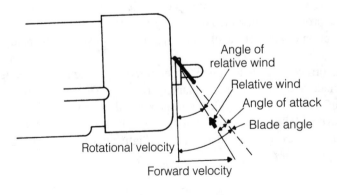

Angle of relative wind

Relative wind

Angle of attack

Blade angle

Rotational velocity

Forward velocity

Fig. 5-4. *Propeller aerodynamic factors.*

The term *pitch* refers to the geometric pitch illustrated in FIG. 5-1, which is the theoretical distance the aircraft moves forward during a single revolution of the propeller; however, the two are directly related. Within limits, as the blade angle increases, so too does the geometric pitch. The bigger the bite of air taken by the blade, the farther forward the aircraft moves in one revolution of the propeller.

Blade angle is controlled from the flight deck and is changed as necessary to get maximum efficiency from the propeller during all phases of flight. The *angle of attack* is the relationship between the chord line of the blade and the *relative wind*. Because the propeller rotates, it creates its own relative wind apart from the relative wind that interacts with the aircraft's wing. Though they are functionally similar, they are two different situations. With the propeller, the relative wind flows parallel and opposite to blade rotation; with the wing, the relative wind flows parallel and opposite to the direction of the flight path of the aircraft.

Forces acting on a propeller

Five basic forces act on a rotating propeller (FIG. 5-5):

- Centrifugal force
- Torque bending force
- Thrust bending force
- Aerodynamic twisting force
- Centrifugal twisting force

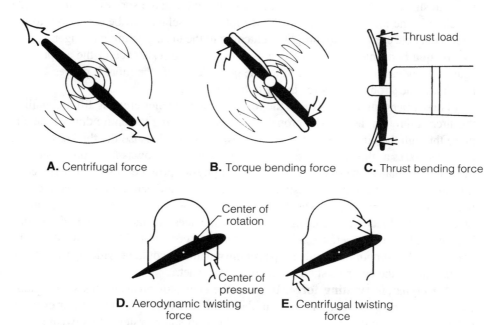

A. Centrifugal force   **B.** Torque bending force   **C.** Thrust bending force

D. Aerodynamic twisting force   **E.** Centrifugal twisting force

Fig. 5-5. *Forces acting on a rotating propeller.*

135

Knowledge of how these forces interact with the propeller is a key element in accident investigation. When logically applied to evidence obtained from post-crash engine analysis, examination of the propeller can tell investigators several important things:

- Power, if any, that was being produced at time of impact
- Rotational speed of the engine
- Propeller blade angle
- Ground speed of the aircraft

Centrifugal force. The faster the propeller rotates, the more that centrifugal force tries to pull the individual blades out of the hub and launch them like missiles. With a turbine engine compressor, for example, centrifugal force is so tremendous that on the very rare occasion a turbine engine compressor "throws" a blade for some reason, the blade has been known to cut right through the engine casing, cowling or nacelle, and even fuselage. Propellers only turn at a fraction of the speed of a compressor, but the mass of a propeller blade is significantly greater than that of a small compressor blade. For instance, when a propeller begins to accumulate ice and the crew activates the prop deicing system, centrifugal force pulls the ice off the blade (after the ice is loosened by the deice system).

Propeller-driven aircraft that frequently operate in icing conditions are characterized by the chipped paint on the fuselage directly in line with the propeller's plane of rotation. Ice is slung outward off the prop blade; as a result, the blade closest to the fuselage slings ice against the aircraft in essentially the same spot every time. Inside the aircraft, the loud "whoomp" as the ice hits the fuselage can be eerie even to the knowledgeable crew and downright frightening to the unsuspecting passenger.

Torque bending force. This force is caused by the resistance of the air pushing against the advancing blade. The faster the propeller turns, the more it pushes against the air and the stronger the force trying to hold it back.

Thrust bending force. This is the result of the propeller creating lift and pulling the aircraft through the air. The propeller is literally pulling the entire drag of the airplane through the air, causing the prop to bow forward. Because the thickness of the blade decreases as the distance from the hub increases, the bow is more pronounced at the blade tips than at the shank. This little bit of physics provides a significant piece of information during accident investigation. If the blades are bent forward near the tips, that is a clue that the engine was developing high power at time of impact. If, on the other hand, the blades are bent slightly aft, this indicates that the propeller was turning but with little or no power, essentially a windmilling condition. This test would not be conclusive, but supported by other types of information it could provide significant understanding of the engine's status at the time of impact.

Aerodynamic twisting force. Because the center of pressure lies substantially ahead of the center of rotation on each blade, the resulting force-moment tries to increase the blade angle. As will be discussed later, in some systems this twisting moment is used to move the propeller into feather under certain conditions.

Centrifugal twisting force. This is the opposing force to the aerodynamic twisting moment. Because this force is greater, it tries to move the blades toward a reduced blade angle.

A propeller is designed to withstand the effect of these forces, but the forces are nonetheless important factors in design and operation. The effect of these forces accumulates across the length of the blade with the greatest stress at the hub. As the rotational speed of the propeller increases, so too do the stresses acting upon it. Given the various forces acting upon a propeller, it is not difficult to understand the serious problem associated with even small nicks or scratches that could weaken the integrity of the propeller.

Propeller aerodynamics

To understand how a propeller moves an aircraft through the air, it is necessary to look at it from an aerodynamic rather than a mechanical perspective. Figure 5-6 depicts the side view of a propeller detailing the blade path, blade chord, and relative wind. The illustration reveals two types of motion associated with the propeller blades: *rotational* and *forward*. As a blade moves downward, it simultaneously moves forward. This has a significant effect on the relative wind making it strike the blade at an angle that is between straight ahead and straight down. This angle that the relative wind strikes the blade is called the *angle of attack*. The relative wind hitting the descending blade is deflected rearward causing the dynamic pressure on the engine side of the blade to be greater than the pressure on the back of the blade. Again, within limits, as the blade angle increases, so too does the angle of attack. On a wing, this situation is called lift; on a propeller it is *thrust*. Thrust is the result of the camber of the blade and the angle of attack of the blade.

CONSTANT-SPEED PROPELLER

A propeller *governor*, as part of the constant-speed propeller construction, maintains propeller rpm by automatically varying the blade angle. Propeller rpm—on a reciprocating or turbine engine—is set by the pilot by moving a control lever in the cockpit. The full forward position moves the blade angle to its shallowest position, which yields the maximum rpm possible, consistent with the available power, and is designated the normal takeoff and initial climb prop setting. The shallow pitch is also used on landing for two reasons: in case the pilot must abort the landing (and needs maximum performance to initiate a climb) and to prepare for reverse pitch (if needed) after landing. Engagement and operation of *reverse thrust* is accomplished with the thrust levers, but the prop levers should be in the full-forward position.

Moving the propeller lever to its full-aft position places the blade angle in a very steep pitch situation that results in the minimum rpm possible. Minimum rpm is consistent with maximum long-range economy cruise; however, the actual propeller rpm operating range of turboprops is very narrow. In the Beech King Air C90A, for instance, the normal operating range is 1800 to 2200 rpm. In most turboprop aircraft, the

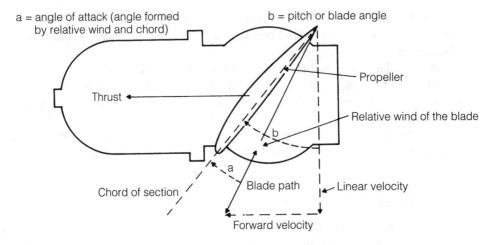

Fig. 5-6. *Propeller forces.*

most aft position of the prop lever is just forward of a detent that prevents accidental movement of the lever farther aft.

Different aircraft have different methods for moving farther aft of that position, such as lifting the lever or shifting it to one side, but movement past the low rpm position will cause the propeller to move into the feather position. Some aircraft use feathering buttons rather than aft movement of the prop lever; the result is the same. To review:

- Low (shallow) pitch = high rpm
- High (steep) pitch = low rpm

The option to select a specific blade angle allows the pilot to select the most efficient propeller blade angle of attack for any given phase of flight, such as takeoff, cruise, or descent. Similar to the lift-to-drag curves calculated for wings, engineers have determined the same information for propellers. The most efficient propeller angle of attack is between +2° and +4°. This should not be confused with the blade angle required to maintain the angle of attack; the actual blade angle will vary slightly depending upon the forward speed of the airplane. Remember, for any given propeller rotational speeds, the angle of attack will vary as forward speed increases or decreases; angle of attack is the product of rotational speed and airspeed. The propeller's control system is actually divided into two modes: flight operation and ground operation.

Flight operation mode

In the flight operation mode, the constant-speed propeller will make small adjustments in blade angle to assure maximum efficiency under slightly varying conditions, with limitations. By consulting the aircraft operating manual, the pilot is able to select the desired speed of the propeller given the conditions of flight. It is important to remember, however, that unlike reciprocating engines, the turboprop operating speed range is very limited. Typically, on takeoff, the propeller levers are set full forward for maximum rpm and are kept in that position until cruise level-off. A second, slightly slower rpm setting of 1900 is used to handle all cruise conditions. To vary performance from maximum cruise to maximum range power, for instance, the pilot checks the appropriate performance chart and sets the *torque* as indicated; the rpm remains unchanged.

Torque. The measured amount of shaft horsepower absorbed by the propeller—torque—is the primary control of aircraft performance in flight. The pilot sets engine torque with the *power lever*, which is the turbine engine's equivalent to the throttle. Moving the power lever for each engine sends a signal from the flight deck to the fuel control unit requesting a specific amount of engine power. The fuel control unit and the propeller governor are interconnected and operate together to establish the precise combination of rpm, fuel flow, and propeller blade angle to provide the requested power. As the power levers are moved forward, the torquemeters and the exhaust temperature indicators will reflect an increase. The governor maintains a constant propeller rpm even as the engine power increases.

To maintain a constant rpm, the blade angle automatically increases to take a bigger bite of air. It is the increased air load that prevents the propeller from speeding up as the torque increases. The net result is that the increased torque goes to processing a greater mass of air per second, which results in more thrust. So, in the flight operation mode, the propeller blade angle and fuel flow for any given power lever setting are governed automatically according to a predetermined schedule.

Power vs. engine speed. To the experienced reciprocating engine pilot, it is a somewhat confusing characteristic of the turboprop engine that a change in power does not relate to engine speed, but instead to turbine inlet temperature. During flight, the propeller maintains the engine speed at a constant rpm known as *100 percent rated speed*. This is the design speed of the engine that provides maximum efficiency and power. If the pilot desires more power and pushes the power lever forward, the fuel control unit simply schedules an increase in fuel flow.

The result of more fuel is an increase in the turbine inlet temperature, which effectively means more energy is available to turn the turbine. The turbine is forced to absorb the extra energy that is transmitted to the propeller in the form of torque. If there were no governor, the propeller speed would increase; instead, the blade angle increases to maintain a constant engine rpm and the propeller is able to take a larger bite of air.

Negative torque signal. Thus far, the operational condition discussed has been the engine turning the propeller. Occasionally, conditions can occur when the propeller attempts to turn the engine. There are several possible causes for a negative torque situation, but the engine is protected from an actual occurrence by a *negative torque signal* (NTS) system. Potential negative torque situations are:

- Temporary fuel interruption
- Air gust load on the propeller

- Normal descents with lean fuel scheduling
- High compressor air bleed loads at low power settings
- Normal engine shutdown

So, two ways of inducing a negative torque situation are something trying to turn the propeller faster than scheduled, or something causing engine speed to suddenly slow down to fewer than scheduled rpm. In either case, the NTS system immediately signals the propeller governor to increase the blade angle until the negative torque situation ends.

Propeller feathering. The term feathering refers to the ability to turn the propeller blades in such a manner as to make the blade leading edges face directly into the ambient air flow; essentially a 90° blade angle. The result is that the propeller creates no aerodynamic forces. When an engine fails, the forward motion of the aircraft continues to create a relative wind. The relative wind will continue to turn the propeller of the failed engine—which is also called *windmilling*. Do not allow this to happen for two reasons.

The first reason is safety. A windmilling propeller creates a tremendous amount of drag that, when combined with the thrust being created by the operating engine, causes a very strong yawing movement toward the inoperative engine. The slower the airspeed, the more pronounced the problem and, if uncorrected, this condition of *asymmetric thrust* will result in loss of aircraft directional control. In addition, it is important to realize that because of the tremendous drag imposed by the inoperative engine and windmilling propeller, very few twin-engine aircraft are even able to hold altitude with a windmilling propeller. Perhaps even more revealing, most twin-engine aircraft lose more than 80 percent of their climb capability even with a feathered propeller. It is imperative that the multiengine pilot be well trained and current in the proper single engine procedures for the aircraft being flown. Under any condition other than cruise, an engine failure requires prompt, positive action.

The second reason to not allow an engine to windmill is damage control. If the engine has failed by itself, there must be a reason. If the reason is either a mechanical or foreign object ingestion problem, continuing to allow the propeller to turn could result in significant engine internal damage. In most systems when the engine is inoperative the oil pump is also not operational. The result of a windmilling prop is that the shaft continues to turn in bearings that are not being lubricated. Any of these situations can rapidly lead to major engine damage. For all these reasons and more, concern about a windmilling propeller is so serious, especially during takeoff and initial climb, that manufacturers incorporate an autofeather system.

Autofeather systems. Depending upon the manufacturer, there are different types of autofeather systems. The *thrust-sensitive signal system* (TSS) arms itself whenever propeller-positive thrust is in the takeoff range. A subsequent loss of thrust on that engine causes the system to automatically feather the propeller.

Other autofeather systems are pilot selectable. Most aircraft operating handbooks call for the activation of autofeather during the critical takeoff and initial climb segments. When activated, this type of system typically compares the torque of one engine against the other. If one engine's torque drops below a preset value, the system automatically feathers that engine's prop. This system has a safeguard built in that precludes accidental feathering of a propeller when the pilot has retarded one thrust lever for whatever reason. Most autofeather systems will completely feather a propeller in 5–15 seconds.

Ground operation mode

In the ground operation mode, things are somewhat different. If the power levers are brought all the way back to the flight idle position, it is possible to physically lift the levers out of their flight idle safety detent and move them back farther into what is known as the *beta* range. In this area of operation, the coordinated rpm/blade angle schedule that governs normal flight operations is no longer appropriate. As a result, whenever operating in the beta range, the propeller blade angle is no longer being controlled by the propeller governor, but is directly controlled by the position of the power lever itself. (*See* the heading **Turbopropeller assembly** in this chapter for more information about beta range.)

Reverse thrust. When the power lever is full aft and stopped at the flight idle position with the prop levers full forward to maximum rpm, the propeller blades are sitting on the low pitch stop. The low pitch stop is a mechanical device that physically prevents the blades from rotating from the lowest possible blade angle position to reverse. When the power lever is first lifted above the flight-idle safety detent and moved aft, the low pitch stop is automatically opened to allow the propeller blade angle to rotate into full reverse position. This provides reverse thrust for slowing down the aircraft after landing, or for taxiing. In some aircraft, it is possible to slowly back up the aircraft.

As the power lever continues to move farther aft, the blade angle stays the same but the engine, which has been at idle, begins to accelerate causing an increase in reverse thrust. It is important to recognize that similar to reverse thrust in the turbojet engine, the term reverse thrust does not imply any reversal of the engine. In the case of the propeller, the blade angle rotates through the lowest normal flight blade angle setting past the low pitch stop and continues beyond until the blades have rotated far enough around to create thrust opposite to the direction of aircraft movement. The propeller is now trying to push the aircraft backward. For most propellers, reverse-thrust blade angle is between $-5°$ and $-15°$.

All turboprop operating handbooks specify a minimum speed on landing roll below which full reverse should no longer be used. This is the speed below which the propellers will move foreign objects forward faster than the forward speed of the aircraft. Use of full reverse when slower than the published speed increases the potential for foreign object ingestion or prop damage, and reduced forward visibility because sand and dust will be kicked up in front of the aircraft. Use of reverse during taxi should be done with caution; however, reverse does minimize brake wear.

Be particularly careful about using reverse thrust to back up the aircraft. While it is very impressive to those watching you back your pretty, new cabin-class turboprop into a parking space, you look pretty dumb when you hit the brakes and the airplane sits down on its tail.

PROPELLER SYSTEM OPERATION

The turboprop combination provides one of the most flexible yet efficient systems for converting power to thrust. Turboprop system operations vary based upon the manufacturer of the system; however, each system accomplishes the same task, which is moving the aircraft through the air in the most efficient manner possible. The following general description of how a constant-speed, feathering, reversible propeller operates is based upon a Hartzell system.

Reduction gear assembly

The output shaft of the engine drives a reduction gear assembly that reduces the fast engine rotational speed to a more suitable range for propeller operation while maintaining a constant engine rpm. Depending upon the manufacturer, the assembly might also incorporate other additional features (FIG. 5-7). The illustrated assembly, for instance, incorporates an NTS system, propeller brake, and an independent dry-sump oil system. The illustration also shows the torquemeter assembly (upper right-hand corner).

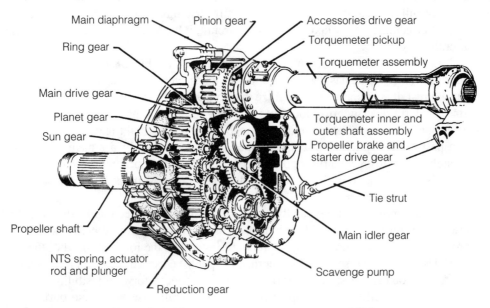

Fig. 5-7. *Reduction gear assembly.*

Propeller brake. The prop brake system serves two purposes. After an in-flight shutdown—even after the propeller has been feathered—it is still possible for the prop to rotate slowly due to variable air loads buffeting the propeller blades and the relatively low friction associated with the propeller-to-turbine shaft. Regardless, the shaft would be turning on bearings that are not being lubricated because the engine is inoperative. The propeller brake assures the shaft will not turn. The second reason is to slow down prop rotation quicker after engine shutdown on the ground. This is particularly beneficial on a quick turnaround when dropping off or picking up passengers; it is wise to shut down the engine on the passenger door side before allowing passengers to approach or leave the aircraft. Without a prop brake the slowdown time can take several minutes. The brake also secures the prop on the ground when the aircraft is unattended.

The propeller brake is a *friction-cone* system composed of a stationary inner member and a rotating outer member. When the engine is operational, oil pressure holds the outer member away from the inner member and prevents brake application. When the engine is shut down, or the prop feathered, the oil pressure drops and a spring automatically moves the outer member against the inner member, applying brake pressure.

Turbopropeller assembly

Part of the propeller assembly (FIG. 5-8) is the control assembly that contains the oil supply and necessary equipment to supply the prop pitch-changing mechanism with necessary hydraulic power and to appropriately direct the supply to vary the blade pitch for the condition of flight. The primary task of the control assembly is to assure a constant engine rpm throughout the *alpha*, or normal, flight operations range. The alpha range—approximately +14° to +90° blade angle—is that range in which the propeller governor controls blade angle to maintain the requested prop rpm. During ground handling and reverse (beta range), the pilot can operate the propeller at either zero thrust or reverse thrust. Beta range, approximately 0° to +14°, is the nongoverning range because blade angle is controlled by the power lever position. When the power lever is moved below the start position, blade angle moves into reverse. Numerous subsystems such as the prop control oil system including oil reservoir and the brush housing that distributes the electric power for the deicer rings are contained within the control assembly.

The items on the left side of the *afterbody assembly* in FIG. 5-8 rotate with the propeller; items on the right side are stationary. The front and rear spinners are secured to the propeller assembly and shaped to keep parasitic drag to a minimum by enclosing the dome and barrel. The front spinner also funnels ram air to provide cooling for the oil used in controlling propeller pitch. The afterbody assembly is fixed to the engine gearbox and used to enclose the control assembly. This also aids in streamlining the airflow to reduce drag.

Constant-speed system. The large steel dome (*spinner*) at the front of the propeller covers the hub, which contains the mechanism that causes the individual blades

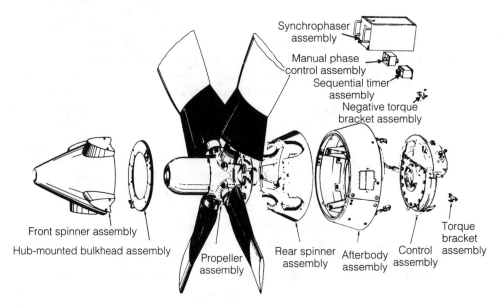

Fig. 5-8. *Propeller assembly and associated parts.*

to rotate according to regulation by the governor control; the hub also contains a retention system that secures the blades to the hub against centrifugal force. Within the hub is a central spider gear that rotates the individual blades.

Constant-speed propellers utilize two forces to increase or decrease blade angle. The forces are *fixed* and *variable*. Centrifugal force is the fixed force, named because it is a constant by-product of rotation and always exists during propeller operation. Centrifugal force is used as a constant that attempts to move the prop counterweights into the plane of rotation, which increases the blade angle (FIG. 5-9). The variable force, which opposes fixed force, is oil pressure. The governor varies the oil pressure as necessary to counterbalance centrifugal force and subsequently decrease the blade angle. The governor accomplishes this task with a hydraulic piston inside a movable cylinder that is located within the propeller dome. Oil pressure, which is greater than the centrifugal force acting on the counterweight, is directed by the governor to flow into the cylinder. The oil fills the cylinder causing the cylinder to move outward away from the fixed piston; that cylinder movement pulls the blades that rotate on the spider gear and decreases the blades' angles.

Governor. The governor, in conjunction with the control assembly, manipulates oil pressure as necessary to adjust blade angle to maintain the rpm selected by the pilot. A set of flyweights inside the governor rotate in direct relationship with the propeller. When in operation, two opposing forces are acting upon the flyweights: centrifugal force and spring pressure. The centrifugal force component acts upon the rotating flyweights—this centrifugal force is similar to, but separate from, the centrifugal force acting upon the propeller blades. The faster the flyweights rotate, the greater the tendency for them to move outward (FIG. 5-10).

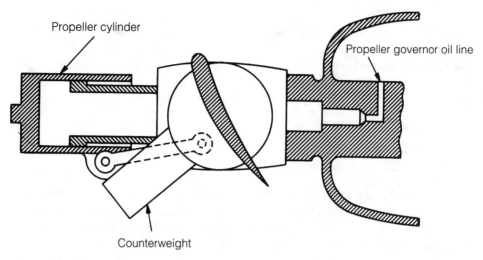

Fig. 5-9. *Pitch change mechanism for a counterweight propeller.*

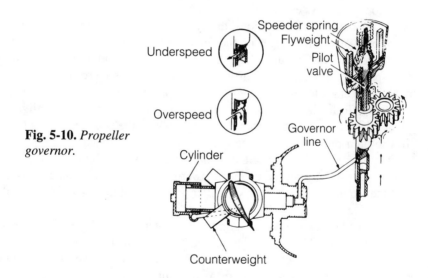

Fig. 5-10. *Propeller governor.*

The opposing force, *speeder spring* pressure, is simply a spring that attempts to pull the flyweights inward by applying pressure to them. The spring is designed to receive inputs from the propeller lever that will increase or decrease the spring pressure. For instance, the pilot selects 1900 rpm with the prop lever. Because it is connected to the speeder spring it schedules its tension to an appropriate value. The tension value allows the flyweights to maintain the appropriate balance between propeller centrifugal force and oil pressure to maintain 1900 rpm. The flyweights are able to control the flow of oil by moving a *pilot valve* up and down.

In one position the pilot valve permits oil from the oil pump to flow into the propeller hub to counteract the effect of centrifugal force on the propeller blades. Oil pressure attempts to move the blade angle toward low pitch. Another valve position drains oil out of the hub, allowing centrifugal force to dominate and move the blade angle to high pitch. As prop speed increases above the preselected value, which is a condition called *overspeed*, the following sequence of events occurs:

1. The flyweights move out.
2. The pilot valve moves up.
3. Oil drains from the dome.

As a result, the centrifugal force on the prop blades dominates and the blade angle moves toward high pitch/low rpm. As the prop speed decreases below the preselected value (an *underspeed* condition), the reverse occurs. The flyweights move in, the pilot valve down, oil flows from the pump into the hub, and the prop blades move toward low pitch/high rpm. When the prop is turning at the preselected speed (an *on-speed* condition), the centrifugal force on the flyweights is in balance with the force of the speeder spring. In that situation, the pilot valve is positioned in such a manner that no oil moves in any direction and the system remains in a status quo condition.

Under normal operating conditions, the system operates so smoothly that it is impossible to detect any rpm changes on the flight deck. If, for instance, the aircraft enters a steep descent that would result in an overspeed condition, the flyweights instantly respond by appropriately shifting the pilot valve, which results in a readjustment of the blade angle to prevent any noticeable rpm fluctuation.

Feathering. To feather a propeller, the pilot activates the appropriate flight deck control that releases oil pressure from the hub, allowing the counterweights to move the blades to high pitch. To assist the counterweights, and speed up the process, a heavy feathering spring is also installed in the hub that, when released by the feathering initiation, aids in pushing the cylinder toward the piston to evacuate the oil. The length of time required to feather the propeller is dependent upon the diameter of the oil passage; typically it takes 3–10 seconds.

An often asked checkride question on this type of system is "What happens if the system loses oil pressure?" The answer, of course, would be the prop goes to feather. In general, if there is a propeller system failure, it is better to have the prop go into feather than into an overspeed condition. As a result, most propeller systems are designed to do so in the event of normally anticipated failures.

SYNCHROPHASERS

If you have ever been in any propeller-driven aircraft—with more than one engine and prop—and felt, as much as heard, a type of throbbing sound, chances are the props were not synchronized (in *sync*). For some people, extended exposure to props out of sync causes headaches, anxiety, and general discomfort. More often than not, the people are unable to identify a cause because it is a sound that falls only slightly above the human threshold of awareness.

Most multiengine aircraft have a prop *synchronizer* system that fine-tunes each propeller to assure duplication of exact rpm. Doing so significantly reduces vibration and the audible beat. Operational specifics vary among systems; typically, one engine is designated as the master and any other engine is a slave. A frequency generator for each engine translates the propeller's rpm into a specific, discrete frequency. A comparator unit compares the frequencies to the master and sends a command to each prop governor that controls a propeller that does not have the same frequency as the master. The command directs a specific governor to either slow down or speed up as appropriate. The span of control for this type of system is limited; therefore, before activating prop sync it is important that all props be within 100 rpm of the master.

Synchrophasing is a more sophisticated version of a prop synchronizer and is designed to virtually eliminate differential prop noise. In addition to synchronizing the propellers to the same rpm, this system also allows the pilot to adjust the phase angle of the slave propellers to assure that no two blades continuously move through the same position at the same time. Take, for instance, a four-engine turboprop aircraft. If the ascending blades of all four engines simultaneously passed through the horizontal position every rotation, the flight deck occupants would sense a vibration or beat. In a manner somewhat similar to the synchronizer system, the synchrophaser interrogates each propeller, compares it to a designated master propeller, and interacts with the individual prop governor to eliminate the problem. On the flight deck, the pilot is able to adjust a control that permits a limited variability in phase control so it can be fine-tuned to yield the minimum vibration.

PREFLIGHT AND MAINTENANCE

Propellers involved in accidents, especially when the prop is stopped by striking the ground or other object, are always suspected of internal damage. In such a case, the entire propeller system and shaft should be removed from the engine, disassembled, and have all parts inspected. Depending upon the nature and severity of the accident, it might also be prudent to do a nondestructive evaluation, such as a magnetic inspection, to assure that no part has sustained fatigue cracks or other abnormalities.

All constant-speed propellers require periodic scheduled maintenance. Different manufacturers have different recommended maintenance procedures and schedules; however, all props have common considerations. Major overhauls also include complete disassembly of the entire system. All the parts are carefully inspected for wear, fatigue, and adherence to manufacturer's tolerances. A nondestructive evaluation is usually conducted on certain high-stress and high-wear parts.

Preflight

During each preflight the pilot should visually inspect blades, hubs, and the like, for security and general condition. Look for oil on the ground underneath the propeller, running along the bottom of the nacelle, or pooled inside the engine nacelle. Oil leaks can reflect a serious problem and are usually easy to spot if the airplane is kept clean. Next, check the individual propeller blades.

Many pilots have developed bad habits regarding blade examination, with only cursory observations. Considering the tremendous forces acting upon the propeller, and the fact that a prop is the only thing moving you forward, it is surprising how quickly it gets passed by. If microscopic nicks and flaws can eventually lead to blade failure, it's a pretty good idea to give each blade a thorough examination. For instance, run a fingernail along the leading and trailing edges. If the fingernail catches on something, carefully investigate why. Small nicks, such as dings from gravel and the like, are tolerable within limits and even those will be filed out by the mechanic during the next scheduled maintenance. Even a small nick might be unacceptable if located farther out on the blade toward the tip. If in doubt, always consult your mechanic.

Propeller vibration

Propeller vibration is often very elusive. It is something we can feel but are not certain about the cause. One helpful test is to observe the prop spinner while it is turning in the 1,200 to 1,500 rpm range. If the spinner moves in an elliptical orbit, however small, the vibration originates with the propeller (verify that the spinner is squarely attached to the hub); otherwise it is probably an engine-related problem. Remember that turbine engines by their design have minimal vibration. If prop vibration is strong enough to cause engine vibration the reason is probably one of three possibilities: *blade tracking*, *differential blade angle settings*, or *blade imbalance*. All three are not pilot correctable and must be corrected by a mechanic.

Blade tracking. This procedure checks to assure that each blade tip on a given propeller rotates in the same plane. By way of clarification, let's suppose we took a piece of white paper and put it on top of a small table. We then place the table directly under a propeller so that the blade tips just lightly touch the paper. Now daub some black ink on each of the tips and turn the propeller through 360°. If the blades are tracking true, there should only be a single line drawn on the piece of paper after all the blade tips have passed by. A single line is unlikely; however, the manufacturer does specify maximum deviations for a fundamentally similar test. The greater the difference among blade tips, the greater the resultant vibration.

Blade angle setting. Propeller blade angle is measured from a manufacturer's designated blade station. If properly measured, all blade angles should be the same at any given time. If they are not, the result can be significant vibration and loss of propeller efficiency.

Propeller balancing. Two types of propeller imbalance are *static* and *dynamic*. Static imbalance is the result of the propeller's center of gravity being off the center of rotation. The test to determine if a propeller has static imbalance is to put it on a test stand that supports the propeller on each side so that the prop sits on the stand as if attached to the aircraft. The stand allows the propeller to freely rotate about its normal axis. The test should be conducted in a room free of drafts.

The mechanic rotates the propeller so that one of the three blades is pointing straight down with the remaining two pointing upward, as if forming the letter Y. Each blade, in turn, should be placed on the bottom. A perfectly balanced propeller will stay in that position. Four-blade props are set with two blades horizontal and two vertical with the same results. Any tendency for the propeller to rotate indicates static imbalance. Within limits, small weights can be added or removed to balance the propeller.

Dynamic imbalance results when the individual centers of gravity of similar parts of the propeller are not tracking in the same plane of rotation; for instance, misaligned counterweights. This fault is typically minimal in propellers when tracking is within the manufacturer's specifications.

Cleaning propeller blades

The few rules pertaining to cleaning the propeller are based upon common sense. Because propellers are made of aluminum and steel, they can be easily washed with soap and water or a number of commercially available cleaning solvents using a soft bristle brush or cloth. Avoid anything that will, or potentially could, scratch the surface: no steel wool, no metal bristle brushes, nothing of the like. Washing propellers might seem a luxury to some, but it is absolutely necessary if the aircraft has been exposed to saltwater or even saltwater spray, which is corrosive to metal. It is very important to thoroughly wash the propeller and all exposed parts as soon as possible to assure that all traces of the saltwater have been removed. After the propeller dries completely, a thin coat of clean, lightweight engine oil should be applied.

Some individuals prefer a high-gloss polish on the prop. Aside from the elbow grease required, it's easy enough with excellent commercially available metal polishes. Once finished and rubbed out, the blades should be lightly coated with very clean engine oil to preserve the finish and protect the metal.

MANUFACTURER DOCUMENTATION

(The following information is extracted from Beech Aircraft Corporation pilot operating handbooks. The information is to be used for educational purposes only. This information is not to be used for the operation or maintenance of any aircraft.)

King Air C90

Propeller system. Each engine is equipped with a McCauley four-blade, full-feathering, constant-speed, counterweighted, reversing, variable-pitch propeller mounted on the output shaft of the reduction gearbox. The propeller pitch and speed are controlled by engine oil pressure, through single-action, engine-driven propeller governors. Centrifugal counterweights, assisted by a feathering spring, move the blades toward the low rpm (high pitch) position and into the feathered position.

Governor-boosted engine oil pressure moves the propeller to the high rpm (low pitch) hydraulic stop and reverse position. The propellers have no internal, low rpm (high pitch) stops, which allows the blades to feather after engine shutdown.

Low-pitch stops. Low-pitch propeller position is determined by the low-pitch stop, which is a mechanical/hydraulic stop. This mechanism allows the blades to rotate beyond the low-pitch position into ground fine and reverse when selected during ground operation. Beta and reverse blade angles are provided by displacing the governor beta valve controlled by the power lever in the ground fine and reverse ranges.

Propeller governors. Two governors, a constant-speed governor, and an overspeed governor, control the propeller rpm. The constant-speed governor, mounted on top of the gear reduction housing, controls the propeller through its entire range. The propeller control lever controls the propeller rpm by means of this governor. If the constant-speed governor should malfunction and the propeller exceeds 2200 rpm, an overspeed governor cuts in at 2288 rpm, and dumps oil from the propeller mechanism.

A solenoid, actuated by the PROP GOV TEST switch, is provided for resetting the overspeed governor to approximately 1900–2100 rpm for test purposes.

If the propeller sticks or moves too slowly during a transient condition, causing the propeller governor to act too slowly to prevent an overspeed condition, the power turbine governor, contained within the constant-speed governor housing, acts as a fuel topping governor. When the propeller reaches 2420 rpm, the fuel topping governor limits the fuel flow to the gas generator, reducing N_1 rpm, which in turn prevents the propeller rpm from exceeding approximately 2420 rpm. During operation in the reverse range, the fuel topping governor is reset to approximately 95 percent propeller rpm before the propeller reaches a negative pitch angle. This ensures that the engine power is limited to maintain a propeller rpm somewhat less than that of the constant-speed governor setting. The constant-speed governor therefore will always sense an underspeed condition and direct oil pressure to the propeller servo piston to permit propeller operation in beta and reverse ranges.

Autofeather system. The automatic feathering system provides a means of dumping oil from the propeller servo to enable the feathering spring and counterweights to start the feathering action of the blades in the event of an engine failure. Although the system is armed by a switch on the pilot's left subpanel, placarded: AUTOFEATHER - ARM - OFF - TEST, the completion of the arming phase occurs when both power levers are advanced above 90 percent N_1, at which time L AUTOFEATHER and R AUTOFEATHER annunciators indicate a fully armed system. The system will remain inoperative as long as either power lever is retarded below the 90 percent N_1 position. Should torquemeter oil pressure on either engine drop below a prescribed setting, the oil is dumped from the servo, the feathering spring starts the blades toward feather, and the autofeather system of the other engine is disarmed. Disarming of the autofeather of the operative engine is indicated when the AUTOFEATHER light for that engine extinguishes.

Propeller synchrophaser. The propeller synchrophaser matches rpm of the two propellers, and also positions the propellers at a preset phase relationship. This phase relationship decreases cabin noise.

Signal pulses are obtained from magnetic pickups, one located at each propeller hub. The pickup is mounted on a bracket on the engine case, while the "target" for the pickup is mounted on the back of the propeller spinner bulkhead, so that it rotates with the propeller. In this way, one pulse is produced for each revolution of the propeller. Electric pulses generated by the target, passing each magnetic pickup, are fed into the control box. An electromagnetic coil for rpm trimming is mounted in each propeller governor close to the flyweights. Any difference in the pulse rates will cause the control box to vary the governor coil voltages until the propeller rpms match, due to control by the governors.

Propeller rpm is a function of the position of the propeller control lever in its quadrant, since linkage from the lever sets the governor flyweight position. The synchrophaser cannot reduce the rpm set by the propeller control lever. It can increase the rpm over a predetermined limited range. The rpm of one propeller will follow changes in rpm of the other propeller, within a limited range. This limited range limits rpm loss to a fixed value on the operative propeller in the event the propeller of one engine is feathered with the synchrophaser ON. In no case will the operative propeller rpm fall below the rpm set by the propeller lever.

The propeller synchrophaser may be used on takeoff and landing at the pilot's option. (The limited range of the synchrophaser will be reduced near maximum propeller rpm.) For all other operations, the synchrophaser should be switched OFF before adjusting the propeller rpm. Adjust the propeller levers to obtain synchronization and then switch ON the synchronizer. This will keep the synchrophaser within its limited range. **Note:** If the synchrophaser is ON but does not synchronize the propellers, it has reached the limit of its range. Switch the system OFF, adjust the propeller levers to obtain synchronization and then switch the synchrophaser ON.

6
Hydraulic and pneumatic systems

NEXT TO ENGINE FAILURE, HYDRAULIC FAILURE IS PROBABLY THE MOST serious problem a pilot can encounter in a large, high-speed aircraft. Hydraulic failure means loss of flight control function. So critical is the requirement for hydraulic power on large, turbine-powered aircraft that redundancy has become a significant engineering concern. The McDonnell-Douglas, four-engine, C-17 transport, for instance, utilizes four totally independent hydraulic systems. Each system is powered by a different engine with no transfer of hydraulic fluid between any of the four systems. Even the fluid lines for the four systems are totally separated within the fuselage. And yet, while it is clear that hydraulic systems are a major concern to the pilot, very few pilots know much about how the system works.

The purpose of the hydraulic system is to assist the operator in accomplishing a mechanical task that would otherwise be impractical or impossible because of the amount of work required. On some light aircraft, for instance, flaps are directly connected to a flap handle that the pilot pulls. The pressure felt by the pilot when deploying the flaps is the weight of the control surface combined with the air load on the surface, plus any constraints inherent to the mechanism. On the ground, there is little resistance and the flaps are very light. In the air, the faster the aircraft is moving, the greater the strength required to extend the flaps. In a large or fast-moving aircraft, the air load can easily be too great for the pilot to physically deploy the flaps. To solve that problem, the necessary work is done by moving a fluid in such a manner that it gives the pilot a mechanical advantage. Such a system is called *fluid power*.

Fluid is often a misunderstood term; there are many forms of fluid. The word fluid often makes one think of water, which is probably the best known fluid. (The word hydraulic is based upon the Greek word hydros, which means water.) There are, however, numerous examples of fluids besides water. Any gas or liquid is a fluid, including air, nitrogen, gasoline and even a soft drink. Fluid power, therefore, takes one of two forms: *pneumatics* that are made up of compressible gases and *hydraulics* that are incompressible liquids. Perhaps the most graphic example of pneumatics is the tremendous damage that can be caused by a tornado. An excellent example of hydraulics is the awesome power of water in the making of the Grand Canyon. For all its negative potential, clearly the ability to harness fluid power is a significant benefit to technology as can be demonstrated by the windmill and waterwheel.

When the wind turns a windmill or water turns a waterwheel, time is also a critical element. If, for instance, a miller wants to grind wheat in a windmill but the wind is very light, it might take all day to accomplish the task. On the other hand, if the day is very windy that same task might take only an hour. The problem is that both the windmill and waterwheel are *open systems* and the user has little or no control over system pressure beyond the natural pressure of the source of the fluid. To be of maximum mechanical benefit, fluid power should be in a *closed system* because it allows more work to be accomplished with a fixed amount of fluid. Common examples of such systems include brakes, automotive and construction jacks, gas station lifts, and bearing presses. In aircraft, hydraulic power is used for brakes, landing gear, and wing flaps. In high-speed aircraft, where control surface air loads are excessive, hydraulics are also used to move the flight controls and speed brakes. With limited exceptions, pneumatics are typically limited to aircraft pressurization and instrumentation usage.

SYSTEM THEORY

A hydraulic system's mechanical advantage is based upon fundamental physics. To understand why a hydraulic system works it is necessary to review a few terms.

Volume. Any container that can contain a fluid has a definable volume regardless of the shape of the container. Generally, hydraulic reservoirs are cylindrical, so the volume is determined by taking the area of the bottom of the container and multiplying it by the height of the container. The result is cubic units, such as cubic feet (FIG. 6-1).

Area. Area is the two-dimensional size of the surface of an object measured in square units, such as square inches or square feet. The area of a rectangle, for instance, is the result of multiplying its length by its width. With hydraulic systems, the area be-

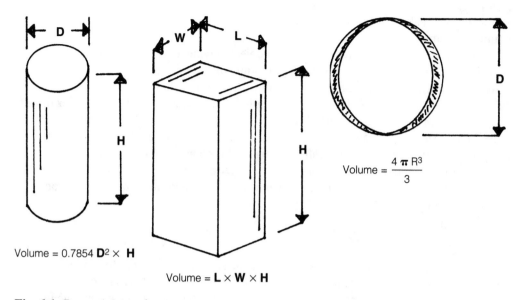

Volume = 0.7854 **D**² × **H**

Volume = **L** × **W** × **H**

$$\text{Volume} = \frac{4 \pi R^3}{3}$$

Fig. 6-1. *Determining volume.* Susan M. Lombardo

ing measured is commonly the circular surface of a hydraulic piston. A circle is simply a closed curve in which every location on the circumference is the exact same distance from the center as every other location. Because there is no length and width of a circle, the formula for determining its square is:

$$0.7854 \ \yen \ D^2$$

Where
   D = circle diameter

Force. Force is the result of energy being exerted. A concentration of force on an object might result in a motion. An example of a force that results in motion would be when a tug pushes an aircraft backward. It is also possible to exert a force with no resultant motion, as would be the case when a pilot pushes against the nose of a very large aircraft. Even though the aircraft does not move, the pilot is still exerting a force and will soon become tired. Force is often measured in pounds.

Pressure. When the area over which a force is applied is taken into consideration, the result is pressure. Pressure is measured in units per square inch such as pounds per square inch. In the case of the hydraulic piston, when the area of the piston surface is determined, it becomes easy to accurately calculate how much pressure can be generated by the hydraulic system's piston-driven actuator to the affected component.

Distance. Hydraulic system actuators operate in a linear manner—a straight line. For example, the piston in an actuator will move back and forth. The amount of movement is called distance and is typically measured either in inches, feet, centimeters or meters.

Work. When a given force results in an object moving a distance, the result is that work is accomplished. Work is measured in units of distance per force; *foot pounds.* Work is a misleading term because it implies the accomplishment of a given task. It does not indicate how long it took to accomplish the task. For instance, consider the tug pushing the aircraft backward. It is one thing if you want to move the aircraft 20 feet into a parking space and something totally different if you want to move it 3 miles across an active airport. Theoretically, the tug could accomplish the task, but it makes more sense to taxi the aircraft under its own, greater power.

Power. Power, then, is the result of work divided by the time required to accomplish the work.

$$\text{Power} = \text{force} ¥ \text{distance}$$

The most common measure of power is *horsepower*, which is 33,000 foot-pounds of work accomplished in one minute (550 foot-pounds per second). Measuring power in hydraulic systems is a bit more complicated in that it is necessary to compute the hydraulic fluid flow rate in gallons per minute (1 gallon = 231 cubic inches) and the pressure exerted in pounds per square inch. For instance, if the flow rate of a given hydraulic system is 1 gallon per minute with a fluid pressure of 1 pound per square inch, the equivalent power is 0.000583 horsepower.

Fluid power and conservation of energy

Certain fundamental principles govern fluid dynamics as they apply to hydraulic systems. Underlying them is the basic law of the conservation of energy that states that energy can neither be created nor destroyed. The importance of that statement cannot be overstated as it pertains to energy transmission. Within certain limitations, it means that a force applied at one location in one direction can be transmitted to another location in a different direction. Through the application of basic physics, it is also possible to increase the force exerted through a mechanical advantage. For instance, pressing forward on brake pedals will have the effect of pressing disk brakes against pads in another location and with significantly greater pressure than that exerted on the rudder pedals; however, even the most efficient mechanical system appears to lose some energy. In reality, the energy is not lost, but because of friction, some of it is converted to heat, a different form of energy.

Energy exists in two fundamental forms: *potential* and *kinetic*. Potential energy is the result of the position of the object or its ability to store energy mechanically. A good example of potential energy as a result of position is the water stored above a dam. Until the dam is opened, the water sitting behind the dam is calm but full of potential energy. By opening the dam, the water spills over and can be used to turn a turbine that powers components to create hydroelectric energy. As the water spills over, the water becomes kinetic energy. An example of potential energy as a result of mechanical storage would be the spring in a mechanical stopwatch. Until the timing device is activated, the spring is calm and unmoving. When activated, the gradual uncoiling of the spring produces kinetic energy that operates the stopwatch.

Kinetic energy is the energy that results from motion; as in the above examples, kinetic energy is created when the water falls over the dam or when the spring unwinds. When considering fluid power, it is easier to think of potential energy as the pressure of the fluid and kinetic energy as the velocity that the fluid moves through the system. It is then possible to divide fluid dynamics into two categories of laws: *static* and *dynamic*.

Pascal's law. This law dealing with a static fluid explains how we are able to transmit a force using a fluid moving through hoses and tubes. Pascal discovered that a change in pressure to a fluid in a closed system will be transmitted uniformly throughout the system and acting at a right angle to the walls of the system (FIG. 6-2). Because the pressure acts uniformly throughout the system, it is possible to increase or decrease the force exerted at a given location by increasing or decreasing the surface area upon which the fluid pushes. For instance, one pound of pressure acting on a piston with a surface of 1 square inch will produce a pressure of 1 pound per square inch.

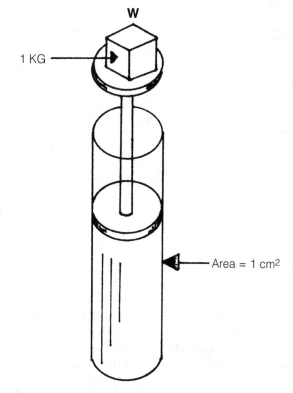

Fig. 6-2. *Uniform pressure transmission.* Susan M. Lombardo

If the fluid is connected to a cylinder that contains a 10-square-inch piston, it will balance a load of 10 pounds (FIG. 6-3) because the pressure is the same in both containers: 1 pound per square inch. If the 1-square-inch piston moves 10 inches (10 pound inches), the 10-square-inch piston will move 1 inch (10 pound inches). The same amount of work (force ¥ distance) done by one side is exactly balanced by the other. It is through this method that a lesser pressure applied in one area can gain a mechanical advantage and perform work in another area with significantly greater force.

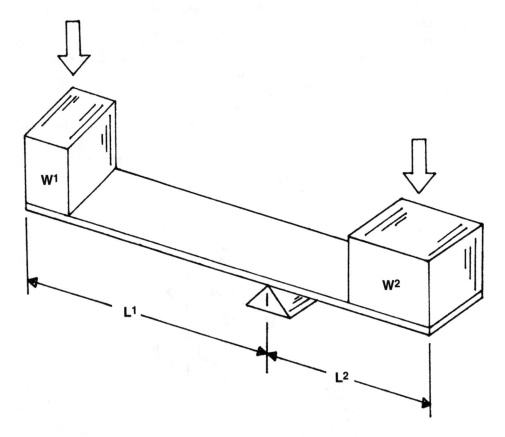

Fig. 6-3. *Hydraulic mechanical advantage.* Susan M. Lombardo

Hydrostatic paradox. Confusing as it might seem, the pressure exerted by a fluid is the result of the height of the column of fluid and not the tank's volume (FIG. 6-4). For example, two side-by-side tanks of different volumes are connected at the bottom. Both tanks will seek a common fluid level, stabilizing the pressure between them.

Bernoulli's principle. Pascal's law and the hydrostatic paradox pertain to fluid when it is in a static, nonmotion, state. When in operation, hydraulic systems move fluid, thereby bringing another set of rules into consideration. Bernoulli's principle

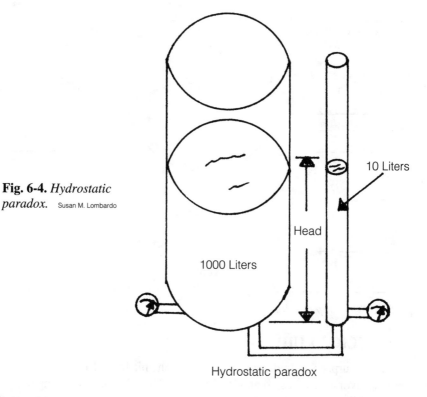

Fig. 6-4. *Hydrostatic paradox.* Susan M. Lombardo

10 Liters

Head

1000 Liters

Hydrostatic paradox

provides the explanation of what is happening in a closed system, one in which the pipes and tubes are sealed, preventing energy increase or decrease (loss). It is assumed that the fluid in the system is incompressible, such as hydraulic fluid or water. Bernoulli's principle essentially says that the faster a fluid flows, less sideways pressure is exerted.

The fluid, traveling through a tube at a constant velocity will exert a constant pressure against the walls of the tube (FIG. 6-5). The energy of the fluid is a combination of a constant potential energy (pressure) and constant kinetic energy (movement). If the tube narrows, the speed at which the fluid travels must increase to allow it to pass through the narrow area; remember, the pump continues to push the fluid through the closed pipes at a constant rate. Because energy can neither be created nor destroyed, the increased kinetic energy must be offset by an equivalent decrease in potential energy (pressure).

As a result, the fluid pressure in the narrow portion of the tube decreases. As the fluid passes through the restriction and enters an area of normal pipe width, the fluid will slow down and the fluid pressure will appropriately increase to the same level prior to the restriction. In effect, the fluid only has a certain amount of energy that must be divided between speed and sideways pressure. If the speed increases, the sideways pressure must decrease; it's that simple.

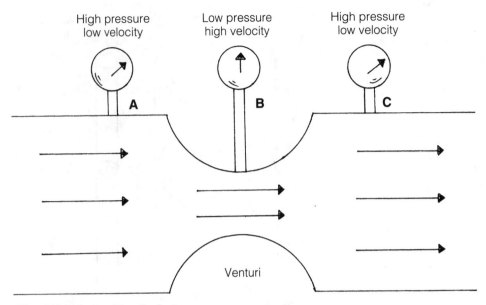

Fig. 6-5. *Relationship of velocity on pressure.* Susan M. Lombardo

HYDRAULIC FLUID

The primary purpose of hydraulic fluid is to transmit force from one place, through nonmoving hydraulic tubes, to another location. The advantage of using a liquid is its incompressibility. Except for minor friction losses as the fluid passes through the tubing, all applied force is transmitted throughout the system, according to Pascal.

Hydraulic fluids are different and generally speaking they cannot be mixed. The manufacturer determines the best composition of hydraulic fluid for the specific application and hardware, then limits system usage to that specific fluid. Many considerations go into determining the specific composition of a hydraulic fluid, including the nature of the system, the type of pressure at which the system will operate, potential for corrosion, anticipated operating temperatures, and the type of seals used within the system. The manufacturer selects a specific hydraulic fluid to be used in a system based upon four primary considerations:

- Viscosity
- Chemical stability
- Flash point
- Fire point

Viscosity. A measure of a fluid's internal resistance to flow: Viscosity is one of the most important selection criteria. To understand viscosity, think of a fluid in a pipe. Water, which has a very low viscosity, flows readily through a pipe. Mud, on the other hand, has a very high viscosity and might not pass through the pipe at all. Other considerations regarding viscosity include temperature. The colder a fluid gets, that fluid

160

becomes more *viscous*—resistant to flow. Other considerations include the ability of the fluid to lubricate and seal the system, transmit power efficiently, and flow easily enough to prevent the pump from having to work too hard. The correct fluid for a given system is a balance between all these and other considerations. For instance, a very thin fluid might transmit power easily and provide a little resistance to the pump, but might not do a good job of preventing corrosion or lubricating moving parts adequately, which leads to premature failure.

Chemical stability. Whenever a fluid is worked and heated, a chemical change takes place. Chemical stability is a measure of a fluid's ability to resist such changes. Temperature is particularly a catalyst and the anticipated system operating temperature is a primary consideration in selecting the appropriate fluid. Unfortunately, the normal operating temperature of the fluid in the reservoir is not always a good indicator of system operating temperature. Hot spots are usually located throughout the system in areas of high friction or pressure. As fluid constantly passes through such areas, the high temperature might slowly cook the fluid and turn it to sludge via a process called *carbonization*; the more viscous the fluid, the greater the potential for carbonization.

Another, similar consideration is exposure of the fluid to metals that react in such a manner as to lead to undesirable chemical reactions in the fluid. Ultimately, improper chemical stability can cause a fluid to turn to sludge and form deposits that will block fluid flow through the system. This, in turn, will cause the pump to work harder, overheat, and accelerate the chemical breakdown of the fluid ultimately leading to system failure. This process, even under ideal conditions, causes the fluid to darken in color and become more viscous; therefore, it is imperative to change hydraulic fluid at the frequency prescribed by the manufacturer.

Flash point. Flash point is the temperature at which the fluid produces enough combustible vapor that it will ignite momentarily or flash when a flame is applied. A fluid with a high flash point can get very hot before it becomes susceptible to flashing. Looking at it from another perspective, a fluid with a high flash point has minimal evaporation under normal operating conditions; therefore, a high flash point is a desirable characteristic of hydraulic fluids.

Fire point. This is the next step up the temperature spectrum from flash point. A fluid's fire point is that temperature at which sufficient vapor is given off that the fluid will burn when a source of ignition exists. From the perspective of hydraulic fluid, the higher the fire point the better.

Types of hydraulic fluid

Choosing the best hydraulic fluid for a given system can be a major task. When a component or system manufacturer determines the exact type of hydraulic fluid to be used, future refinements to that system are predicated on the use of that fluid. Proper system operation and the very health of the system depend upon the fluid. Never mix hydraulic fluids or use a fluid other than that specified by the manufacturer. Due to the importance of keeping these fluids separate, a color-coded system similar to aircraft fuels is employed; hydraulic fluids are blue, red, or clear purple. Three types of hydraulic fluid are:

- Vegetable base
- Mineral base
- Synthetic or phosphate ester base

Vegetable base. This blue fluid, identified by *military specification* (*milspec*) number MIL-H-7644, is a combination of foul-smelling caster oil and alcohol. Of the three types of hydraulic fluids, this is the oldest and least used. The fluid resembles automotive blue fluid, but the two are not interchangeable. The major drawbacks to vegetable-base fluid systems are extreme flammability and a requirement for natural rubber seals that react negatively with other fluids.

Mineral base. Identified by milspec MIL-H-5606 and colored red, this fluid is a flammable petroleum product and smells like penetrating oil. Systems using mineral base fluids incorporate synthetic rubber seals and, as with all hydraulic systems, you should never use vegetable-base or phosphate ester-base fluids.

Phosphate ester base. In the late 1940s, as the number of commercial aircraft began to increase, a new hydraulic fluid was needed. After World War II, for the first time in history, there was general confidence among the public in the use of air travel. The demand was met by increasing the passenger carrying capability of aircraft and that meant larger, high-performance piston engines, and an increasing reliance on the new turbine engines. The increased size, weight, speed, and complexity of these new transport aircraft placed significantly higher demands upon their hydraulic systems. Of considerable concern was the flammability of existing hydraulic fluids. As a result, significant research and development resulted in a phosphate ester-base fluid called Skydrol that is produced and marketed by Monsanto Company.

Skydrol fluids have been upgraded and the most commonly used are Skydrol 500B and in transport category aircraft Skydrol LD. Both are clear purple liquids. The LD version has a lighter weight per gallon for weight savings in consideration of the tremendous quantity of fluid on board large aircraft. They share the common characteristics of good performance even at low temperature, minimal effect on most common aircraft metals such as aluminum and stainless steel, and excellent flammability characteristics; Skydrol will flash at very high temperatures upon direct application of a source of ignition, but when the ignition source is removed burning stops. Skydrol will not support combustion, so it is not capable of spreading fire.

Skydrol is generally safe to handle, has low toxicity, and poses minimal health hazard, but can be an eye and respiratory irritant and should be handled carefully. Because it is a phosphate ester base, Skydrol does react with thermoplastic resins, such as oil- and lacquer-base paints and asphalt. The effect is short term and a quick cleanup of the spill should minimize, if not eliminate, any problem. Aerospace industry standard epoxies and polyurethanes are not affected by Skydrol. Intermixing is not acceptable.

Remember, one of the most severe consequences of using an improper fluid is the potential deterioration of the hydraulic seals, which leads to total system failure.

HYDRAULIC SYSTEM COMPONENTS

Reservoirs

The primary purpose of the reservoir is to act as a storage tank for the system's fluid. It assures a constant supply of fluid is available to replace any lost by system leaks. Two types of fluid reservoirs are *integral* and *in-line*. An integral reservoir is built into the working component, rather than having a separate housing with pipes connecting it to the component. This is very common in small, dedicated systems, such as automotive brakes. An in-line reservoir is used in larger systems that have several components, all requiring hydraulic fluid.

In-line reservoirs are separate components that supply fluid to various system components through narrow pipes referred to as *hydraulic lines*. In addition to the task of fluid storage, these reservoirs serve several other important functions. The airspace above the fluid level provides room for the fluid to expand as its temperature increases (FIG. 6-6). The airspace also provides space for air trapped in the system to collect, rather than continue to circulate through the lines. A dipstick or glass sight-gauge is usually incorporated into the reservoir to provide a method of checking fluid level.

Depending upon the manufacturer's intended mission of the aircraft, two system venting approaches are available. For aircraft that operate at lower altitudes, hydraulic reservoirs are typically open and vented to the atmosphere. In that case, normal ambient air pressure is sufficient to provide enough pressure to assure that hydraulic fluid flows from the reservoir to the hydraulic pump. Turbine-powered aircraft, because they routinely operate above 18,000 feet where the air pressure is very low, are always equipped with closed, pressurized reservoirs. This assures sufficient pressure on the fluid to supply the hydraulic pump under all conditions.

Reservoir pressurization. Two basic methods of positively pressurizing hydraulic system reservoirs are air pressure and hydraulic pressure. Air pressure is a very common method of pressurizing hydraulic systems. Air, diverted from either the cabin pressurization system or tapped from a turbine engine compressor, goes directly to the airspace located above the fluid in the reservoir. This produces a positive pressure on the fluid that assures reliable flow through the lines to the hydraulic pump.

Because turbine compressor air pressure is high—compared to the needs of the hydraulic system—it is necessary to reduce the air pressure before entering the reservoir with an air pressure regulator. It is important to understand that while air is being used to exert pressure on the fluid, the air does not actually mix with the fluid and flow through the system.

Hydraulic pressurization employs an additional hydraulic pump upstream from the main pump near the reservoir. The additional pump boosts fluid pressure to assure a positive flow to the hydraulic pump regardless of ambient pressure.

Fluid integrity. Another function of the reservoir is to assure that air does not mix with hydraulic fluid and move throughout the system. This is accomplished by strategically placing baffles inside the reservoir. The baffles sharply reduce fluid sloshing as the result of turbulence and shifting flight attitudes associated with normal maneuver-

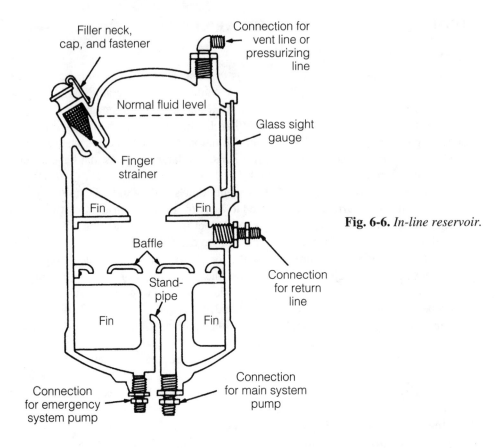

Fig. 6-6. *In-line reservoir.*

ing. It is highly desirable to reduce, if not totally eliminate, foaming because air in the lines is a source of trouble. Air is compressible and the primary advantage of hydraulic fluid is that fluid pressure can be transmitted readily from one system point to another because hydraulic fluid is not compressible.

Air also causes pump *cavitation*—the pump keeps turning, but accomplishes no work because the pump is filled with air, not fluid. Air in the fluid also compromises the fluid's ability to lubricate the hydraulic system's moving parts; air does not lubricate and premature pump failure can be expected.

Most hydraulic systems on turbine engine aircraft have a redundant *backup* hydraulic system. Typically, the backup system uses the same reservoir as the primary system. Because a major reason for a backup system is the potential loss of the primary system due to fluid leakage, the reservoir incorporates a safeguard to prevent that possibility. The primary system receives fluid through a standpipe that stands several inches above the bottom of the reservoir (FIG. 6-6). The backup system takes fluid from the bottom of the reservoir, which assures at least several inches of fluid in the reservoir even if a major leak in the primary system drains all available fluid.

HYDRAULIC PUMPS

The pump is the hydraulic system's equivalent of a heart. A pump assures the proper amount of fluid is available throughout the system at the appropriate pressure. (Aircraft design is simultaneously demanding more hydraulic power and fewer weight increases; therefore, the trend is to design smaller pumps that weigh less and provide higher pressure. The *fly-by-wire* aircraft is increasing this demand because flight control surfaces are being constantly adjusted by commands from redundant computer systems.)

Two primary methods of operating hydraulic pumps are engine power and electrical power. Most commonly, hydraulic pumps are powered off the accessory section of the turbine engine. A second pump might be operated off the other engine as a backup system, but certain aircraft utilize an electric pump backup system.

Two hydraulic system pumps are *constant displacement* and *variable displacement*.

Constant displacement. Also known as a *fixed-delivery* or a *constant-volume* pump, the constant displacement pump is the least sophisticated and simplest design of the two types of pumps. A common example is the geared pump that is typically limited to smaller aircraft and very simple hydraulic systems (FIG. 6-7). A constant displacement pump delivers the exact same amount of hydraulic fluid every revolution, regardless of the system's requirements; therefore, fluid flow, which is the number of gallons per minute that the pump puts out, is dependent upon the rpm of the pump. Because this has no direct relationship to the system's hydraulic fluid requirements, all constant-delivery pumps must incorporate a pressure regulator to maintain the required system fluid pressure.

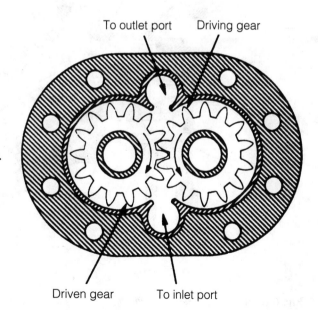

Fig. 6-7. *Gear-type pump.*

To outlet port   Driving gear

Driven gear   To inlet port

Variable displacement. The more sophisticated variable displacement pump includes a method to compensate for variable system demands when one or more subsystems are activated. Hydraulic fluid output varies to meet the system's fluid pressure requirements, regardless of the number of subsystems activated. The methodology used to achieve the variable displacement differs among various manufacturers and models, but typically incorporates a piston, power-driven design (FIG. 6-8). The primary exception to the piston pump is the cam pump (FIG. 6-9). One commonality is a regulated supply of hydraulic fluid that meets the system's varying demands.

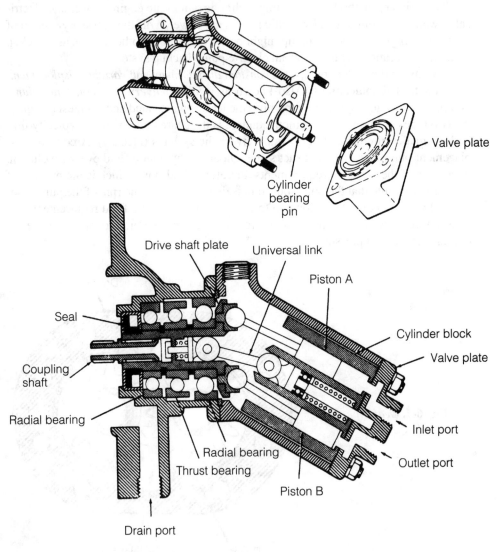

Fig. 6-8. *Piston-type, angular pump.*

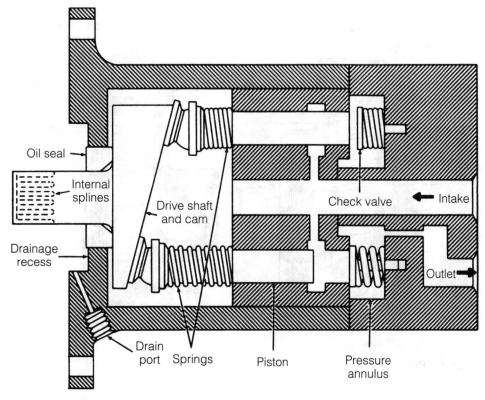

Fig. 6-9. *Typical cam-type pump.*

Hydraulic valves

Two categories of hydraulic valves are *flow control* and *pressure control*.

Flow control valves. Fluid is routed throughout the system by flow control valves. One type of flow control valve is a *selector valve* that operates like a door, permitting the user to selectively channel hydraulic fluid to a component to accomplish a specific task, such as extending flaps. Another type is a *sequence valve* that automatically performs an operation in the same sequence every time. For example, a sequence valve assures that during extension of the landing gear, the gear doors open first, then the gear extends through the open doors. *Hydraulic fuses*, a third type of flow control valve, are simply safety valves that prevent fluid flow in the event of a serious system leak. The primary purpose of the fuse is to assure that sufficient fluid remains in the reservoir to assure backup system operation in the event of a primary system failure. (A *check valve* only prevents fluid from backing up in the system.)

Pressure control valves. The *pressure relief valve* is the most fundamental of the pressure control valves; the relief valve will release hydraulic fluid, usually back to the

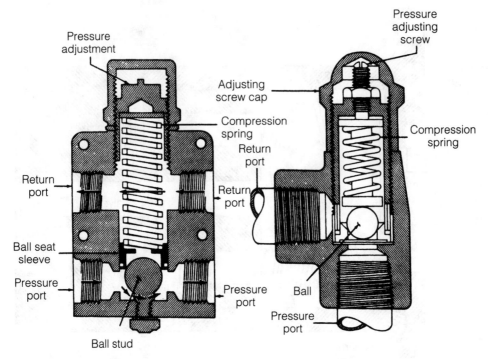

Fig. 6-10. *Pressure relief valve.*

reservoir, if system pressure increases beyond a prescribed level (FIG. 6-10). This helps reduce the possibility of hydraulic line rupture or component failure due to excess system fluid pressure. These valves should not be confused with pressure regulators. Relief valves are spring operated and preset for a specific pressure that is higher than the system's designed operating pressure; they are an escape valve. Pressure regulators are incorporated into systems that use a constant-delivery pump. *Pressure regulators* divert fluid flow from the system as necessary to maintain system pressure within a designed range.

ACCUMULATORS

The accumulator is the shock absorber of the hydraulic system. The most common types of accumulators are the *diaphragm* version and the *bladder* version; both are very similar in appearance and design. These accumulators are most commonly found on light aircraft, but their operation is very easy to understand and is functionally the same as the type used on turbine-powered aircraft. The diaphragm accumulator consists of a hollow steel sphere roughly divided in half internally by a rubber diaphragm (FIG. 6-11). Hydraulic fluid at system pressure is on one side of the diaphragm; com-

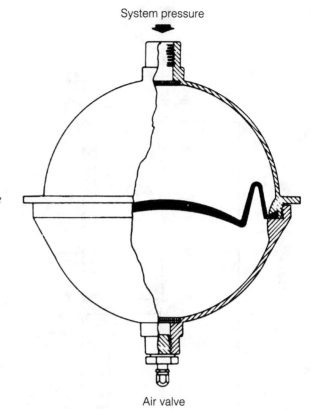

System pressure

Fig. 6-11. *Diaphragm-type accumulator.*

Air valve

pressed air or, more commonly, nitrogen (an inert gas) is on the other side. System hydraulic fluid is allowed to flow freely into and out of the accumulator.

The nitrogen, constantly exerting pressure on the diaphragm, compresses and decompresses in response to diaphragm movement when the amount of fluid in the accumulator increases or decreases. The increasing volume of fluid in the accumulator is the result of the system's demand for fluid. As the demand increases, the fluid moves out of the accumulator under the pressure of the nitrogen to flow to whatever actuator needs fluid. When an actuator puts fluid back into the system, the fluid then flows back into the accumulator, pushing against the diaphragm and compressing the nitrogen.

Warning: It is important to note that only an experienced mechanic should ever work with hydraulic systems because the pressure charge is continuously maintained by the accumulator—even when all aircraft systems are shut down. Accumulators store a dangerously high amount of fluid pressure and must be depressurized properly before any part of a closed system can be safely opened.

Amongst several types of accumulators, the most common type found on turbine-powered aircraft is the *piston accumulator* (FIG. 6-12). Constructed slightly differently

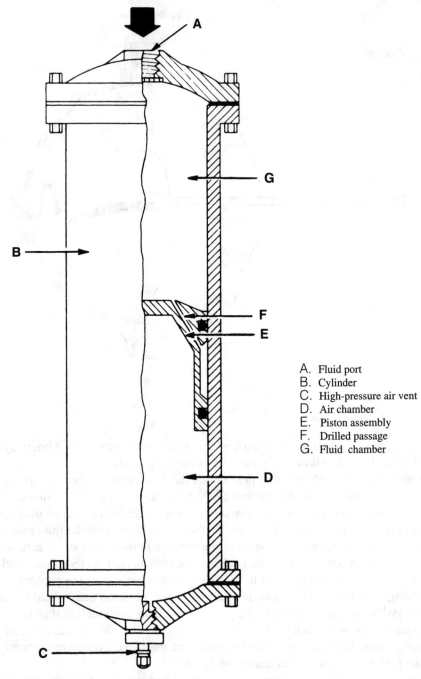

Fig. 6-12. *Piston-type accumulator.*

A. Fluid port
B. Cylinder
C. High-pressure air vent
D. Air chamber
E. Piston assembly
F. Drilled passage
G. Fluid chamber

than its diaphragm and bladder cousins, the piston-type accumulator serves the same functions. The accumulator has four functions:

- Reduce pressure bumps
- Provide supplemental system pressure
- Maintain system pressure
- Serve as a pressure storage unit

Reduce pressure bumps. When a component served by the hydraulic system is activated, an immediate demand for pressurized fluid is felt throughout the system. The instantaneous movement of fluid results in a drop of system pressure that causes pressure-sensing switches to activate the pump to maintain fluid pressure. In a closed system, with no accumulator, the result would be a shock wave or "bump" throughout the system—because the fluid does not compress—that would shake the hydraulic lines momentarily. Each bump would cause the lines and fittings to jiggle slightly, which could eventually cause line and fitting failures. The bump is absorbed by the accumulator's nitrogen cushion on the other side of the diaphragm; thus, the accumulator acts as a shock absorber, preventing the lines from shaking.

Provide supplemental system pressure. When a component of the system is activated that requires a significant amount of hydraulic fluid, or when several smaller components are activated, the demand for fluid might momentarily exceed the capability of the pump. To prevent a lag in the system, the compressed nitrogen in the accumulator assists the pump by maintaining a constant pressure on the fluid flowing out of the accumulator into the system. This supplemental pressure helps the pump maintain a constant system pressure until the pump can catch up to demand. When that happens, fluid again flows into the accumulator, recompressing the nitrogen on the other side of the diaphragm.

Maintain system pressure. When the hydraulic system is dormant and the pump is not operating, the system must remain pressurized to maintain a state of readiness. In the ideal closed system, that is not a problem because no fluid leaks out, but few, if any, hydraulic systems are leak-free. As a result, most hydraulic systems are constantly losing a small amount of fluid through undesirable leaks. The compressed nitrogen maintains constant pressure on the remaining fluid, thereby maintaining constant fluid pressure throughout the system, even as the fluid quantity decreases. Eventually the fluid quantity would decrease to the point that the nitrogen charge would no longer be able to maintain system pressure. At that time, only activation of the pump would bring fresh fluid from the reservoir into the system and increase fluid pressure back to normal.

Pressure storage unit. The accumulator also acts as a pressure storage unit to permit limited operation of the system when the pump is not activated, for example, flap deployment for preflight inspection prior to engine start. Sufficient compressed nitrogen pressure is in the accumulator to allow system fluid to flow into the flap actuator system without the use of the pump.

FLUID LINES AND FITTINGS

Hydraulic fluid lines are classified as rigid or flexible. Rigid lines might be made of either aluminum tubing on low-pressure systems or stainless steel for higher pressure systems and oxygen systems. The major advantages of rigid lines are the ability to handle high pressures, no kinking or twisting after installation, and long service life. The disadvantages are they can be accidentally bent, which might block fluid flow, lack resilience to bumps and other abuse, and are not flexible, so they cannot move with the actuation of a component. For instance, one quirk of rigid lines is that regardless of how short they might be, all rigid lines must have at least one bend. The bend allows for line expansion and contraction as the temperature changes.

Flexible lines, actually hoses, are the answer to the disadvantages of the rigid lines. Flexible lines are either constructed of synthetic rubber or Teflon and are black in color. Hoses are also typically wrapped with stainless steel braid for added strength. The primary disadvantage of hosing is its ability to twist, causing fluid flow stoppage and/or putting a strain on the hose that can lead to failure during pressure surges. As a result, flexible hoses have a yellow lay-line running like a stripe on the outside of the hose. This is a visible confirmation that the hose is not kinked; therefore, the lay-line should run essentially straight and not twist around the hose.

Some hoses are green with a white line. They are high pressure hoses that can only be used with Skydrol hydraulic fluid. Teflon hoses in general have become very common on aircraft because they can be used in hydraulic, fuel, pneumatic, and oil systems.

Amongst many types of hydraulic fittings, some are not interchangeable. A proper fit between hose or tube connectors is essential; even the smallest leak can dramatically spray fluid when the system is pressurized. Again, remember that only a qualified mechanic should loosen a hydraulic fitting because the system must be discharged to prevent possible injury.

FILTERS

If every system has a weak point, contamination is the Achilles' heel of the hydraulic system. Specifically, contamination of the hydraulic fluid causes problems. Two system contaminants are abrasives and nonabrasives.

Abrasive contaminants. This type of contamination includes any solid, hard objects that infiltrate the fluid, most typically the result of a lack of quality control when pouring fluid into the system. Of particular concern are contaminants that can circulate in the air and be drawn into the system as you pour the fluid. Examples include sand, dirt, or even rust. Bear in mind that hydraulic system components are machined very smooth and to extremely close tolerance. The viscous hydraulic fluid will hold contaminants such as grit in suspension and will act like sandpaper as it continuously passes through the system eroding internal surfaces.

Nonabrasive contaminants. Soft particulate matter, such as particles from worn internal seals and oil oxidation, are not as critical as abrasive contaminants, but they can eventually lead to system failure.

A pilot can do very little to determine a level of hydraulic system contamination. Contaminants that can cause major harm are often microscopic or near microscopic in size and not conducive to visible inspection. A pilot inspection might turn up large impurities, such as bits of seal or traces of metal indicating the deterioration of a major system component, but exposing the fluid to visible inspection would subject it to picking up serious airborne contamination in the form of abrasives such as sand and dirt. Rather than pilot inspection, it is more important to know that maintenance on an aircraft is conducted by a reliable, professional organization that prevents contamination from occurring as a result of thorough maintenance practices and procedures.

Careful maintenance procedures will go a long way toward reducing system contamination, but complete elimination is impossible. As a result, the filter is a key element of system health. Contaminants, held in suspension by the fluid, flow through the system until they encounter the filter. The exact location of the filter varies according to the manufacturer, but a common location is near or even within the fluid reservoir. In any case, all fluid is forced to pass through the filter as it cycles through the system.

In-line filter assemblies have a bypass valve in the event the filter becomes so clogged with contaminants that the filter would severely restrict or even prevent fluid passage. Bypass valves are spring-loaded and allow unfiltered fluid to flow through the system when system pressure reaches a preset value. While the situation is a result of a clearly significant problem—components might be disintegrating—the bypass valve does permit continued system operation despite the possibility of rapid deterioration. Activation of the bypass valve should be recognized for the problem it is: Potential short-term reduced system effectiveness and long-term permanent damage. Regular inspection of the hydraulic system by qualified maintenance personnel is an important part of assuring the health of an aircraft.

Three filters used in hydraulic systems are *magnetic*, *micron*, and *porous metal*. Micron filters (FIG. 6-13) consist of a bowl-like housing that contains a treated paper filter held in place with a spring. Fluid enters the filter bowl inlet at a system pressure fewer than the 50 psi and goes into the bowl that contains the filter. The porous paper micron filter, which is rated in microns (1000th of a millimeter), allows the fluid, and anything else smaller than 20 microns, to pass through while stopping anything larger and trapping it on the outside of the filter (FIG. 6-14).

No significant amount of matter larger than 20 microns should be in the system. If sufficient matter exists to cover the filter and stop the flow of fluid, continuing system demands will cause system pressure downstream from the clogged filter to decrease. Eventually, the lower pressure will exceed the bypass relief valve spring pressure. When that occurs, the spring will yield and the fluid will flow through the bypass and continue through the system.

ACTUATING CYLINDERS

Actuators are the muscle of the hydraulic system; they do all the work. According to the definition, work is accomplished when a given force results in an object moving a

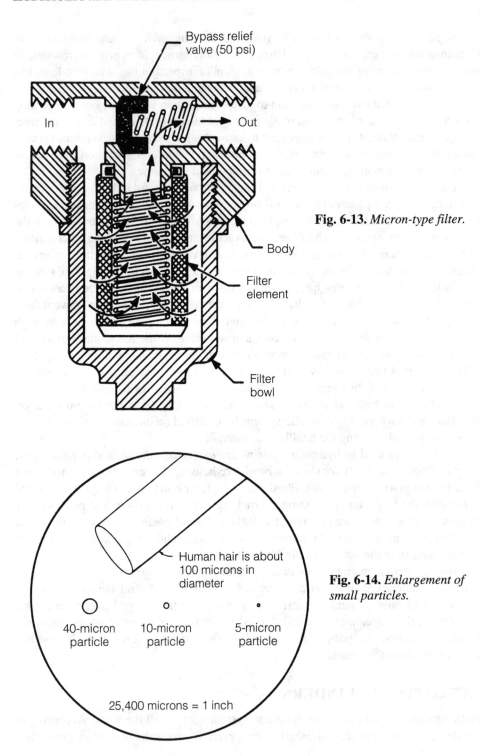

Bypass relief valve (50 psi)

In

Out

Fig. 6-13. *Micron-type filter.*

Body

Filter element

Filter bowl

Human hair is about 100 microns in diameter

Fig. 6-14. *Enlargement of small particles.*

40-micron particle

10-micron particle

5-micron particle

25,400 microns = 1 inch

distance. Actuators translate the force of system fluid pressure into either linear or rotary movement.

Linear actuators. This type of actuator is made up of a piston inside a cylinder. The cylinder is permanently attached to a portion of the aircraft's structure in such a way that there is no relative movement between the structure and the cylinder. As a result, the piston moves in relationship to the cylinder and airframe and is connected to the device, which is intended to be moved. Three primary types of linear actuators are *single-acting*, *double-acting balanced*, and *double-acting unbalanced* (FIG. 6-15).

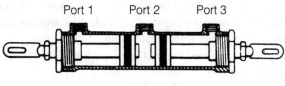

Port 1   Port 2   Port 3

Three-port actuating cylinder

Fig. 6-15. *Actuating cylinders.*

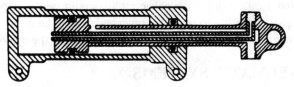

Actuating cylinder having ports in piston rod

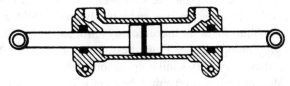

Double-action actuating cylinder
having two exposed piston rod ends

Single-acting. The piston only moves in a single direction as the result of system pressure. The piston returns to its original position as a result of spring pressure when system pressure is reduced.

Double-acting balanced. Balanced actuator shafts move in both directions as a result of system pressure. Both sides of the piston have an equal amount of area upon which the fluid can act. The result is a balanced force on both sides. Movement of the piston is the result of selecting one side to receive a flow of fluid under pressure while the other side is allowed to drain itself of the fluid as the piston moves in that direction.

Double-acting unbalanced. Though the piston is round, one side has a rod, or actuating arm, attached that results in less surface area than the side without the actuating arm. This permits more work to be done in one direction than the other. Landing gear extension is an example of where a double-acting, unbalanced actuator might be used. Gear extension requires relatively little work as the weight of the gear and the force of the airstream will help the gear extend. As a result, gear extension would be accomplished by pressurizing the side of the actuator with the reduced piston surface area. Gear retraction, on the other hand, requires significantly greater work because weight of the gear alone, coupled with airstream resistance, creates tremendous total resistance that the hydraulic system must overcome. Gear retraction would be accomplished by pressurizing the side of the actuator that had the greatest piston surface area.

Through the use of cushions, snubbers, shuttles and other flow control devices it is possible to adapt these three types of actuators to accomplish a wide array of lateral movements. For instance, it is possible to move an actuating arm in one direction, slowly at first, then with greater speed toward the end of its travel; or, to move with minimal force for the first half of travel but with significantly greater force during the latter half of travel.

Rotary actuators. This actuator causes rotary motion and would replace an electric motor in certain applications. The greatest arguments for using a hydraulic rotary actuator rather than an electric motor would be the demand for significantly higher torque, the ability to instantly reverse direction, and the elimination of the concern of potential fire if the load freezes up and prevents the motor from turning.

PNEUMATIC SYSTEMS

Pneumatics provide a limited, but necessary source of energy in the aircraft. Pneumatic systems operate in principle very much like hydraulic systems. Both systems use a confined fluid as the motivating force. Hydraulic fluid, a viscous liquid, is virtually incompressible; air, or more commonly nitrogen, is also a fluid but unlike hydraulic fluid it is very compressible. Regardless, because both are fluids trapped in a confined space, they share very similar operational traits.

Most aircraft rely on either the electrical system or the hydraulic system as the primary means of accomplishing the heavy work in an airplane. Jobs such as raising and extending landing gear, moving major control surfaces, and operating the brakes are typically left to hydraulics. Electrical motors might also move control surfaces, occasionally operate the gear on certain smaller business aircraft, and frequently move minor control surfaces, such as flaps, spoilers, and slats. Despite the popularity of hydraulic and electrical systems, however, pneumatics can, and occasionally do, perform all these tasks and more.

Compressed air does have some significant advantages over the more popular hydraulic and electrical systems. For one thing, air is everywhere and if the system happens to leak, the hangar floor remains clean. Compared to high-pressure hydraulic systems that require very solid components, heavy fluid and return-line requirements, pneumatic systems are relatively lightweight. Compressed air, which can get very hot, does not present the temperature-related problems associated with hydraulic fluid, and, finally, no fire hazard is associated with compressed air.

Types of pneumatic systems

Depending upon the work to be done, pneumatic systems fall into two broad categories: high-pressure and low-pressure. High-pressure systems are used as primary power sources for heavy-duty applications instead of hydraulics. Also, a pneumatic system used as a backup for a hydraulic system would also be high pressure. Low-pressure systems drive the flight instruments and perform miscellaneous tasks.

High pressure systems. Emergency and backup pneumatic systems normally use metal bottles to store air compressed to the range of 1000–3000 pounds per square inch. The air bottle will have two valves, one for charging the bottle and the other, a shutoff valve, to activate and deactivate the system. The bottle may be charged, through the charging valve, by a ground-based compressor. The major drawback to this system is the limited amount of air available in the bottle. Such a system could only function as a backup.

Some aircraft use a similar system as a primary source of fluid power by adding permanently installed air compressors. When leaks result in decreasing system air pressure, the compressor automatically activates and brings the system pressure back up to full charge. Figure 6-16 depicts the schematic of a two-stage air compressor. This particular system uses three check valves. Similar to the check valves used in hydraulic systems, they permit the flow of air to occur in one direction only.

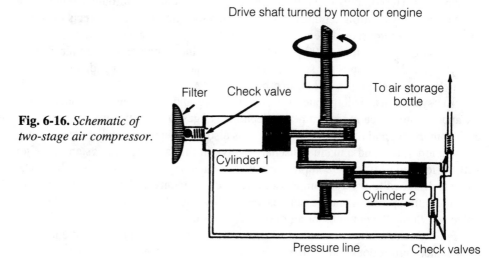

Fig. 6-16. *Schematic of two-stage air compressor.*

The drive shaft of the compressor might be powered directly off the accessory section of the engine or might use an electric motor with a pressure sensing on/off switch. In either case, as the shaft rotates, the pistons move back and forth in their respective cylinders. As that happens, the piston in cylinder one draws ambient air through the filter into the cylinder as the partial pressure created by the piston moving back offsets the spring tension on the one-way check valve. When the piston reaches bottom dead center and begins to move forward, the partial pressure ceases and the spring-activated check valve closes the cylinder intake. The piston then compresses the air that goes through the pressure line and the second check valve.

The partially compressed air is now drawn into cylinder two, where it is further compressed. The check valve directly upstream from the cylinder prevents the air from backing up the pressure line and escaping out the intake filter. The check valve directly downstream from cylinder two allows the newly pressurized air to flow to the storage bottle but prevents air in the storage bottle from backing up into cylinder two.

Despite its rather limited popularity as a primary system, the Fairchild Hiller FH-227B (FIG. 6-17) is one example. The Rolls-Royce Dart-powered turboprop aircraft has long been out of production, but is still used successfully by select regional air carriers. The automatically controlled pneumatic power system provides air pressure for operation of the main wheel normal and emergency brakes, landing gear normal extension and retraction, landing gear emergency extension, main landing gear operation as a drag brake, nosewheel steering and centering, propeller brakes, and passenger door retraction.

The system includes two engine-driven compressors, one driven by the accessory gear box in each nacelle, and three compressed air reservoir bottles. Air, supplied by a duct routed from the cowling on top of the engine is compressed by the four-stage compressors. The compressed air is routed via an intercooler line through a relief valve to an electrically controlled unloader valve. The valve automatically directs the compressed air as appropriate depending upon the status of the pneumatic system. If system pressure is fewer than 2700 pounds per square inch, the compressed air charges the system to 3100 pounds per square inch; if the system is fully charged, the unloader valve channels the compressed air overboard.

The system also incorporates various purifying and regulating components, and the appropriate indicating instruments. Separate gauges indicate the pressure output of each compressor and reservoir bottle. There is also an external ground charging valve, an isolation valve, and a manual discharge valve in the landing gear system. A relief valve is located next to the compressor outlet in each engine accessory gear box. The valve is designed to protect compressor circuit components in the event of excessive pressure buildup. If system pressure exceeds 4200 pounds per square inch, the relief valve will dump the compressor air output overboard.

For the sake of illustration, follow this molecule of air from the time it enters the intake duct until it does some work, such as activate the nose steering. The molecule enters the intake duct above an engine and is rushed down to the intake port of the compressor. It is compressed one step at a time through the four stages of the compressor. If, for any reason, excessive pressure builds up in the compressor, relief valves are located between stages one and two, and also between stages two and three that can vent the excess pressure overboard.

When fully compressed, the molecule flows through the unloading valve to the *moisture separator*, which removes approximately 98 percent of the moisture of the

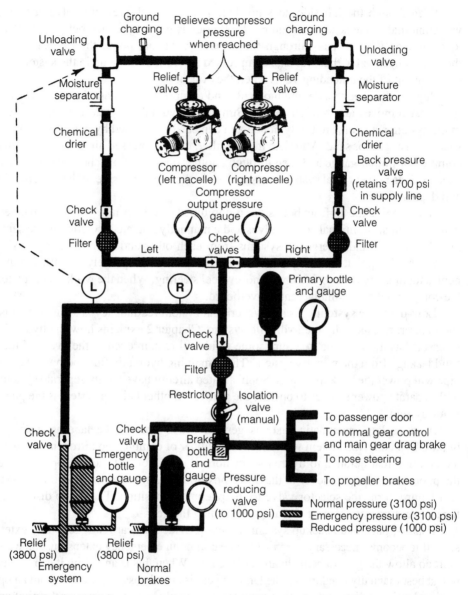

Fig. 6-17. *FH-227B Pneumatic power system schematic.*

air. After the separator, the air is routed through a chemical dryer that effectively removes all remaining moisture. At that point, the dry air on the right side travels through a back-pressure valve. This valve prevents the air from going through the system until the supply-side pressure reaches at least 1700 pounds per square inch, the minimum pressure required to effectively operate the moisture separator.

Even though the left side does not incorporate a back-pressure valve, a spring valve internal to the separator requires 1500 pounds per square inch before it will allow the supply air to pass through the separator. At that point, the dry air passes through a check valve that prevents compressed air from backing up in the system and escaping out of the unloading vents when they open.

Beyond the check valve, the molecule and compressed air pass through a micron filter that removes any foreign matter. Similar to hydraulic system actuators, pneumatic system actuators are finely honed and can easily be ruined by particulate matter such as dust, grit, or sand. After leaving the filter, the compressed air is registered on a compressor output gauge and then through another one-way-flow check valve. These check valves prevent total system pressure loss in the event one side of the system fails and depressurizes.

At this point, the air from both sides is combined and flows into primary and emergency systems. Additional check valves and a manually operated isolation valve separate the primary and emergency systems from each other and from the compression system. At this point, the compressed air is channeled appropriately into the component actuating device associated with nosewheel steering. When the work is completed the compressed air is simply vented overboard.

Low-pressure systems. As always, critical system redundancy is a significant issue in aircraft, especially the hydraulic system. Chapter 2 explains how the hydraulic reservoir has a stand pipe to assure some fluid will remain even in the event of total fluid leakage from the primary system. The remaining hydraulic fluid below the stand pipe will power the backup system. Sophisticated aircraft have totally separate systems with separate power sources to operate the pumps. Another backup system is the pneumatic system.

As a backup, pneumatic systems generally will operate the landing gear and brakes. Typically, the system will include a cylinder of compressed nitrogen with a selector valve that defaults to hydraulic for normal operation. At the appropriate time, the pilot moves the gear lever to the down position and, in the event of hydraulic system failure, turns the selector valve to the pneumatic position. The bottle discharges the air pressure into the gear actuators, extending the gear.

Usually, pneumatic backup systems do not provide for gear retraction, only extension. If it becomes necessary to retract the gear after an emergency extension, some aircraft do allow the gear to be manually cranked up. When this is an option it can be said that at best manually cranking up the landing gear is an exhausting process. From an operational point of view, pneumatic gear extension should be used only after the pilot has a very high degree of confidence that the approach will result in a successful landing.

In instrument meteorological conditions, when runway visual sighting is not assured—particularly if there is a power-related problem, such as engine failure—pneumatic gear extension must be very carefully considered. If the runway environment does not appear upon the conclusion of the approach, it might not be possible to successfully retract the gear in sufficient time to prevent undesirable contact with the ground or loss of directional control due to excessive airspeed reduction. Always carefully read and understand all aspects of emergency gear operations for an aircraft as outlined in the approved pilot's operating handbook.

Low-pressure pneumatic systems. A very common use of pneumatics in all aircraft is to drive the gyros in the flight instruments. In addition, sophisticated aircraft such as the Beechcraft Super King Air 200 also use low-pressure pneumatics to do several other housekeeping chores, such as operate the surface deicing equipment, inflate the door seal to reduce noise and allow efficient pressurization, operate the rudder boost and bleed-air warning systems, and pressurize the aircraft (FIG. 6-18). With turbine-powered aircraft such as the King Air 200, the most common method of supplying low-pressure air is tapping the engine bleed-air system.

The King Air 200 is powered by two 850-shaft horsepower Pratt & Whitney Canada PT6A-42 turboprop engines. Each engine features a three-stage axial flow, single-stage centrifugal-flow compressor. High-pressure bleed air is tapped from the third stage of the compressor, also known as *P3 air*. As P3 air leaves the compressor, its pressure is approximately 120 pounds per square inch and its temperature is 650°F. It is interesting to note that the high temperature of the compressed air is solely the result of compression.

The P3 air is then routed through a normally open (N.O.) firewall shutoff valve. The purpose of the firewall shutoff valve is to be able to stop the flow of 650°F air in front of the firewall in case of a leak on the other side. The area aft of the firewall contains the fuel tank. After passing through the firewall, the P3 air goes through a check valve that prevents pressurized air from backing up in the case of an engine failure. If, for instance, there were no check valves and the right engine failed, the P3 air from the operating left engine would pass through the left firewall shutoff valve across to, and through, the right firewall shutoff valve and back into the compressor of the left engine where it would spin the compressor and the rest of the engine backwards. The check valves prevent that from happening.

MAINTENANCE

Hydraulic and pneumatic system maintenance consists of servicing, troubleshooting, removal and installation of components, and operational testing; all of which should be accomplished by appropriate maintenance personnel. The hydraulic fluid level or air compressor oil level should be checked daily. Most systems provide either a sight gauge or dipstick for the pilot to check during preflight. If the fluid level is low, it should be brought to the attention of maintenance personnel. Hydraulic and pneumatic systems are susceptible to contamination with potentially very serious consequences.

Periodically, maintenance personnel will purge both systems to remove moisture, contamination, and in the case of the pneumatic system, any accumulation of oil. After a pneumatic system is purged and cleaned, the air bottles will also be drained until empty to remove moisture and impurities. Finally, the bottles will be recharged with nitrogen or clean, dry compressed air. With either system, the final step is to test the fully charged system to assure system integrity and operation.

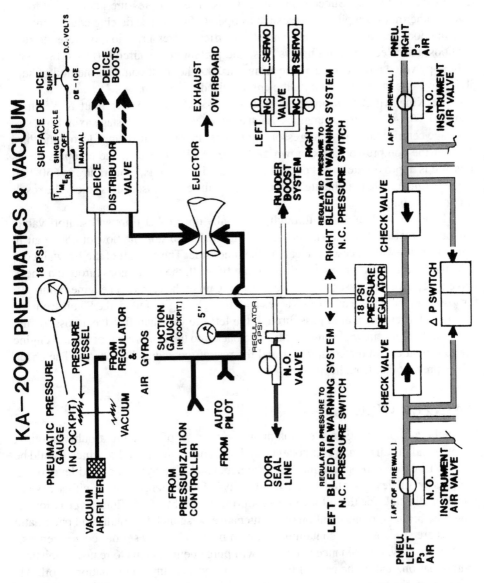

Fig. 6-18. *Beechcraft King Air 200 pneumatics and vacuum system schematic.* Flight Safety International Inc.

MANUFACTURER DOCUMENTATION

(The following information is extracted from Beech Aircraft Corporation pilot operating handbooks and Falcon Jet Corporation aircraft and interior technical descriptions. The information is to be used for educational purposes only. This information is not to be used for the operation or maintenance of any aircraft.)

Beechjet 400A

Hydraulic system. The airplane is equipped with a closed-center hydraulic power system that has two pumps, a hydraulic package, ground connectors, and associated electrical components (FIG. 6-19). The system supplies hydraulic power to the flaps, speed brake, power brake/antiskid, thrust reversers, and landing gear. The system operating pressure is 1500 psig.

Pumps. A variable displacement, rotary plunger type hydraulic pump is installed on and driven by each engine accessory gearbox. The displacement varies from zero at an output pressure of 1500 psig to 3.9 gpm at about 1,400 psig. This pump not only varies the volume of fluid pumped but also regulates the system pressure; therefore, a pressure regulator is not needed.

Hydraulic package. The hydraulic package consists of a reservoir, an air filter, vacuum and air pressure relief valves, check valves, fluid filters, shutoff valves, a hydraulic relief valve, a bypass valve, low-pressure switches, a high-pressure switch, a pressure transducer, and lines. The reservoir provides hydraulic fluid for system operation. Its capacity is approximately 1.1 U.S. gallons of MIL-H-5606 hydraulic fluid. The reservoir is pressurized by filtered engine bleed air that is regulated to 15 psi. Air pressure in the reservoir is controlled by an air supply check valve, a vacuum relief valve, and a pressure relief valve. The reservoir is equipped with a sight gauge (window) that is used to check proper fluid level. A float type switch installed inside the reservoir warns the flight crew when the reservoir fluid quantity reaches 0.6 U.S. gallons or less. Activation of this switch will illuminate the HYD LEVEL LO annunciator.

Shutoff valves are mounted at the fluid supply ports of the reservoir to stop fluid flow to the engine driven pumps. These valves are normally open and are closed in case of an engine fire or overheat condition. During an engine fire procedure, depressing the LH or RH ENG FIRE PUSH button electrically closes the fuel and hydraulic shutoff valves.

Three filters extract foreign particles from the fluid to prevent damage to moving parts in the system. One filter assembly is installed in each pressure passage from the pumps and the third filter is in the fluid return passage to the reservoir. If a filter element becomes clogged, a bypass valve will allow fluid to bypass the element for continued system operation. After passing through the filters, fluid passes through a check valve to prevent reverse flow of pressure in case of engine shutdown.

The system relief valve dumps abnormally high pressure to the return side when the system pressure rises excessively. The system can be depressurized through the

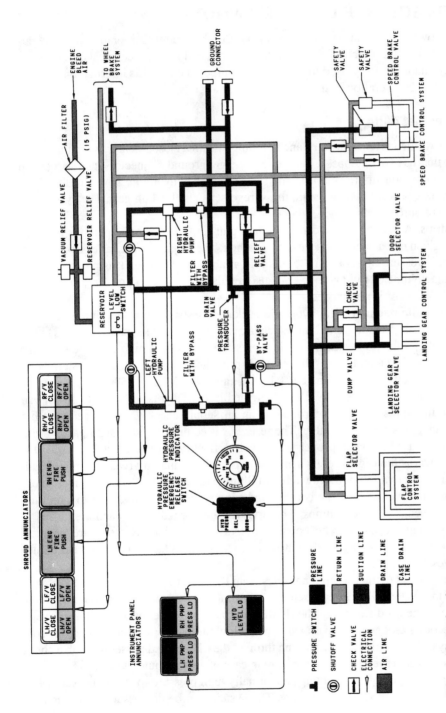

Fig. 6-19. *Hydraulic system.* Beech Aircraft Corporation

bypass valve that is electrically energized by the HYD PRESS REL switch to prevent excessive fluid temperature rise caused by the sustained flow through the system relief valve. The switch should be set in the NORMAL position. When moved to the HYD PRESS REL position, the bypass valve is opened to bypass the hydraulic fluid to the reservoir.

A low-pressure condition in the pump outlet lines is sensed by the low pressure switches. If one or both pumps fail (pressure below 750 ±100 psig for 15 seconds or more), the L or R H PMP PRESS LO annunciator illuminates. The same indication is provided for excessive leakage due to any hydraulic component failure. The system pressure can be monitored by the hydraulic pressure indicator. After fluid passes through all the indicating and controlling devices, it is ported through the hydraulic valve to the systems utilizing hydraulic pressure for actuation. Systems utilizing hydraulic pressure for activation are the landing gear, flaps, speed brakes, power brake/anti-skid, and thrust reversers.

Falcon 900B

Hydraulic system. Each engine drives a hydraulic pump that supplies hydraulic power to the two independent hydraulic systems (FIG. 6-20). The No. 1 system is powered by hydraulic pumps driven by the side engines while the No. 2 system is powered by a hydraulic pump driven by the center engine and by an electrically driven standby pump. The two hydraulic systems provide for operation of the following services:

	No. 1	No. 2
Primary flight controls	*	*
Landing gear (normal & emergency)	*	
Slats (normal extension)	*	
Outboard slats (automatic extension)	*	*
Outboard slats (emergency extension)		*
Inboard slats (automatic extension)	*	
Flaps		*
Airbrakes		*
Brakes	Normal	Standby
Nose wheel steering		*
Parking brake		*
Thrust reverser		*
Elevator Arthur-Q unit	*	
Aileron Arthur-Q unit		*

The hydraulic systems are designed for operation with MIL-H-5606 fluid to a working pressure of 3000 psi, except that the electric standby pump delivers a pressure varying between 1600 and 2150 psi.

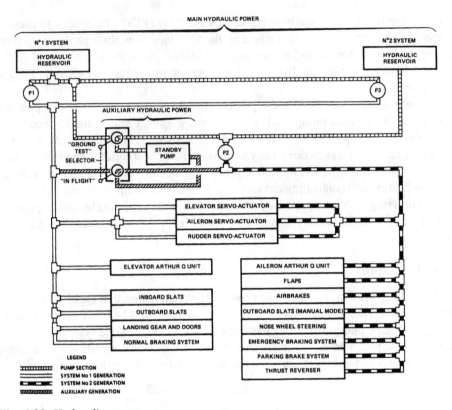

Fig. 6-20. *Hydraulic power.* Falcon Jet Corporation

The hydraulic reservoirs are located in the rear compartment and are fitted with a gauge for checking the fluid level. A control panel and fluid pressure and quantity indicator are provided in the flight deck.

Each hydraulic system is fitted with easy-to-reach ground connectors. A mechanically operated valve allows the standby pump to be used on the ground for maintenance operation of either the No. 1 or No. 2 system.

Hydraulic pressure lines are made of titanium except in certain areas such as the nacelles and wheel wells where they are made of steel.

All filters have replaceable paper cartridges.

7
Fuel systems

Aꜰᴜᴇʟ ɪs sᴏᴍᴇᴛʜɪɴɢ ᴛʜᴀᴛ ʙᴜʀɴs ᴀɴᴅ ᴘʀᴏᴅᴜᴄᴇs ʜᴇᴀᴛ ᴡʜᴇɴ ɪɢɴɪᴛᴇᴅ. ɪᴛ does so because it contains chemical energy that is released during combustion as heat energy. When a fuel is mixed with air prior to ignition and used to power an aircraft internal combustion engine, the heat energy is converted to mechanical energy. The mechanical energy is then used to produce thrust either directly as the result of hot gas exhaust, or indirectly as the result of a propeller's motion.

Three major classifications of fuels are separated into categories according to the physical state: solids, gases, and liquids.

Solid fuels. These fuels are characterized by their slow burning rate and low heat value. Examples of solid fuels include coal, wood, and charcoal. Storage and distribution are major problems associated with this type of fuel and when coupled with other problems, solid fuels are predominantly limited to external combustion engines such as the steam engine, and to area heating such as a furnace or fireplace.

Gaseous fuels. To a limited extent, gaseous fuels, such as natural gas and liquefied petroleum gas, are used in selected internal combustion engines. The most common application has been in the automotive area with some taxi companies equipping their entire fleet. The major problem for aircraft use is that to carry enough fuel for extended flight, it would require very large storage containers.

Liquid fuels. These are the best fuels for internal combustion engines because liquid is easily stored and dispensed. Liquid fuels are divided into two categories: volatile and nonvolatile. Nonvolatile fuels are heavy oils used in diesel engines. Some of the more common volatile petroleum fuels include gasoline, alcohol, and kerosene. Petroleum, an organic chemical from a family of hydrocarbons, is a compound composed of hydrogen and carbon.

To convert the chemical energy to heat energy, it is necessary to produce a chemical reaction by first mixing hydrocarbon fuel and oxygen together. Then the temperature must be raised to the appropriate kindling point, or that specific temperature at which the combination will combust. While the kindling temperature varies among compounds, when it is reached, *oxidation* (the union of a substance with oxygen) occurs. Prior to that point, you simply have two substances coexisting side-by-side; after oxidation you have a single, new substance burning.

This type of fuel is generally dispensed into internal combustion engines with a fuel metering device that sprays it in a partially vaporized condition directly into a cylinder or combustion chamber. The two most common forms of aviation fuel are aviation gasoline (avgas) for reciprocating engines and kerosene for turbine engines.

TURBINE ENGINE FUELS

Distillate fuel, more commonly known as jet fuel, is also made up of hydrocarbons, though with a few more carbon molecules and a higher sulfur content than its aviation fuel counterpart. A few additives are also included to discourage corrosion and reduce oxidation. Another common additive, sometimes included by the refiner, sometimes added at the dispensing point, is an anti-icing compound.

Three types of jet fuel are Jet A, Jet A-1, and Jet B. Unlike avgas, the letter and number designators used with jet fuel are strictly arbitrary and have no relationship to performance. Jet A, a kerosene-grade turbine fuel, is referred to as JP-8 by the military after additives are injected. Jet JP-5, which has a higher flash point and lower freezing point than commercial kerosene, is nonetheless very similar. JP-5 also has a very low vapor pressure, which means minimal fuel is lost over time to evaporation or to boiling off at high altitudes. Jet A-1 is a derivative of Jet A except that it is designed to operate at lower temperatures. A-1 is the most commonly used fuel among air carriers outside of the United States while the use of Jet A predominates within the United States.

Jet B has all but vanished from use, except in some parts of Canada. JP-4 is the military designation for Jet B with anti-icing and static additives. JP-4 is slowly being phased out with scheduled elimination in 1997. Jet B, a blend of gasoline and kerosene, is very similar to Jet A with a slightly different relative density (*specific gravity*). While most turbine engines are capable of burning either, the difference in relative density can be enough to require fuel control adjustments prior to use. As a result, Jet A and Jet B should not be considered readily interchangeable.

Caution must be used when refueling turbine-powered aircraft to assure the proper type of fuel because, unlike avgas, jet fuels are not dyed. It is impossible to visually tell one fuel from the other. Even the same type of jet fuel will have color variations from one batch to another ranging from colorless to amber depending upon the characteristics of the petroleum source used to make it. Even the fuel's age can change the color.

Volatility. This is one of the most important characteristics of any fuel and varies considerably among different fuels (FIG. 7-1). Fuel with a high volatility has the desirable trait of being easy to start under low temperature conditions, such as during the winter and operating at high altitude. While the latter is particularly important for in-flight restart attempts at high altitude, high volatility is also the cause of vapor lock and the reason why fuel evaporates.

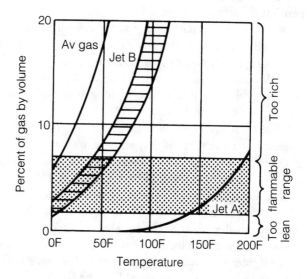

Fig. 7-1. *Vaporization of aviation fuels at atmospheric pressure.*

In fact, at normal temperatures, a closed tank of avgas can produce and trap so much vapor that the fuel/air mixture that fills the airspace above the liquid avgas can be too rich to burn. If the next airplane on the line has fuel tanks that contain Jet B, though the conditions are identical, the fuel/air mixture in those tanks can be highly explosive. Jet A, on the other hand, has such a low volatility that under those same conditions it will not give off sufficient vapor to become flammable.

FUEL SYSTEM CONTAMINATION

The intricacies of metering, mixing, and delivering fuel and air to the combustion chamber, combined with the wide operating temperature range associated with turbine engines, leaves jet fuel very susceptible to numerous forms of contamination. While these same contaminants are of concern with avgas, the greater viscosity and relative density of jet fuel gives it the tendency to hold dispersed contaminants in suspension. This makes detection and elimination extremely difficult, if not impossible. The most common contaminants include:

- Accidental mixing with improper fuels
- Water
- Rust
- Dirt

Water. While there are many differences between jet fuel and avgas, one of the most significant operational differences from the pilot's point of view is water contamination. Jet fuels are a blend of heavy distillates that result in a relative density that is much closer to water than is avgas. As a result, jet fuel tends to hold water in suspension for a much longer time, making it difficult to remove, or, for that matter, to even detect visually. Given enough time, without agitation or other motion, eventually water will separate but it is not a practical solution to the problem.

Two types of water contamination in jet fuel are *dissolved* and *entrained*. Jet fuel naturally contains some dissolved water, much like humidity in the air. You cannot see it or easily detect it in any way. So long as dissolved water remains dissolved, it poses no serious operational threat.

Entrained water, on the other hand, is often visible and is a potential threat. Entrained water in excess of 30 parts per million can result in severe degradation of engine performance and even cause flameout. Though they are held in suspension as tiny droplets, the nature of water is to reflect light; therefore, contaminated jet fuel will appear dull, hazy, or cloudy. But be cautious about forming opinions because fuel can appear cloudy for two reasons. One is water contamination, in which case the cloud will disappear at the top as the water slowly works its way toward the bottom. The other reason is air mixed in the fuel that can be detected if the cloud disappears at the bottom as the lighter air slowly works its way to the top. Another reason for preventing water contamination of fuel is the health of the fuel pumps and fuel control unit. Both units depend upon fuel as the sole source of lubrication; they are internally lubricated. If an appreciable amount of water is allowed to enter either unit, permanent damage occurs very quickly.

Most pilots at one time or another have observed avgas contaminated with water. If not, it is worth your effort to take a small glass jar and fill it with avgas. Then take a tablespoon of water and dump it in. The water will sink right to the bottom of the container and sit there as a clear bubble. With jet fuel it does not work so easily. Fill the same jar three-quarters full of jet fuel, add a tablespoon of water, then shake vigorously. The picture will be significantly different.

If water contamination of 30 parts per million or more can cause flameout, fewer than 30 parts per million is not much better. Free water can easily cause icing throughout the fuel system. The most common place for jet fuel to become contaminated with water is in the airport fuel storage facility. As jet fuel and water are pumped into the aircraft fuel tank, both are mixed together thoroughly by the flowing action. As a result, water droplets are held in suspension in the fuel. The most common areas for icing to disrupt the fuel flow include boost-pump screens and low-pressure filters. When low temperatures are encountered at high altitudes the water droplets freeze into minute slivers or needles and block filters by piling up on the filter like a haystack.

This is a significantly more serious problem than avgas water contamination because the water in jet fuel does not readily drain to the low point of the system and collect together for easy draining. In addition, water might short-circuit electrical fuel quantity probes, making fuel quantity readings erratic or incorrect. Finally, if the water contamination happens to be saltwater, there are all sorts of concerns regarding corrosion of fuel system and engine components.

Foreign particles. This category is a catch-all for any solid object contamination. Between refining and being sprayed into a combustion chamber, fuel changes hands quite a few times and is stored or flows through many environments. As a result, there is ample opportunity to pick up rust, sand, aluminum and magnesium compounds, brass shavings, and even rubber.

Contamination with other types of fuel. With all the handling safety precautions that are followed, this category should not exist, but it does. Improper fueling of aircraft is still a problem with catastrophic implications. Engines are designed to operate with a specific type and grade of fuel. The use of any other can seriously erode engine performance, obviate aircraft performance charts, and potentially result in immediate engine failure when the fuel is introduced to the combustion process. Reciprocating engines are touchier than turbines about what you feed them; recips simply quit if you fill up with jet fuel. Turbines are a bit more tolerant because you can mix jet fuels and even avgas in the 80–145 octane range with turbine fuel if you are in a pinch. The problem is that avgas contains tetraethyl lead (TEL) to increase its critical pressure and temperature. When avgas is mixed with jet fuel, the result is that the lead ends up sticking to the turbine blades and vanes, which, in the long run, might cause a loss in engine efficiency.

Microbial growth. Microbial growth is the result of several different types of microorganisms that make their home in the murky area where water meets jet fuel. As they propagate, they form a slime that looks very much like that found in swampy, still water. Microbial color varies between red, brown, gray and black. If left unchecked, the organisms multiply rapidly. They are an interesting lot because they feed on hydrocarbons, which are plentiful in jet fuel, but free water is necessary to multiply. The water, being heavier, gravitates to low points in tanks or sumps and is not pumped out with the fuel. When airborne spores enter tanks through vents, they migrate to the water where they live while feeding on the hydrocarbon fuel.

When the process begins, the microorganisms attract each other and form ever enlarging blankets, usually brown in color. The growth rate of microorganisms is so rapid they can form colonies numbering millions within a few days. These colonies seemingly act as a magnet drawing the existing water into them and thereby accelerating the growth of more microorganisms. It is a self-perpetuating cycle as long as water and hydrocarbons are present.

Microbiological contamination can be an expensive and potentially dangerous condition in turbine aircraft fuel systems. The waste products of these organisms can produce matted, fibrous formations and thick slimes. If they are not cleaned out, these wastes can eventually plug fuel lines, attack fuel tank coatings, cause fuel quantity indicators to be inaccurate or nonfunctional, and expose structural metal surfaces to corrosion.

Sediment. When taking a fuel sample, the liquid should essentially be free of all visible sediment, though occasionally some trace might be seen. Recurrent instances of sediment or particulate matter is indicative of a problem. Sediment includes organic and inorganic matter that might appear as dust, powder, fibrous materials, grains, flakes, or stain. Typically, the sediment that is usually seen is a very fine powder. Individual specks or granules large enough to see are at least 40 microns large (FIG. 7-2), and more than just a very few such particles indicates a problem.

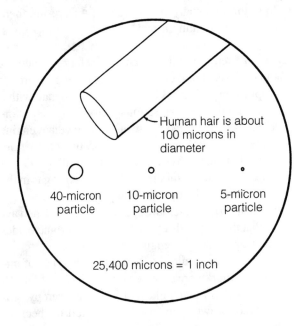

Fig. 7-2. *Enlargement of small particles and comparison to human hair.*

Most likely it is either a malfunction of a filter/separator, or something producing the particles downstream of the filter/separator. Fibrous materials in the fuel indicate a filter element is coming apart and beginning to shed into the system. Any metallic sediment or metal shavings indicates a potentially serious problem, most probably a mechanical failure.

Two categories of sediment or solid contamination are coarse and fine. *Coarse sediment* can be easily visually detected because it is anything greater than 10 microns (FIG. 7-2). This type of sediment is usually easily removed by normal system filtration. Coarse particles that are not removed from the fuel system can easily clog orifices and sliding valve clearances that might result in system malfunction. The particles will also adversely affect fuel controls and metering equipment. Also, fuel filter screens located in fuel nozzles and elsewhere are very susceptible to this size of sediment.

Fine sediment is anything smaller than 10 microns. The great majority of fine sediment in fuel can be removed by settling and filtration. Individual particles are not visible to the naked eye, but they will reflect light and can appear as small flashes in the fuel or as a slight haze within the fuel. Because of the small size, they can be held in suspension for a relatively long time, easily more than an hour (FIG. 7-3); therefore, any

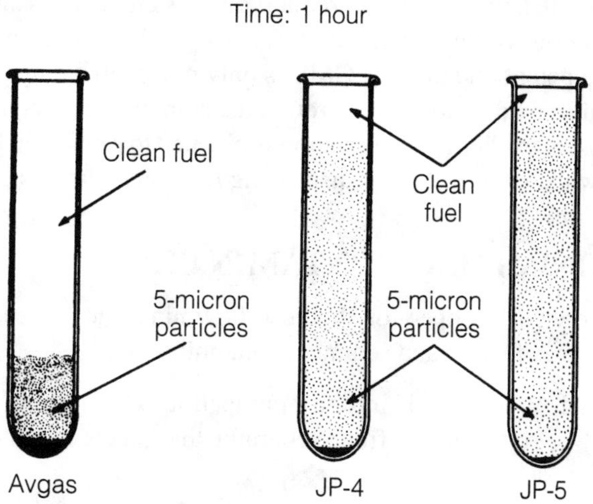

Time: 1 hour

Fig. 7-3. *Comparison of particle's rate of settling in three types of fuel.*

Clean fuel

Clean fuel

5-micron particles

5-micron particles

Avgas

JP-4

JP-5

water or sediment in the fuel means waiting as long as necessary for the foreign material to settle prior to engine operation.

Contamination detection. Due to the pervasiveness of water in turbine fuel, it is important to test for it as you are refueling the aircraft. Commercially available kits, such as Exxon's Hydrokit, make it a simple process. That particular kit includes an indicator powder prepackaged in a 10-milliliter evacuated test tube. Simply take a fuel sample and if there is 15 parts per million or more undissolved water, the sample will turn pinkish tan indicating no go. Otherwise, even a visual inspection is better than none at all.

The fuel should be clean, shiny, and contain no perceptible free water. Cloudy fuel indicates the presence of water. It is important to understand that as much as three times the volume of water considered acceptable can be contained in the fuel with no visible trace, so cloudy fuel is a definite problem.

Contamination control. Several fuel additives are available to control contamination. Anti-icing additives prevent entrained water from freezing within limits. All such additives are chemically identical. Flight into temperatures lower than the protection allows, will require the use of fuel heat.

Microbiocidal additives kill microbes, fungi, and other types of growth bacteria. Certain fuels do have these additives premixed at the refinery, many do not. Virtually all airport operators who offer jet fuel also have additives available that can be mixed in during refueling. Prist, probably the best known and most used additive, is designed to be added at the point of sale. Similar to other fuel additives, it is designed to be an anti-icing and biocidal agent.

All additives should be handled as a combustible liquid and stored away from direct sunlight and other sources of heat. The user should be cautious to prevent inhaling the fumes because the primary ingredient of Prist is ethylene glycol monomethyl ether (EGME), which is a central nervous system depressant. Symptoms include headache, drowsiness, blurred vision, weakness, lack of coordination, tremor, unconsciousness, and potentially death. EGME is only mildly irritating to the eyes and skin; however, it can be readily absorbed through the skin in toxic amounts with the same symptoms. A less toxic, diethylene variation is planned to replace EGME. As with all chemicals, follow the manufacturer's dispensing recommendations carefully.

MINIMIZING CONTAMINATION

Every airport operator should follow nine rules when handling jet fuel to reduce the hazard of water and sediment contamination:

1. When fuel is initially pumped into the airport's main storage tanks, it should go through a filter-separator that meets the latest edition of the API Bulletin #1581.
2. A minimum of two tanks for storage of each type of turbine fuel—fuel in one tank can be settling while the other tank is dispensing. Fuel should be allowed a minimum settling time of one hour per foot of fuel depth before dispensing.
3. After the fuel has settled, the sump should be tested for water. After all water has been drained, the hydro kit may be used to test the fuel. During periods of heavy rain, fuel should be periodically checked regardless of the length of time since delivery.
4. Fuel suction lines inside the tank should be of the floating type because they remove the fuel from the top, rather than the bottom. This type of system allows maximum settling time and ensures that the most water-free fuel will be dispensed.
5. Fuel being dispensed from the storage tank should pass through an API 1581 filter/separator.
6. If the fuel is transferred to a fuel truck, rather than dispensed directly from the original storage tank, extreme care must be exercised to prevent airborne dust, dirt, rain, and the like, from entering.
7. All fuel storage tanks or fuel truck tanks should be constructed of either stainless steel, nonferrous material or steel-coated with an approved, inert material, such as epoxy paint.

8. Fuel being pumped into the aircraft should be filtered with another API 1581 filter/separator, or a filter monitor, that is known to have met the performance of API 1581.

9. Maintain maximum quality control procedures including regular and frequent inspection and changing of the filter and frequent fuel quality check.

FUEL HANDLING SAFETY

When refueling any aircraft, personnel should follow standard refueling techniques used for all types of fuel. Jet fuel requires additional caution. When pumping avgas into an aircraft fuel tank, the vapor-to-air mixture above the fuel level in the tank is likely to be so rich that ignition is difficult if not unlikely. The vapor above jet fuel, on the other hand, could easily be in the ignition range on a hot day; one spark could set the whole tank off. The problem is that the act of fuel flowing causes an electrostatic charge to develop that can discharge in the form of a spark and cause a fire if fuel vapors are present.

For many years, the standard practice for refueling aircraft has included a three-part safety procedure: grounding the fuel truck, grounding the aircraft, and bonding the fuel truck to the aircraft. The theory behind this practice was that the primary contributor to the electrostatic charge was the fuel flowing through the pipes, valves, and fuel hose; however, it was discovered that the primary culprit is the action of the fuel flowing through the fuel truck's filter, monitor, and filter separator. This caused the National Fire Protection Association to reissue NFPA Bulletin 407 on August 17, 1990, which significantly changed the recommended procedure. NFPA recommended that the truck be bonded to the aircraft, but that neither truck nor aircraft be grounded. To understand the reason for the amendment, it is necessary to understand the nature of an electrostatic charge.

An excellent description of the process appears in the Gamgram Bulletin #40, *Bonding vs Grounding*, published by Gammon Technical Products, Inc. In part, the bulletin explains that an electrostatic charge is basically pluses (+++) that are physically separated from minuses (– – –). If you connect a wire or any other conductor between those two places, a current flows and the pluses immediately cancel the minuses so that no net charge remains.

Think about what happens during fueling. Flow begins through the filter with a separation of the pluses and minuses. Pluses or minuses go along with the fuel into the aircraft and the opposite (pluses or minuses) stays in the filter, creating a large voltage difference. If the fuel were a conductor, this would not happen.

The charges in the filter eventually travel to the filter vessel casing and into the truck frame. The charges that are carried away in the fuel to the aircraft will migrate to the airframe. And now what do you have? You have the truck all charged up with minuses and the airplane all charged up with pluses, or vice versa. You could ground the truck and get rid of that charge. You could ground the aircraft and get rid of that charge. When there is no voltage difference, there is no charge.

A better way to deal with the whole problem is simply to connect the truck and the airplane together with a bonding wire—the pluses on the airplane go back to the truck to cancel the minuses, yielding no net charge. The practice in the United States was to bond and ground, but tests proved that if the aircraft and the truck are bonded, a grounding wire carries no charge at all. This is why NFPA 407 no longer specifies grounding for safety during aircraft fueling.

NFPA 407 does not mean that an aircraft should not be grounded for electrical reasons or for maintenance. Personnel who have those responsibilities must make their own decisions about grounding and provide the proper size of cable for their purposes. A typical cable used for electrostatic grounding is usually far too small to satisfy electrical grounding needs—some have simply melted right under the aircraft where fuel vapors could be ignited. This can happen when the aircraft electrical system or the ground power unit/generator malfunctions. This was a further reason for NFPA 407 to delete the electrostatic ground wire.

You might be thinking that the bonding can be eliminated if the fuel has been made conductive by adding a conductivity improver. That is not a good idea because a metal wire is always going to be a better conductor than liquid. You might also argue that now that NFPA 407 specifies hose with a conductive cover, this constitutes a bond between the truck and the aircraft. NFPA 407 specifically forbids this because the wire makes a superior bond. The conductive cover of the hose provides an added factor of safety, just in case; therefore, you absolutely must bond the aircraft to the source of fuel where the final filter is located. The source might be a refueler truck, hydrant servicer, or a cabineted fueling station. The main points to remember about aircraft refueling are:

- Grounding can become a hazard if the cable is insufficient to carry the power from an electrical apparatus such as a ground power unit or generator.
- The charge difference is between the truck and the aircraft, not between the aircraft and ground.
- Conductivity improver or conductive cover hose is not a proper solution to the problem.
- Grounding only the aircraft without bonding to the refueler creates a new hazardous situation because the refueler will be left with a substantial charge.

Personnel handling jet fuel should be aware that the fuel is poisonous if taken internally and fuel is a skin irritant. All refueling personnel should be aware that the effect of fuel upon humans will vary among individuals. Some people will have more drastic reactions than others; therefore, it is a good idea to be aware of the following recommendations:

- First and foremost, know good first aid procedures and have an emergency medical team telephone number readily available. An emergency shower should be adjacent to every fueling facility. Do not underestimate the toxicity of jet fuel.

- Do not let fuel contact a body in any way, especially not skin; wear long sleeves and long legged pants. Jet fuel is poisonous and can cause severe skin problems, blindness, and other maladies. Particularly avoid inhaling the fumes that can permanently damage lung and nasal tissues.
- If you do get fuel on your skin, wash it off immediately with soap and water. If fuel soaks your clothes, take them off, shower, and put on something else. Do not wash those clothes in a load with other clothes.
- Do not use jet fuel as a solvent to remove paint, oil, grease, road tar, and the like, because the fuel is truly a toxic chemical, not a household cleaner. Never allow jet fuel to run off onto lawns or into free water such as rivers or lakes. You should even keep it out of storm sewers because they end up dumping into open water. Always appropriately clean up any spills.

FUEL SYSTEM THEORY

For the reciprocating engine pilot, the turbine engine fuel system is perhaps one of the most baffling. Its wide range of operating conditions demand a very sophisticated and responsive system, but the criticality of high-altitude operation also requires that the system be almost foolproof. Control of turbine engine power is achieved by varying fuel flow, but in the case of turbopropeller engines, the propeller is the additional consideration. In that situation, thrust control is through two variables: fuel flow and propeller blade angle.

Turbine-powered aircraft have a very wide vertical operating range that spans from sea level to more than 45,000 feet. This encompasses drastic temperature and pressure changes for which the fuel system must continually and automatically adjust. Several critical problems can result from improper fuel flow. For instance, given the ambient conditions, the fuel flow becomes too high for the mass airflow through the engine. A major problem that can result from this condition is called *rich blowout*. Rich blowout refers to the condition where the amount of oxygen in the air becomes insufficient to support combustion. Extra fuel that is added cools the fuel/air mixture below the combustion temperature and literally puts out the fire. If the fuel quantity is reduced too much, that might cause the turbine blade temperatures to exceed the maximum permissible temperature. If the lean condition continues, it might become insufficient to support combustion, given the airflow, and the flame will extinguish. This condition is known as *lean die-out*.

In addition to delivering fuel to the combustion chamber in the proper amount, the fuel system must also deliver fuel in the proper condition—the fuel nozzles spray vaporized fuel into the combustion chamber. Not only must the fuel system take all these factors into account, but must also make adjustments to assure a satisfactory engine start either on the ground or in flight. All these considerations and more are within the realm of the fuel control unit.

JET FUEL CONTROLS

Fuel control units (FCU) fall into two categories: hydromechanical and electronic. The electronic system, which is not commonly found on aircraft, is a combination of hydromechanical and electronic. The methodology varies somewhat between the types of units, but both accomplish the same goal: produce thrust appropriate to conditions. To accomplish the task, depending upon the type of FCU, one or more of the following items are taken into consideration:

- Power lever position
- Engine rpm
- Compressor inlet temperature
- Compressor inlet pressure
- Burner pressure
- Compressor discharge pressure

By far the most common FCU is hydromechanical.

Hydromechanical. The hydromechanical FCU is responsible for determining the appropriate quantity of fuel flow given the ambient conditions and the demand scheduled by the power lever position. It must provide fuel to the combustion chamber at the proper pressure and volume, while assuring that it never allows so great a fuel flow to exceed safe operating limitations.

For instance, take the JFC (jet fuel control) 12-11 used on a Pratt & Whitney turbojet engine (FIG. 7-4). The unit precisely meters fuel to the engine to assure maximum efficiency at controlling rpm, and simultaneously prevents overheating, surging, rich blowout, or lean die-out. The unit interprets engine rpm and burner pressure signal inputs and adjusts fuel flow as necessary to maintain the power requested by the power lever while staying within safe operating parameters.

Two flight deck input controls schedule the JFC: fuel shutoff lever and power lever. The fuel shutoff lever governs engine start and shutdown by activating, in the proper sequence, a fuel shutoff valve, windmill bypass feature, and manifold dump valve signal. The power lever is used to set engine rpm for all forward- and reverse-thrust modes. For any given position of the power lever, a specific amount of thrust is anticipated from the engine. Recall that thrust is the result of accelerating a mass of air through the engine, but as ambient air density differs, so too does the thrust because it is directly related to the density of the incoming air.

For any given steady-state condition, for instance cruise power resulting in a constant mass airflow through the engine, a specific amount of fuel is required. One method of measuring the flow of air through the engine, albeit somewhat rough, is burner pressure; changes in ambient air density will immediately be reflected. Because of that, burner pressure is used to control fuel flow to prevent lean die-out during engine deceleration; however, during acceleration, fuel flow is metered as a function of both engine rpm and burner pressure. Metering by this more comprehensive measurement aids in preventing rich blowout, surging, or overheating.

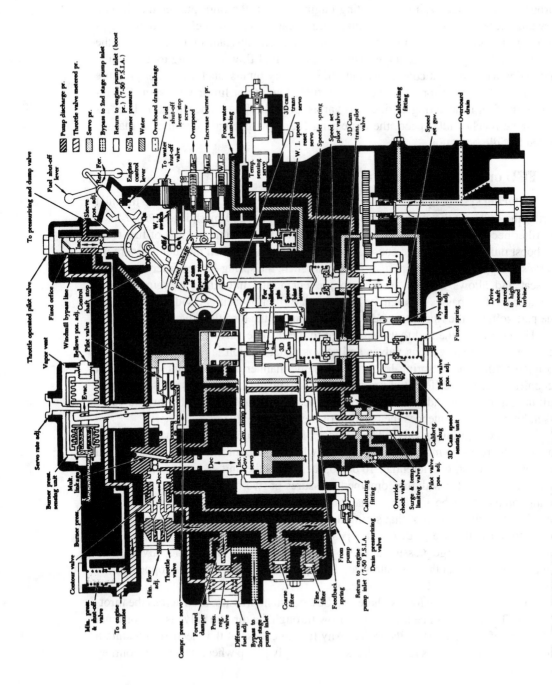

Fig. 7-4. *JFC12-11 schematic operating diagram.*

199

From a fuel control logic perspective, an FCU is significantly different than the carburetor associated with reciprocating engines; when the pilot pushes the throttle forward, there is essentially an immediate and linear increase in fuel flow to the reciprocating engine. FCUs do not operate in that manner. Instead, FCU logic uses the power lever's request as the target while metering fuel flow according to a schedule that varies with ambient conditions and other sensory inputs, such as burner pressure and engine rpm; therefore, the increase in fuel flow is not linear with rpm. For instance, the fuel flow during engine start varies, but follows a set acceleration line. As the engine reaches idle speed, the increasing fuel flow becomes constant at that required for idle operation.

FCU operating description

Metering system operation. Fuel leaves the boost pumps and flows into the inlet port of the FCU. At that point, all fuel is routed through a 200-mesh filter (FIG. 7-5). If the filter becomes clogged or otherwise inoperative, the continuous flow of fuel from the boost pumps will begin to cause a rise in fuel pressure. When the fuel pressure increases by approximately 25 pounds per square inch, the spring-loaded filter moves off its seat and allows the fuel to pass to prevent a fuel interruption engine failure. Whether the fuel passes through, or bypasses the filter, it then is divided in two with one part going to the fine filter and the other going to the throttle valve.

The portion of the fuel that does not go to the throttle valve will be high-pressure fuel used to operate various valves and servos. That portion is channeled through a 35-micron filter that is spring-loaded in the same manner as the 200-mesh filter. If it becomes clogged, a differential pressure of approximately 10 pounds per square inch will unseat the filter and allow the fuel to bypass. This fuel goes to the throttle valve assembly where it influences the position of the contour valve.

Pressure-regulating valve. The purpose of the pressure-regulating valve is to assure that the fuel flowing between the coarse filter and the throttle valve is maintained at a constant pressure differential. To do so, fuel enters the unit and encounters a valve that pushes against it with the combined pressures of two opposing sources. One is a spring force that has been preset to maintain the desired pressure drop across the throttle valve. It is the tension of that spring that can be adjusted by the maintenance technician to allow the use of different types of fuel. The second opposing force is the throttle valve discharge pressure as fuel is routed down from the throttle valve and allowed to push against the opposite side of the pressure-regulating valve, opposing the high pressure fuel from the coarse filter.

Due to the balancing forces, there will be a constant pressure drop across the throttle valve. The result will be that the fuel flow through the throttle valve is proportional to the size of the opening of its orifice. Any fuel in excess of that needed to maintain the pressure differential is rerouted back to the supply pump where it starts the journey all over again.

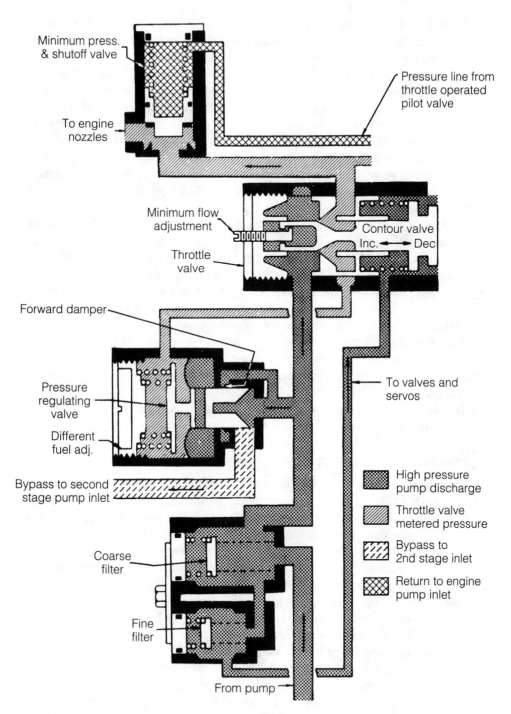

Minimum press.
& shutoff valve

Pressure line from
throttle operated
pilot valve

To engine
nozzles

Minimum flow
adjustment

Contour valve
Inc. — Dec

Throttle
valve

Forward damper

To valves and
servos

Pressure
regulating
valve

Different
fuel adj.

Bypass to second
stage pump inlet

High pressure
pump discharge

Throttle valve
metered pressure

Bypass to
2nd stage inlet

Return to engine
pump inlet

Coarse
filter

Fine
filter

From pump

Fig. 7-5. *Metering system.*

Throttle valve. This main metering valve of the fuel system controls the main fuel flow from the engine-driven fuel pump to the fuel nozzles. The contour valve is controlled by two opposing forces. The force that attempts to move the throttle valve open, increasing the flow of fuel to the nozzles, is the combined input of the compressor pressure servo and the governor servo. Neither of these two are depicted in the illustration, but do act upon the contour valve. The opposing force, which tries to move the valve to the closed position to decrease fuel flow to the nozzles, is the contour valve spring. The contour valve is spring-loaded to the closed position, but a minimum-flow adjustment screw prevents the valve from completely seating; therefore, the contour valve never completely closes. This, combined with the minimum setting of the pressure regulating valve, assures idle fuel flow when the power lever is placed in the idle position.

In addition to supplying the engine nozzles with metered fuel, the throttle valve also routes a portion of fuel to the spring side of the pressure regulating valve as described previously. Finally, the metered fuel can also flow in the direction of the throttle-operated pilot valve. When the fuel shutoff lever is moved into the cutoff position and the engine shuts down, the pilot valve allows the fuel to divert and drain rather than be trapped in the system where it could drip out of the engine nozzles and collect in the burners if the shutoff valve does not properly seat.

Shutoff and minimum pressure valve. At the very top of FIG. 7-5 is the minimum pressure and shutoff valve. Again, two opposing forces are acting upon a valve. On one side is the throttle valve discharge pressure; a spring force and an additional force that reflects the mode of operation are on the other side. The purpose of this valve is to stop fuel flow through the nozzles during engine shutdown and provide a fuel drainage route.

Fuel scheduling system

Speed-set governor. The speed-set governor senses the rpm of the engine high-speed rotor (N_2). The governor is a centrifugal, permanent-droop system that influences the fuel flow to the nozzles via the throttle valve. The power lever is connected through various linkages to the speeder spring of the governor. Movement of the power lever varies the tension on the speeder spring, which attempts to hold the system's flyweights inward. Centrifugal force, which increases as a function of increasing N_2, opposes that force and attempts to move the flyweights outward. The system is fundamentally the same as a propeller governor system.

As N_2 speed increases faster than target speed as scheduled by the power lever (overspeed), the governor senses the increase because a set of flyweights move outward causing the speed-set pilot valve to lift (FIG. 7-6). When the pilot valve goes up, it results in a decrease in the fuel/air ratio, which causes N_2 to decelerate back to the selected rpm. If N_2 decreases (underspeed), the centrifugal force acting upon the flyweights is insufficient to overcome the tension of the opposing speeder spring and the flyweights move inward. This results in the pilot valve moving downward, which in-

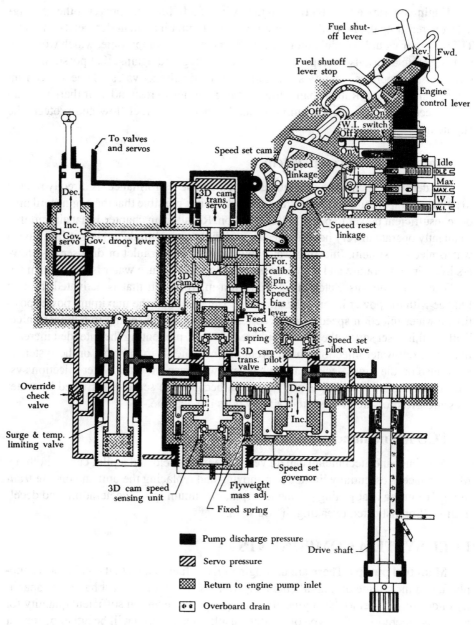

Fig. 7-6. *Scheduling system.*

creases the fuel/air ratio, and N_2 increases back to selected engine rpm. When N_2 and the power-selected engine rpm are the same, an onspeed condition exists and the system is in equilibrium.

Engine overspeed protection. In the event of a failure that prevents the rpm signal from reaching the FCU, the governor would interpret it as an underspeed condition. The governor would schedule more fuel to the nozzles to compensate, which would, in reality, create a serious overspeed condition. To safeguard against that possibility, the governor incorporates a safety system that automatically activates whenever the rpm signal is interrupted. Rather than allowing the governor to respond as if there were an underspeed condition, the system prevents the increase in fuel flow and protects the engine from overspeed.

Water injection reset system

Air density, which has a dramatic effect on engine performance, is directly related to ambient temperature and pressure. On hot days, it is possible that the engine will produce insufficient power to permit safe operation. Recall from chapter 1 that aircraft that frequently operate in very hot climates or at high density-altitude airports might have a water injection system. Injecting water into the compressor inlet or diffuser case lowers inlet air temperature and increases inlet air density because water is denser than air.

The FCU controls water injection with a microswitch that is activated through linkage with the power lever. When the lever is moved into the maximum power position, a water injection speed reset servo resets the N_2 speed setting to a higher value. Without this reset servo, the FCU would reduce rpm, preventing the intended increase in thrust from water injection. In the event that no water is in the tank, or the system is not armed on the flight deck, nothing happens with respect to the water injection system, including the resetting of the N_2 speed, because the servo is activated by water pressure. No pressure: The system simply does not activate.

FCU maintenance

A maintenance technician can do very little in the field and a pilot can do nothing. Maintenance is essentially limited to removing and replacing the unit and on-site *trimming*. Trimming is adjusting engine idle rpm, maximum rpm, acceleration, and deceleration. (Sophisticated tweaking, if you wish.)

FUEL SYSTEM COMPONENTS

Main fuel pumps. These are the engine-driven fuel pumps that continuously supply fuel to the engine during all engine operations. The pumps must have the capability of delivering fuel to the engine at the proper pressure and in sufficient quantity for all engine operations. The type of positive displacement pump will be either piston- or gear-driven. Two categories of this type of pump are *constant displacement* and *variable displacement*. The determining factor in selecting which will be used is based upon the type of system used to regulate fuel flow to the FCU. If a pressure-relief-valve system is utilized, then the choice would be a constant-displacement, gear-driven pump. With that type system, the pressure relief valve will make the necessary pressure adjustments to the constant flow of fuel. The other choice would be a feedback method for regulating pump output which would allow use of a variable-displacement, piston-driven pump.

Constant-displacement pump. The gear-driven pump puts out a constant flow of fuel without regard for any variables or ambient conditions. This pump puts out more than enough fuel for normal operation; therefore, a pressure-relief system is necessary to relieve the system of the excess fuel.

Figure 7-7 is a drawing of an engine-driven, constant-displacement fuel pump. Fuel pressure is increased from 15 to approximately 45 pounds per square inch as fuel is drawn through the pump by the action of the *impeller* (a boost element). The exact pressure varies somewhat with engine speed because the engine drives the impeller. Whatever the speed, the impeller turns at a higher rate than the high-pressure gears, ensuring that the gears receive an adequate fuel supply. Both high-pressure gear elements discharge their fuel out of the same fuel outlet. The combined effect of the gears is to process approximately 50 gallons per minute at an output pressure of 850 pounds per square inch.

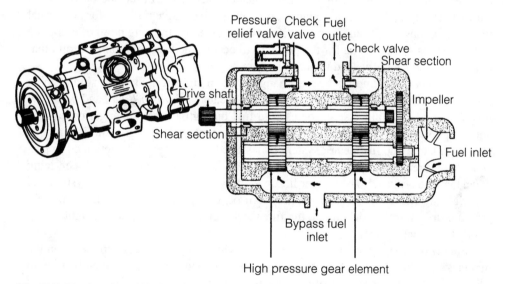

Fig. 7-7. *Engine-driven fuel pump.*

Each element has independent shear sections so that if one gear fails, the remaining gear will continue to operate normally. To prevent the operating gear from forcing fuel through the failed gear, a one-way check valve is placed downstream from each gear. The effect of a single element failure on overall performance is significant, but not critical. One element is capable of producing sufficient fuel flow to support engine operation through moderate aircraft speeds. Because a normally operating pump is capable of producing significantly higher pressure than required by the FCU, a relief valve is incorporated in the pump. The valve unseats at approximately 900 pounds per square inch and is fully open at approximately 960, or roughly the total fuel flow. By-passed fuel is rerouted to the inlet side of the high-pressure gears.

Variable-displacement pump. The positive-displacement, variable-stroke pump differs from the constant-displacement pump in that fuel flow is accurately varied to meet the requirements of the engine. Pump output is controlled directly by the FCU, which assesses the current engine demands and varies the pump output accordingly. The problem with this type of system is that pump failure means no fuel. As a result, variable displacement systems incorporate two, parallel operating pumps. Either pump is capable of sustaining normal engine operation and, under normal conditions, both pumps are operated at reduced, approximately equivalent loads. If one pump fails, the remaining pump automatically increases its output to compensate with no significant interruption of fuel flow.

Fuel heater. Given the problems associated with water in turbine engine fuels, it is no surprise that fuel-filter icing is a serious problem whenever the fuel temperature is fewer than 32°F. Considering the operational characteristics of turbine-powered air-craft and the normal temperature lapse rate, this is a pervasive problem for all normal flight operations. While anti-icing additives help lower the freezing point of entrained water, there is still potential for the problem to occur. The best solution is a combination of anti-icing additive and the use of a fuel heater.

A fuel heater is simply a heat exchanger system and is functionally similar to an automobile radiator (FIG. 7-8). Two types of fuel heaters are *air-to-liquid* and *liquid-to-liquid*. The air-to-liquid system routes cold fuel through a heat exchanger containing hot compressor bleed-air. The air-to-liquid system contains hot engine lubricating oil.

In principle, the fuel heater system is designed to be proactively activated to prevent fuel system icing; however, it is sufficiently powerful to thaw fuel screen icing when that has occurred. The danger lies in waiting too long to activate the system and experiencing flameout. To prevent that from occurring, many aircraft incorporate a pressure-drop warning switch into the filter that will illuminate a warning light on the pilot's annunciator panel.

Fuel filters. Two critical places for filters within the fuel system are between the fuel tanks and the engine-driven fuel pump, and between the engine-driven fuel pump and the FCU. A low-pressure filter protects the pump; a high-pressure filter protects the FCU.

Three filters can be used: *micron*, *wafer screen*, and *plain screen mesh* (FIG. 7-9). Of the three filters, the micron filter is the most effective and is functionally the same as the oil system micron filter described in chapter 3. Because of its effectiveness, it is used as the low-pressure filter between the fuel tanks and the engine-driven pump. Micron filters are made of porous cellulose material ranging from 10–25 microns. A micron filter's major weakness lies in its major strength: It is a very effective filter, extremely susceptible to clogging as a result of trapping particulate matter. To protect against potential flameout, a relief valve is included in the filter system that will allow

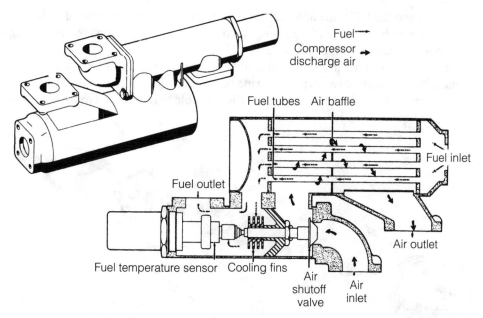

Fig. 7-8. *Fuel heater.*

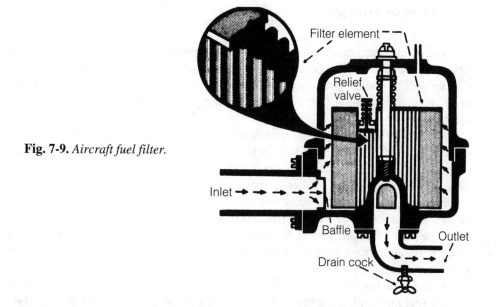

Fig. 7-9. *Aircraft fuel filter.*

fuel flow to bypass a clogged filter. The cellulose also absorbs water, so the micron filter is the last line of defense against entrained water entering the fuel pump.

Also, 200-mesh and 35-mesh filters are composed of multiple layers of metal wire. These wire mesh filters are often used as a final filter between the engine-driven fuel pump and the FCU. Because the screens are made of metal instead of cellulose, they are not only effective at removing micron-sized foreign matter, but are strong enough to be used under high pressure.

The wafer-screen filter (FIG. 7-10), utilizes a replaceable filter element that is formed from multiple layers typically made of bronze, brass, or steel. Somewhat similar to micron filters, the wafer screen filter is also able to remove micron-sized particles while withstanding high pressure.

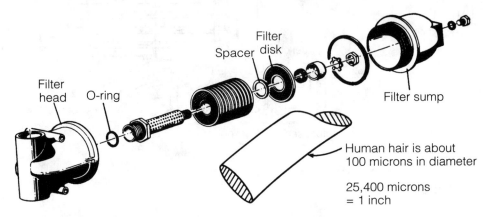

Fig. 7-10. *Wafer screen filter.*

The plain-screen mesh filter is probably the most common type of all filters and has been used in aviation as oil and fuel filters for many years. This filter does not measure up well to micron filters or wafer screen filters, but can be found on turbine engines. Plain-screen mesh filters are predominantly used in low-pressure, noncritical applications.

Fuel spray nozzles and fuel manifolds. The purpose of the spray nozzle is to inject highly atomized fuel into a precise location within the combustion area. Even the proper spray pattern is critical to assure that the burning process occurs in the least possible time and within the smallest possible space. The design of fuel spray nozzles varies among the different combustion chambers for which they are used to prevent hot spots on the chamber walls. Mismatched or misaligned nozzles can literally cause the flame pattern to burn through the combustion liner. In general, nozzles deliver fuel under pressure through a series of small orifices in the nozzle. Two fuel nozzles are *simplex* and *duplex*.

Simplex fuel nozzle. Simplex is the original fuel nozzle design. It is simple, uses a single fuel manifold, and consists of a nozzle tip, insert, and fine-mesh strainer (FIG. 7-11). Very few simplex nozzles are currently in use because they have given way to the far more efficient duplex.

Duplex fuel nozzle. The duplex nozzle is the primary fuel spray nozzle in turbine engines. A duplex nozzle uses a flow divider to produce a highly desirable spray pat-

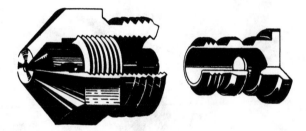

Fig. 7-11. *Simplex fuel nozzle.*

tern that is effective over a wide range of operational pressures (FIG. 7-12). The flow divider splits the fuel into a primary and a secondary flow, spraying them through separate, concentric spray tips. The result is an even, efficient spray pattern through all operational speeds (FIG. 7-13). Fuel enters the fuel inlet port, passes through the screen filter, and moves via a passage through a second screen. The fuel then moves into the primary spin chamber where a change in direction puts a spinning motion on it that establishes the proper spray angle and helps with atomization. At that point, the fuel is discharged from the primary tip.

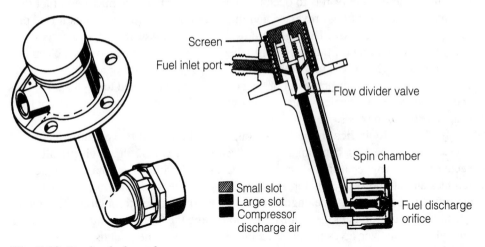

Screen
Fuel inlet port
Flow divider valve
Spin chamber

⬚ Small slot
■ Large slot
■ Compressor
  discharge air

Fuel discharge
orifice

Fig. 7-12. *Duplex fuel nozzle.*

As the engine accelerates from startup, the fuel pressure begins to increase. When the pressure reaches approximately 90 pounds per square inch, it causes the flow divider to open and part of the incoming fuel flow is channeled to the secondary spin chamber. The second chamber performs the same function, but channels fuel to the secondary spray tip (FIG. 7-13).

Fuel pressurizing and dump valves. Duplex nozzle systems require the incorporation of a fuel pressurizing valve to divide the fuel flow between primary and main manifolds. The primary line handles 100 percent of the fuel flow for both engine start and idle at altitude. As the power is increased from either of those two conditions, the

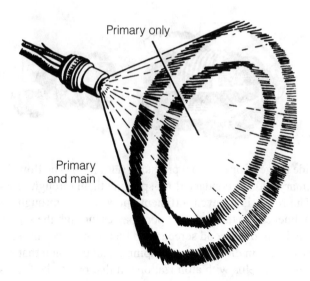

Primary only

Primary
and main

Fig. 7-13. *Duplex nozzle spray pattern.*

fuel pressurizing valve begins to open the main line. By the time maximum fuel flow is achieved, the priority has reversed and the mainline flow now accounts for 90 percent of the total fuel flow. During shutdown, the fuel pressurizing valve serves as a dump valve by cutting off the fuel prior to the manifold to ensure a positive, clean fuel flow stoppage. The excess fuel is then dumped through an overboard drain. This prevents excess fuel from dripping out of the nozzles, which can lead to engine after-fires and carbonization of nozzles, which occurs when unburned fuel is allowed to cool and evaporate on the nozzles.

Fuel quantity indicating units. Among several different fuel quantity indicating devices, one of the most common is the *fuel counter* or *totalizer* that is electrically connected to a flowmeter installed in the engine fuel inlet line. The totalizer, which resembles an automotive odometer, is manually set to the total fuel on board when the aircraft is refueled. When the engine is started, all fuel passes through the flowmeter, which electrically updates the totalizer by subtracting the fuel, in pounds, that has been burned. The totalizer gives the pilot a constant reading of the pounds of fuel remaining. The system is fairly accurate in terms of measuring fuel flow, but does not take into account any fuel that is dumped in-flight, or fuel tank leaks, or any leak upstream from the flowmeter.

Fuel flow can be measured several ways, but the primary method in turbine engines is the *mass-flow*, which is capable of accurately measuring fuel flow between 500 and 2,500 pounds per hour. The system has two cylinders that are placed in the normal fuel stream so that the direction of fuel flow parallels the longitudinal axes of the cylinders (FIG. 7-14). The impeller (upstream cylinder) is driven at a constant angular velocity by a 28-volt dc power supply. As the fuel passes through the rotating impeller, it exits the opposite end with an angular momentum—the fuel is now also rotating. The fuel then flows to the turbine (downstream cylinder), which is not rotating. When the still-rotating fuel enters the turbine, it attempts to turn the turbine. A calibrated restraining spring attached to the turbine measures the angular velocity and translates that into fuel flow. The more pounds of fuel per minute that move through the impeller, the greater the rotational force when fuel enters the turbine and the higher the indication will be on the totalizer.

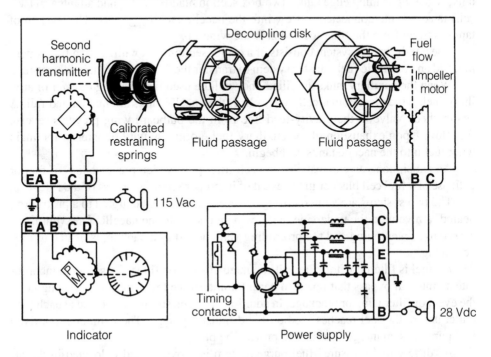

Fig. 7-14. *Schematic of a turbine engine fuel flow indicating system.*

FUEL SYSTEM MAINTENANCE

A flight crew can do practically no fuel system maintenance. Probably the most important thing that can be done is to make sure fuel filters are cleaned and replaced by a maintenance technician on a routine basis in accordance with manufacturer's recommendations. This will help reduce water damage to fuel pumps and control units. Pilots can do their part, for the same reason, by making sure fuel tank sumps and low-pressure filters are drained and checked every day during preflight.

MANUFACTURER DOCUMENTATION

(The following information is extracted from Beech Aircraft Corporation pilot operating handbooks and Falcon Jet Corporation aircraft and interior technical descriptions. The information is to be used for educational purposes only. This information is not to be used for the operation or maintenance of any aircraft.)

Super King Air B200

Fuel system. The fuel system consists of two separate systems connected by a valve-controlled crossfeed line (FIG. 7-15). The fuel system for each engine is further divided into a main and auxiliary fuel system. The main system consists of a nacelle tank, two wing leading edge tanks, two box section bladder tanks, and an integral (wet cell) tank, all interconnected to flow into the nacelle tank by gravity. This system of tanks is filled from the filler located near the wing tip.

The auxiliary fuel system consists of a center section tank with its own filler opening, and an automatic fuel transfer system to transfer the fuel into the main fuel system.

When the auxiliary tanks are filled, they will be used first. During transfer of auxiliary fuel, which is automatically controlled, the nacelle tanks are maintained full. A swing check valve in the gravity-feed line from the outboard wing prevents reverse fuel flow. Upon exhaustion of the auxiliary fuel, normal gravity transfer of the main wing fuel into the nacelle tanks will begin.

An antisiphon valve is installed in each filler port, which prevents loss of fuel or collapse of a fuel cell bladder in the event of improper securing or loss of the filler cap.

The two systems are vented through a recessed ram vent coupled to a protruding heated ram vent on the underside of the wing adjacent to the nacelle. One vent is recessed to prevent icing and the protruding vent, added as a backup, is heated to prevent icing.

All fuel is filtered with a firewall-mounted 20-micron filter. These filters incorporate an internal bypass that opens to permit uninterrupted fuel supply to the engine in the event of filter icing or blockage. In addition, a screen strainer is located at each tank outlet before the fuel reaches the boost and transfer pumps. The main engine-driven fuel pump has an integral strainer to protect the pump.

A "differential pressure" fuel purge system is provided and is located in the aft compartment of each nacelle. The system purges the fuel that is left in the fuel manifolds at engine shutdown by forcing the fuel into the nozzles so that it is consumed in the combustion chamber.

Fuel pumps. The engine-driven fuel pump (high pressure) is mounted on the accessory case in conjunction with the fuel control unit. Failure of this pump results in an immediate flameout. The primary boost pump (low pressure) is also engine driven and is mounted on a drive pad on the aft accessory section of the engine. This pump operates when the gas generator (N_1) is turning and provides sufficient fuel for start, takeoff, all flight conditions (except operation with hot aviation gasoline above 20,000 feet altitude), and operation with crossfeed.

In the event of a primary boost pump failure, the respective red FUEL PRESS light in the annunciator panel will illuminate. This light illuminates when pressure decreases below 10 ± 1 psi. The light will be extinguished by switching on the standby fuel pump on that side, thus increasing pressure above 11 ± 2 psi. **Caution:** Engine operation with the FUEL PRESS light on is limited to 10 hours between overhaul, or replacement, of the engine driven fuel pump.

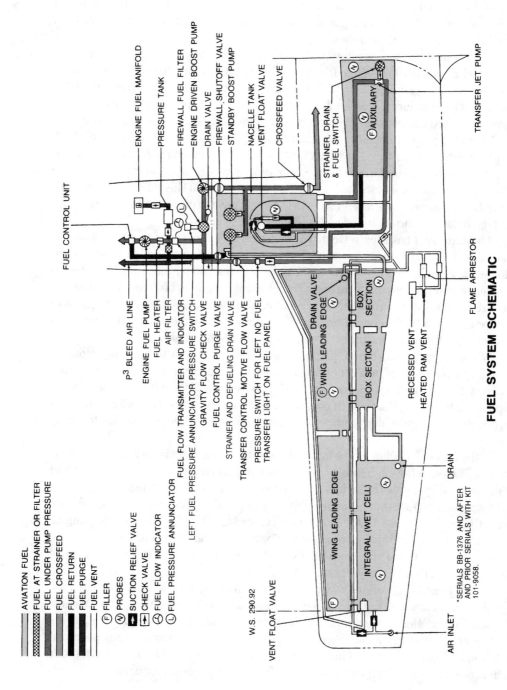

AVIATION FUEL
FUEL AT STRAINER OR FILTER
FUEL UNDER PUMP PRESSURE
FUEL CROSSFEED
FUEL RETURN
FUEL PURGE
FUEL VENT
(F) FILLER
(N) PROBES
SUCTION RELIEF VALVE
CHECK VALVE
(F) FUEL FLOW INDICATOR
(L) FUEL PRESSURE ANNUNCIATOR

FUEL CONTROL UNIT

P³ BLEED AIR LINE
ENGINE FUEL PUMP
FUEL HEATER
AIR FILTER
FUEL FLOW TRANSMITTER AND INDICATOR
LEFT FUEL PRESSURE ANNUNCIATOR PRESSURE SWITCH
GRAVITY FLOW CHECK VALVE
FUEL CONTROL PURGE VALVE
STRAINER AND DEFUELING DRAIN VALVE
TRANSFER CONTROL MOTIVE FLOW VALVE
PRESSURE SWITCH FOR LEFT NO FUEL TRANSFER LIGHT ON FUEL PANEL

ENGINE FUEL MANIFOLD
PRESSURE TANK
FIREWALL FUEL FILTER
ENGINE DRIVEN BOOST PUMP
DRAIN VALVE
FIREWALL SHUTOFF VALVE
STANDBY BOOST PUMP
NACELLE TANK
VENT FLOAT VALVE
CROSSFEED VALVE

STRAINER, DRAIN & FUEL SWITCH

(F) AUXILIARY

TRANSFER JET PUMP

DRAIN VALVE
(N) WING LEADING EDGE
BOX SECTION
FLAME ARRESTOR

RECESSED VENT
HEATED RAM VENT

DRAIN

WING LEADING EDGE
INTEGRAL (WET CELL)

W.S. 290.92

VENT FLOAT VALVE

AIR INLET

*SERIALS BB-1376 AND AFTER AND PRIOR SERIALS WITH KIT 101-9058.

FUEL SYSTEM SCHEMATIC

Fig. 7-15. *Fuel system schematic.* Beech Aircraft Corporation

When using aviation gasoline during climbs above 20,000 feet, the first indication of insufficient fuel pressure will be an intermittent flicker of the FUEL PRESS lights. A wide fluctuation of the fuel flow indicator might also be noted. These conditions can be eliminated by turning on a standby pump.

An electrically driven standby boost pump (low pressure), located in the bottom of each nacelle tank, performs three functions; it is a backup pump for use in the event of a primary fuel boost pump failure, it is for use with hot aviation gasoline above 20,000 feet, and it is used during crossfeed operations. In the event of an inoperative standby pump, crossfeed can only be accomplished from the side of the operative pump.

Electrical power to operate the standby boost pumps is controlled by lever lock toggle switches, placarded STANDBY PUMP-ON-OFF, located on the fuel control panel. It is supplied power from the number 3 or number 4 feeder bus, and is protected by a 10-ampere circuit breaker located on the fuel control panel. This power is only available when the master switch is turned on. These circuits are protected by diodes to prevent the failure of one circuit from disabling the other circuit.

Auxiliary fuel transfer system. The auxiliary tank fuel transfer system automatically transfers the fuel from the auxiliary tank to the nacelle tank without pilot action. Motive flow to a jet pump mounted in the auxiliary tank sump is obtained from the engine fuel plumbing system downstream from the engine driven boost pump and routed through the transfer control motive flow valve. The motive flow valve is energized to the open position by the control system to transfer auxiliary fuel to the nacelle tank to be consumed by the engine during the initial portion of the flight.

When an engine is started, pressure at the engine driven boost pump closes a pressure switch that, after a 30 to 50 second time delay to avoid depletion of fuel pressure during starting, energizes the motive flow valve. When the auxiliary fuel is depleted, a low-level float switch de-energizes the motive flow valve after a 30 to 60 second time delay provided to prevent cycling of the motive flow valve due to sloshing fuel.

In the event of a failure of the motive flow valve or the associated control circuitry, the loss of motive flow pressure when there is still fuel remaining in the auxiliary fuel tank is sensed by a pressure switch and float switch, respectively, which illuminates a light placarded NO TRANSFER on the fuel control panel. During engine start, the pilot should note that the NO TRANSFER lights extinguish 30 to 50 seconds after engine start. A manual override is incorporated as a backup for the automatic transfer system. This is initiated by placing the AUX TRANSFER switch, located in the fuel control panel, to the OVERRIDE position.

Use of aviation gasoline. If aviation gasoline must be used as an emergency fuel, it will be necessary to determine how many hours the airplane is operated on gasoline. Because the gasoline is being mixed with the regular fuel, it is expedient to record the number of gallons of gasoline taken aboard for each engine. Each engine is permitted 150 hours of operation on aviation gasoline between overhauls. This means that if one engine has an average fuel consumption of 50 gallons per hour, for example, it is allowed 7500 gallons of aviation gasoline between overhauls. (Two engines; 15,000 gallons between overhauls.)

Crossfeed. During emergency single-engine operation, it may become necessary to supply fuel to the operative engine from the fuel system on the opposite side. The simplified crossfeed system is placarded for fuel selection with a diagram on the upper fuel control panel. Place the standby fuel pump switches in the OFF position when crossfeeding. A lever-lock switch, placarded CROSSFEED FLOW, is moved from the center OFF position to the left or to the right, depending on direction of fuel flow. This opens the crossfeed valve, energizing the standby pump on the side from which crossfeed is desired, and de-energizes the motive flow valve in the fuel system on the side being fed. When the crossfeed mode is energized, a green FUEL CROSSFEED light on the cautionary/advisory panel will illuminate.

Firewall shutoff. The system incorporates two firewall shutoff valves controlled by two switches, one on each side of the fuel system circuit breaker panel, located on the fuel control panel. These switches, respectively LEFT and RIGHT, are placarded FIREWALL SHUTOFF VALVE-OPEN-CLOSED. A red guard over each switch is an aid in preventing inadvertent operation. The firewall shutoff valves receive electrical power from the main buses and also from the hot battery bus which is connected directly to the battery.

Fuel routing in engine compartment. Just forward of the firewall shutoff valve is the primary engine driven boost pump. From the primary boost pump, the fuel is routed to the main fuel filter, through the fuel flow indicator transmitter, through a fuel heater that utilizes heat from the engine oil to warm the fuel, through the engine driven fuel pump, then to the fuel control unit. Fuel is then directed through the dual fuel manifold to the fuel outlet nozzles and into the annular combustion chamber. Fuel is also taken from just downstream of the main fuel filter to supply the jet transfer pump motive flow.

Fuel drains. During each preflight, the fuel sumps on the tanks, pumps, and filters should be bled to check for fuel contamination. Five sump drains and one filter drain are in each wing, located as follows:

Drains	Location
Leading edge tank	Outboard of nacelle underside of wing
Integral tank	Underside of wing forward of aileron
Firewall fuel filter	Underside of cowling forward of firewall
Sump strainer	Bottom center of nacelle forward of wheel well
Gravity feed line	Aft of wheel well
Auxiliary tank	At wing root just forward of the flap

Fuel purge system. Engine compressor discharge air (P_3) pressurizes a small purge tank. During engine shutdown, fuel manifold pressure subsides, thus allowing the engine fuel manifold poppett valve to open. The purge tank pressure forces fuel out of the engine fuel manifold lines, through the nozzles, and into the combustion chamber. As the fuel is burned, a momentary surge in (N_1) gas generator rpm should be observed. The entire operation is automatic and requires no input from the crew. During engine starting, fuel manifold pressure closes the fuel manifold poppett valve, allowing P_3 air to pressurize the purge tank.

Fuel gauging system. The airplane is equipped with a capacitance type fuel quantity indication system. A maximum indicating error of 3 percent might be encountered in the system. It is compensated for changes in fuel density, which result from temperature excursions.

The LEFT fuel quantity indicator on the fuel control panel indicates the amount of fuel remaining in the left-side main fuel system tanks when the fuel QUANTITY SELECT switch is in the MAIN (upper) position, and the amount of fuel remaining in the left-side auxiliary fuel tank when the fuel QUANTITY SELECT switch is in the AUXILIARY (lower) position. The RIGHT fuel quantity indicator indicates the same information for the right-side fuel systems, depending upon the position of the FUEL QUANTITY switch. The gauges are marked in pounds.

Falcon 50

Fuel system description. The fuel tanks consist of 3 wing and 3 aft fuselage tanks (FIG. 7-16). The approximate capacities are of 1,528 USG (5,787 liters) in the wing tanks and central tank and 787 USG (2,976 liters) in the fuselage tank for a total usable approximate capacity of 2,315 USG (8,763 liters).

Each aft fuselage tank is fed through a transfer pump and a float valve from the respective wing tank; and in emergency, by a differential pressure between forward and aft tanks. Each engine is fed by a booster pump installed in each of the aft fuselage tanks.

A crossfeed system permits operation of two or three engines from the same booster-pump. The tanks (wing and feeder) are pressurized with bleed air, providing differential pressurization backup for the transfer pumps. Feeder tank pressurization backs up the booster pumps.

The fuel system is designed to accommodate all types of fuel approved for the engine. Fuel additives (anti-icing and anti-static) are also acceptable as permitted by the engine data sheet.

Exchange of booster and transfer pumps and most fuel gauges can be performed without draining the tanks. A canister is installed on each booster pump to permit disassembly without defueling the tanks.

Fuel quantity gauging is by a capacitance type indicator system. Total fuel tank and feeder tank contents can be indicated separately.

Single point pressure fueling and defueling is provided as well as two overwing fueling ports with locking tank caps.

216

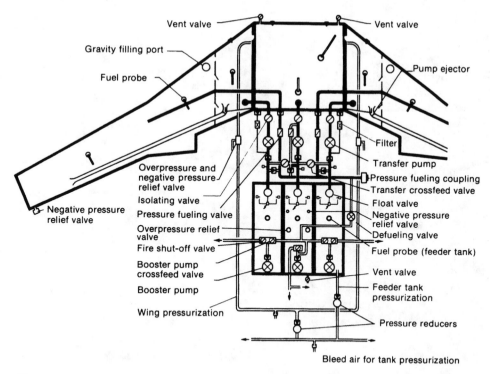

Fig. 7-16. *Fuel system.* Falcon Jet Corporation

Fuel instruments. (The Falcon 50 has) three fuel quantity indicators (with total or fuselage tank quantity selection via a switch on the instrument panel); three fuel-flow indicator totalizers; transfer low-pressure lights; boost pump low-pressure lights; fuel low-level light; and fueling light.

8
Environmental systems

THE HUMAN BODY IS A REMARKABLY ADAPTABLE MACHINE. YET, FOR ALL its versatility the body requires a fairly tightly controlled environment in which to operate. The temperature and atmospheric parameters that will sustain life are relatively narrow, given the extremes found even on our own planet. And while the physical condition of the individual might expand the parameters somewhat, the environmental envelope of extremes for the body's most efficient operation is even smaller. One of the first signs that the quality of the environment is decreasing below that necessary to sustain human life is a gradual diminishing of performance. Logic begins to deteriorate, thought processes become fuzzy, night vision begins to diminish, and worst of all, a feeling of euphoria often begins to set in, which makes all the other symptoms appear to be natural.

NEED FOR OXYGEN

We are all oxygen breathers. Oxygen is so fundamental to our lives that it is an autonomic response; our brain simply will not allow us to hold our breath long enough to cause permanent damage. Unfortunately, our brain cannot control the external environment that surrounds our body, so we must guard against oxygen deficiency. As the body's intake of oxygen decreases, there is a rapid, but subtle, change in bodily functions, thought processes, and degree of consciousness. This oxygen deficiency condition is known as *hypoxia*. Hypoxia can be induced several ways, but the one of major concern to the flight crew is a decrease in the partial pressure of the oxygen in the lungs.

Oxygen is drawn into, and absorbed by, the lungs. To accomplish that task, adequate oxygen pressure must be in the atmosphere. Oxygen exerts approximately ⅕ of the total air pressure at any given altitude, but the total air pressure varies with altitude. For instance, at sea level, the average ambient pressure is 15 psi, of which ⅕, or 3 psi, would be oxygen. That 3 psi is sufficient to saturate the blood through the lungs. As altitude increases, the total pressure decreases, and while the same oxygen ratio of ⅕ is maintained, the actual oxygen pressure is steadily decreasing.

The effect of increased altitude on the human body is practically negligible between sea level and 7000 feet above mean sea level (MSL). As cabin pressure climbs above that altitude, the negative effect of altitude slowly begins to show. Research has revealed that at approximately 8000 MSL, night vision begins to deteriorate and, in some individuals, learning ability and task performance begin to erode. At approximately 10,000 MSL, oxygen saturation of the blood is reduced to 90 percent; extended exposure results in headache and fatigue.

At 15,000 MSL, oxygen saturation is reduced to 81 percent; sleepiness, headache, blue lips and fingernails, impaired vision and judgment, increased pulse and respiration, and some fundamental personality changes occur gradually but relatively quickly. At 22,000 MSL, blood saturation level is 68 percent. Except for those few individuals, such as mountain climbers that are in excellent physical condition and have trained extensively for working in such an environment, the result of a foray at this altitude is gloomy. Without supplemental oxygen, the individual will certainly experience convulsions and death within a fairly short period of time. At 25,000 MSL, the blood saturation is as low as 50 percent; 5 minutes exposure without supplemental oxygen results in unconsciousness and death.

As altitude increases, blood saturation continues to decrease with dramatically shorter times before onset of unconsciousness and death. It is no wonder that Federal Aviation Regulation Part 91 requires at least one pilot at the controls to wear an oxygen mask at all times when the aircraft is above 35,000 MSL. The alternative, and far more common situation, is for both pilots to have the quick-donning type of oxygen mask that can be placed on the face with one hand from the ready position within 5 seconds. At those altitudes, useful consciousness is only a matter of seconds.

ATMOSPHERE'S COMPOSITION

Air, more properly called the atmosphere, is actually a collection of gases, principally nitrogen and oxygen, but with small quantities of other gases including carbon dioxide, water vapor, and ozone (FIG. 8-1). As altitude increases, the amount of the atmosphere decreases, but the relative percentage of its individual components remains the same except for water vapor and ozone. These proportions remain true until a point ap-

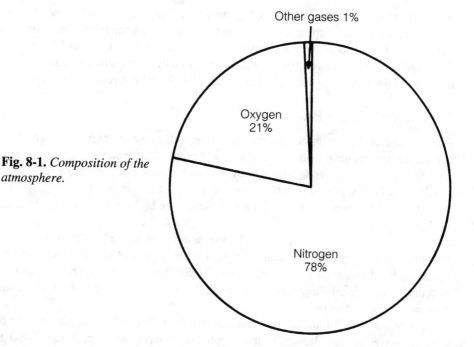

Other gases 1%

Oxygen
21%

Nitrogen
78%

Fig. 8-1. *Composition of the atmosphere.*

proximately 50 miles above the surface of the earth, where fundamental changes in the relative percentages take place, as well as the addition of new gases.

The most common gas, at 78 percent, is the inert nitrogen; inert because nitrogen has a stable atomic structure and has no tendency to unite with other elements to form chemical compounds. Nitrogen is the major component of the atmosphere, but has no direct life-support value in respiration; however, nitrogen is integral to many compounds in biological processes that are required to sustain life.

Even at 21 percent, oxygen is the most critical part of the atmosphere, with respect to sustaining life. Though a relatively small portion of the atmosphere, carbon dioxide is important, used by plants to make the complex substance that humans and other animals consume as food. Carbon dioxide also assists all animals to control their breathing.

One of the most interesting components of the atmosphere is water vapor. Even under the most moist conditions at sea level, water vapor at best accounts for 5 percent of the atmosphere and becomes variable at higher altitudes. Despite its relatively small percentage, water vapor absorbs significantly more of the sun's energy than all other atmospheric gases combined. Vapor is not the only manner in which water exists in the atmosphere because water and ice crystals can also be found almost all of the time. Atmospheric water, in its several forms, plays an important part in the formation of the atmosphere and its weather conditions.

Ozone is important to animal existence, filtering out most of the sun's harmful ultraviolet radiation. Ozone—a permutation of oxygen—is composed of three atoms of oxygen per molecule instead of two. Ozone is found principally in the very high altitudes of the earth's atmosphere. The reason is that ozone is primarily formed by the interaction of oxygen and the sun's rays, though it can also be produced by electrical discharges. Anyone who has been in the vicinity of a lightning strike will have smelled its chlorine odor. Auroras and cosmic rays are also thought to produce ozone.

ATMOSPHERE'S PRESSURE

Despite the fact that the atmosphere is invisible, it does have weight. The weight of a column of air reaching from the surface of the earth into space is known as *atmospheric pressure*. For standardization: A one-inch-square column of air rising to approximately 70,000 feet has been designated as the reference and its average sea level weight is 14.7 lbs; hence the aviation atmospheric standard weight of air at sea level is 14.7 psi.

This same column of air can also be measured against the height of a column of mercury when it is used as a scale to measure the weight of the air. The same one-square-inch column of air will displace 1013.2 millibars of mercury, or 29.92 inches of mercury. These are all simply different methods of measuring the weight of the same column of air at sea level.

The higher the altitude at which you sample the weight of the air, the less the column will weigh because less air is above your position. In fact, approximately 50 percent of the weight of the atmosphere is below 18,000 feet because the weight of the atmosphere at that point is 7.34 psi. A review of FIG. 8-2 will reveal that atmospheric pressure decreases rapidly with an increase in altitude. At 50,000 feet, the atmospheric pressure is almost 1/10 of the sea level value, and at a few hundred miles above the earth the atmosphere is so rarefied that it is considered nonexistent.

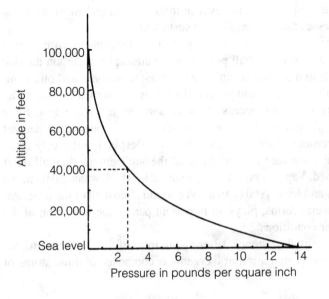

Fig. 8-2. *Weight of the atmosphere vs. altitude.*

TEMPERATURE AND ALTITUDE

Temperature variations on the surface of the earth are well known, but those at high altitude are less variable with a more constant pattern. Virtually all civilian flying takes place within the *troposphere*, which is the earth's atmosphere from sea level to approximately 60,000 feet MSL. In the troposphere, temperature decreases as altitude increases. The actual top of the troposphere, called the *tropopause*, occurs at different altitudes, depending upon the season and the geographical location. The troposphere's maximum altitude is over the equator; the lowest altitude is 30,000 feet over the earth's poles. The tropopause separates the troposphere from the *stratosphere*, which is where temperature no longer decreases as altitude increases; temperature is constant regardless of altitude, stabilized at –67°F (–55°C).

Even a perfunctory review of the atmosphere makes it clearly evident that routine flights above an altitude of approximately 10,000 feet require special consideration. Such flights are best conducted in aircraft that are capable of maintaining a cabin pressure, temperature, and humidity at or below approximately 8000 feet.

PRESSURIZATION

For several reasons, it is desirable to fly at the highest practical altitude based upon the operational parameters of the aircraft and the length of the trip. From an economical perspective, at a given airspeed, turbine engine fuel consumption decreases as altitude increases—the higher the aircraft flies, the less fuel it burns, all other things being equal. From an operational point of view, the higher the flight, the less adverse weather you are likely to encounter.

An aircraft pressurization system has three primary goals:

- Automatically maintain a maximum cabin pressure altitude of approximately 8000 feet at the aircraft's maximum designed cruising altitude.
- Prevent unwanted, rapid changes of cabin altitude regardless of rate of climb or descent.
- Reasonably fast fresh air exchange to eliminate odors and remove stale air.

The concept behind the operation of a cabin pressurization system is very simple. Air is pumped into the cabin under pressure at a steady rate. In turbine-powered aircraft, this pressurized air is provided by bleed air. Because compressed air rises to a temperature of several hundred degrees, it is necessary to cool the air prior to cabin entry. The air circulates throughout the cabin and escapes through a hole (*cabin outflow valve*) in the aft portion of the fuselage. This constant input and outflow assures a constant turnover of the air. Varying the size of the opening of the outflow valve increases or decreases the cabin pressure as desired. In normal operation, the outflow valve is automatically controlled by the aircraft pressurization system; however, in the event of malfunction, the crew can also schedule the outflow valve manually.

A *dump valve* is next to the outflow valve. The dump valve is either fully open or fully closed. When fully open, the cabin pressure will equalize with the ambient pressure.

It is normal for the business-class turboprop and turbojet aircraft cabin, flight deck, and baggage area to be a single *pressure vessel*. Air-carrier aircraft often provide for the isolation of the cargo area from the passenger cabin. Both areas are pressurized, but the separation allows the flight crew to seal off the cargo area in the event of a fire or other problem.

The amount of pressurization that a given aircraft is capable of achieving is predetermined by the manufacturer based upon the designed maximum operating altitude. In turbine-powered aircraft, there is little concern regarding the adequacy of the source of pressurized air; turbine-engine bleed air is more than capable of pressurizing the aircraft. The major limiting factor to the maximum cabin altitude possible for a given airframe is the fuselage's ability to withstand the pressure differential between the inside and the outside of the cabin. The higher the intended cruise operating altitude, the greater the cabin pressure differential, and the greater the fuselage reinforcement required.

Pressurization essentially tries to blow up the fuselage like a balloon with the result that a significant internal stress is placed upon the fuselage's skin. Until the Aloha Airlines incident—in which the top, forward section of the aircraft separated in flight due to metal fatigue—it was generally agreed that the airframe had unlimited life. It is now known that airframes do have limited lives as repeated pressurizing cycles gradually stretch and deform the fuselage skin. The worst scenario is an aircraft that is pressurized for each flight, but each flight is very short, rapidly increasing the number of cycles.

It is a difficult problem from a design point-of-view. If the engineer did not have to be concerned about weight, it would be a simple matter to build a very strong container. But weight is a critical factor in designing aircraft; therefore, it is important to be able to design a fuselage capable of holding a significant amount of air under pressure, yet be light enough to operate efficiently with an acceptable amount of people, bags, and fuel on board. The problem is not trivial when you consider the pressure differentials involved.

Recall that the maximum cabin altitude, under normal operating conditions, should not exceed approximately 8000 feet MSL. With aircraft that are designed for operation at altitudes higher than 25,000 feet MSL, in addition to the 8000-foot requirement, they must also be able to maintain a cabin altitude of 15,000 feet in the event of any reasonably likely failure. For instance, pressurizing the cabin to 8000 feet might require the use of both engines; if one engine is inoperative, thereby losing half of the available bleed air, the remaining engine must be capable of pressurizing the cabin to an altitude not greater than 15,000 feet.

Now consider what kind of pressure differential must exist between the pressurized inside of the fuselage and the ambient air outside. An atmospheric pressure at 8000 feet MSL is approximately 10.92 psi; at 40,000 MSL, the ambient pressure is approximately 2.72 psi. If an aircraft is flying at 40,000 feet with a cabin pressure of 8000 feet, the differential pressure is 8.20 psi (10.92 −2.72 = 8.20). Let's say our aircraft has a cabin vessel internal surface of 10,000 square inches. That means the structure will have 82,000 pounds, or approximately 41 tons, of pressure that it must hold. In addition, regulations require an additional safety factor of 1.33; therefore, the manufacturer will have to construct that aircraft so that it will have an ultimate strength of 109,060 pounds (82,000 ¥ 1.33 = 109,060), or 54.5 tons.

AIR-CONDITIONING AND PRESSURIZATION SYSTEMS

The air-conditioning and pressurization system provides cabin pressurization, heating, cooling, and cooling of certain specific items such as navigation and communication equipment. Two primary methods of operating aircraft cabin pressurization are turbine-engine bleed air and reciprocating engine turbochargers or internal superchargers. By their very nature, both methods rely on compressed air, which is heated by the act of compression. In addition, heat is also produced inside the cabin through ram-air temperature increases, solar heat, electrical equipment heat, body heat, and engine heat, depending upon the engine's proximity to the cabin. As a result, it is typically necessary to cool pressurized air prior to its entry into the cabin. Two primary methods of cooling pressurized aircraft are the air cycle machine (ACM) and the vapor cycle system.

Terms and definitions. Before looking at specific systems, several terms should be defined to be able to understand air-conditioning and pressurization system operation.

Absolute pressure is measured on a scale that uses a complete vacuum as its zero value.

Absolute temperature scale has a zero value that is established as the point where there is no molecular motion (–273.1°C or –459.6°F).

Adiabatic process means no transfer of heat between the working substance and any outside source.

Aircraft altitude is the actual height of the aircraft above sea level.

Ambient temperature is the static, free-air temperature in the area immediately surrounding the object under discussion.

Ambient pressure is the static, free-air pressure in the area immediately surrounding the object under discussion.

Standard barometric pressure is the weight of gases in the atmosphere sufficient to hold up a column of mercury 760 millimeters high (approximately 30 inches) at sea level (14.7 psi). Standard barometric pressure decreases with altitude.

Cabin altitude expresses the cabin pressure in terms of equivalent altitude above sea level.

Differential pressure is the difference in pressure between the pressure acting on one side of a wall and the pressure acting on the other side of the wall—the difference between aircraft cabin pressure and atmospheric pressure.

Gauge pressure is the pressure in a vessel, container, or tube, compared to ambient pressure.

Ram-air temperature rise is the increase in temperature created by the ram compression on the surface of an aircraft traveling at a high rate of speed through the atmosphere. The rate of increase is proportional to the square of the speed of the object.

Temperature scales: Centigrade has 0° representing the freezing point of water and 100° is equivalent to the boiling point of water at sea level; Fahrenheit has 32° representing the freezing point of water, and 212° is equivalent to the boiling point of water at sea level.

Basic requirements. Five basic requirements must be met for an air-conditioning and pressurization system to effectively meet the needs of an aircraft.

- A reliable source of compressed air to pressurize and ventilate the cabin. In turbine-powered aircraft, the source is engine bleed air.
- A pressure regulator to automatically and manually control the cabin pressure by regulating the cabin air outflow through the outflow valve.
- A method to prevent over-pressurization of the cabin. This is accomplished through the use of pressure relief valves, negative relief valves, and dump valves.
- A positive control method for regulating the cabin air temperature. In turbine-powered aircraft this typically means cooling the turbine-engine bleed air used for pressurization.
- An effective sealant system that minimizes uncontrolled leakage of pressurized air. Sealants must be tailored to the normally anticipated pressure differentials.

Major components of an air-conditioning and pressurization system have numerous minor components. Figure 8-3 illustrates a schematic diagram of a typical air-conditioning and pressurization system for a twin-engine turbojet aircraft.

Cabin pressure sources

Bleed air was the most common source of compressed air to pressurize the cabin, but bleed air was not ideal for two reasons: (1) The potential for air contamination if a lubricant or fuel leaked into the intake air, and (2) dependence upon engine performance to keep the cabin pressurized. As a result, another method of pressurizing the cabin was developed that utilizes an independent *cabin compressor* that can be driven by the engine accessory drive gearing, or by bleed air from the compressor. In the latter case, the bleed air does not pressurize the cabin, the bleed air only powers the pump that pressurizes outside air. Cabin compressors are separated into two groups: *positive-displacement* and *centrifugal*. Centrifugal compressors dominate turbine-powered aircraft.

Centrifugal cabin compressors. The fundamental principle behind the operation of a centrifugal compressor cabin pressurization system is similar to the turbojet compressor (FIG. 8-4). A centrifugal unit takes in a volume of air, passes the air through an impeller that is turning at a high rate of speed, and imparts kinetic energy to the air. Because of the impeller's rotation, the induced air is accelerated and, as the air is thrown outward, the induced air is also compressed. The kinetic energy that re-

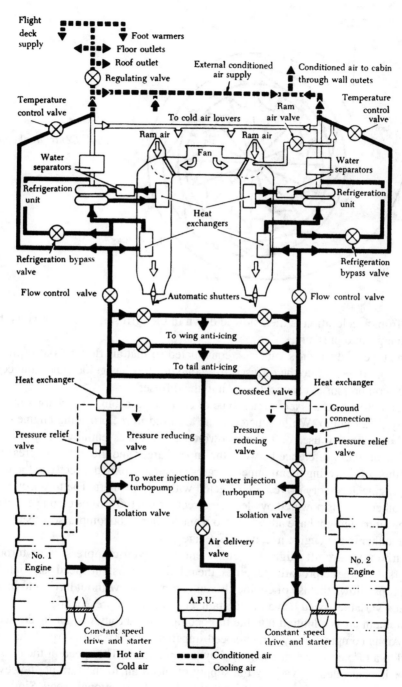

Flight deck supply
Foot warmers
Floor outlets
Roof outlet
Regulating valve
External conditioned air supply
Conditioned air to cabin through wall outlets
Temperature control valve
To cold air louvers
Ram air valve
Temperature control valve
Ram air
Ram air
Fan
Water separators
Water separators
Refrigeration unit
Refrigeration unit
Heat exchangers
Refrigeration bypass valve
Refrigeration bypass valve
Flow control valve
Automatic shutters
Flow control valve
To wing anti-icing
To tail anti-icing
Heat exchanger
Crossfeed valve
Heat exchanger
Ground connection
Pressure relief valve
Pressure reducing valve
Pressure reducing valve
Pressure relief valve
To water injection turbopump
To water injection turbopump
Isolation valve
Isolation valve
Air delivery valve
No. 1 Engine
No. 2 Engine
Constant speed drive and starter
A.P.U.
Constant speed drive and starter
Hot air
Cold air
Conditioned air
Cooling air

Fig. 8-3. *Typical pressurization and air conditioning system.*

Fig. 8-4. *Centrifugal cabin compressor.*

sulted from acceleration is then slowed down in the diffuser, which converts the kinetic energy into pressure.

One type of diffuser is vaneless, constructed so that air flowing out of the compressor goes directly into the diffuser. A second diffuser has guide vanes immediately after the compressor to channel the air into the diffuser.

Figure 8-5 illustrates a common type of centrifugal cabin supercharger that is essentially an air pump. While superchargers are used in reciprocating engine aircraft, the turbocompressor used on turbine-powered aircraft is essentially the same component. The primary difference between the supercharger and the turbocompressor is in the method of powering them. Superchargers are either splined directly to the reciprocating engine accessory drive, or connected with a drive shaft. The turbocompressor, on the other hand, is powered by bleed air directly from the turbine engine compressor. A given aircraft might have anywhere from one to four turbocompressors that might be located either in the engine nacelle or fuselage.

Turboprop aircraft might use either engine-driven compressors or turbocompressors. Unlike reciprocating engines, the relatively constant speed associated with turboprops allows the use of engine-driven compressors without having to include a complex variable-speed drive to maintain a constant compressor speed.

Ambient air is admitted into the turbocompressor through an air intake scoop or duct. Air is compressed by a high-speed impeller (20,000–50,000 rpm) that is connected to a turbine wheel that is turned by pneumatic air. The speed of the turbocompressor is controlled by varying the supply of bleed air to the turbine; however, all supercharger systems have a major limitation: impeller rotational speed. Similar to an aircraft propeller, as the impeller tip speed approaches the speed of sound, the impeller

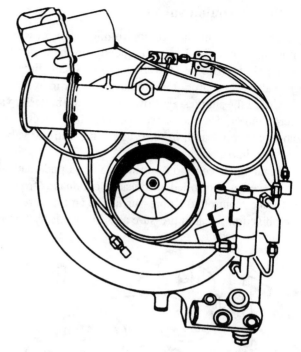

Fig. 8-5. *Pictorial view of a centrifugal cabin supercharger.*

rapidly loses its efficiency as an air pump, and, if allowed to progress, the condition can result in damage to the compressor. Due to this major limitation, most turbocompressor units incorporate an overspeed control that works on the same flyweight principle as a propeller governor. When an overspeed condition occurs, the flyweights close the pneumatic duct shutoff valve.

Another limitation to the turbocompressor's output capability is air duct back pressure. When air pressure output increases, the air has a tendency to "back up" in the ducting, which makes it difficult for the most recently compressed air to flow into the cabin. As this problem gets worse, the impeller might stall or surge.

The pneumatic air supply is bleed air from the turbine engine's compressor. Varying among aircraft, a constant pressure of approximately 45–75 psi is maintained and used as pneumatic air throughout the aircraft. In addition to controlling the turbocompressor, various aircraft use pneumatic air for such tasks as anti-icing, door seal inflation, pneumatic instrument power, and other systems.

The outflow of a turbocompressor is controlled automatically by an airflow control valve and servo-operated inlet vanes. The inlet vanes control the pneumatic (bleed air) system air supply to the turbocompressor turbine. The vanes open or close according to the air pressure signal sensed at the airflow control valve, and turbocompressor speed is increased or decreased to maintain a relatively constant output air volume. Turbocompressor speed will therefore increase with altitude.

Pressurization valves

The primary method of controlling cabin pressure is the *outflow valve*. Located within the pressurized area, frequently on the aft bulkhead, the outflow valve vents cabin air overboard into the atmosphere. Small aircraft might only have a single outflow valve; large aircraft might have three or more.

One of the simplest outflow valves is the butterfly valve, which is controlled by an electric motor. The motor receives amplified electrical commands from the pressurization controller and, in turn, varies the outflow valve opening to maintain the required cabin pressure.

A far more common outflow valve is pneumatically controlled (FIG. 8-6). The valve receives controlled air pressure commands from the pressurization controller in the form of controlled air pressures. The air pressure that operates the outflow valve is a combination of cabin pressure and pneumatic system pressure.

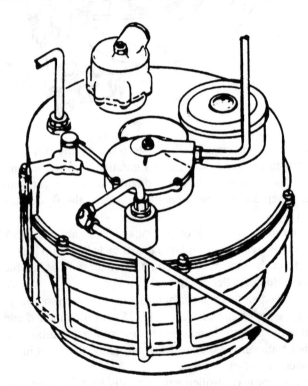

Fig. 8-6. *Typical pneumatic outflow valve.*

In many aircraft, the outflow valve is held in the full-open position while on the ground. This is assured because the system is wired through a landing gear *squat switch*. When weight is on the gear, the outflow valve is open. Once the weight is off the landing gear—detected by the squat switch—the outflow valve is free to follow the commands of the cabin pressurization controller. As pressurizing air flows into the cabin during climbout, the outflow valve slowly closes to assist in holding sea level cabin pressure. In cruise conditions, cabin altitude is directly related to the degree of outflow valve opening.

Another required device on all pressurized aircraft is the automatic cabin pressure relief valve. Depending upon the manufacturer, this valve might be incorporated into the outflow valve or might be a separate valve. The relief valve automatically opens if the cabin pressure differential reaches a preset value. This is simply an over-pressurization safety valve. A negative pressure relief valve, on the other hand, prevents a condition of greater pressure outside the aircraft than inside, which is unlikely. But it is worth noting that an aircraft fuselage is structurally designed to hold high pressure inside, not keep high pressure outside. This valve might also be incorporated in the outflow valve or be a separate unit. On some aircraft, the negative relief valve is simply a hinged flap on the aft cabin bulkhead. During pressurized flight, the internal cabin pressure holds the flap closed. If negative pressure occurs, the higher ambient air pressure would simply push the hinged flap inward and equalize the pressure differential.

A final valve included in all pressurized aircraft is a safety relief valve or manual depressurization valve. Pilots simply refer to it as the *dump valve*. This valve is manually controlled from the flight deck when all other means of control fail. A dump valve is primarily intended to induce rapid depressurization in the event of an emergency, such as smoke or fire in the cabin or in the event of an emergency descent.

Pressurization controls

The pressurization controller illustrated in FIG. 8-7 supplies the control signals that operate the pressurization system. The controller resembles an altimeter with several control knobs. The dial is graduated in cabin altitude increments up to approximately 10,000 feet. Turning the cabin altitude selector knob causes the single pointer to rotate along the scale. The pointer is placed at the desired cabin altitude represented on the scale. More advanced units might have a second pointer that indicates the corresponding aircraft pressure altitude. The knob in the upper right-hand corner adjusts the controller to the existing altimeter setting. It is operated in the same manner as an altimeter. The knob in the lower left-hand corner adjusts the cabin rate of altitude change—how fast the cabin altitude climbs or descends.

The controller adjusts either an electric or pneumatic signaling device inside the unit. The settings are compared to the cabin pressure by an aneroid bellows. Any discrepancy between the requested and the actual cabin altitudes causes an appropriate signal to go to the outflow valve to adjust cabin altitude accordingly. When the bellows notes that the requested and actual cabin altitudes are the same, another signal goes to the outflow valve to readjust appropriately. As factors change, the process is repeated to constantly maintain the requested cabin pressure. This system takes into account variances in actual aircraft altitude, loss of an engine or turbocompressor, and other similar variables.

Three additional instruments are part of the pressurization controller system: cabin differential pressure gauge, cabin altimeter, and cabin rate of climb indicator. The cabin differential pressure gauge indicates the difference between inside and outside

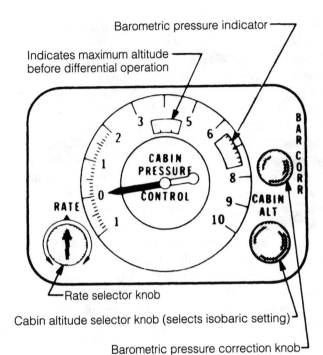

Barometric pressure indicator

Indicates maximum altitude
before differential operation

Fig. 8-7. *Pressurization controller*

Rate selector knob

Cabin altitude selector knob (selects isobaric setting)

Barometric pressure correction knob

pressure. It is important to monitor this gauge to assure that the cabin is not approaching the maximum allowable differential pressure. The cabin altimeter, which indicates the pressure altitude inside the cabin, also helps the crew monitor overall system performance. Certain pressurization controller systems incorporate both functions into a single instrument. The third instrument indicates the cabin rate of climb or descent. A combined pressure differential gauge and cabin altimeter, and a cabin rate of climb indicator are depicted in FIG. 8-8.

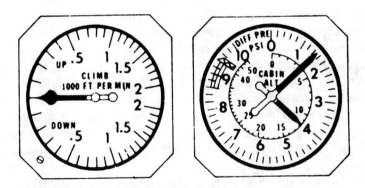

Fig. 8-8. *Instruments for pressurization control.*

Cabin pressure control system

This system provides the crew with a method of regulating cabin pressure, plus provides for over-pressure and vacuum relief, dumping the cabin pressure in an emergency, and a means of selecting desired cabin altitude within the isobaric and differential pressure limitations. The system requires three pieces of equipment to accomplish these tasks: cabin pressure regulator, at least one outflow valve, and a safety valve. A number of different systems are used in aircraft; the following general description is typical for this type of system.

Cabin pressure regulator. Cabin pressure is controlled by varying the position of the outflow valve, which allows cabin air to escape into the atmosphere. Cabin pressure is regulated within the isobaric range selected by the crew and limited to the manufacturer's preset differential value. The isobaric range maintains the cabin at a constant-pressure altitude throughout all normal flight conditions as selected by the crew's inputs to the cabin altitude selector knob on the pressurization controller. This method works until the difference between the inside and outside pressure exceeds the highest differential pressure for which the fuselage is designed. When that occurs, the regulator stops trying to hold the requested cabin altitude and begins to allow it to increase as actual altitude increases. The controller will automatically allow this to happen based upon the maximum differential.

Whether the regulator is constructed as an integral part of the outflow valve, or as a separate unit, the regulator is a differential pressure device and is normally closed. Figure 8-9 is the pneumatically controlled type and incorporates an outflow valve. This type of regulator consists of two principal sections: the head and reference chamber section, and the outflow valve and diaphragm section.

The head and reference chamber section is approximately the upper half of the cabin pressure regulator; the outflow valve and diaphragm section is the lower half. The outflow valve and diaphragm section consists of a base, a spring-loaded outflow valve, an actuator diaphragm, a balance diaphragm, and a baffle plate. The outflow valve rides on the pilot between the cover and baffle plate. The outflow valve is held closed, which is against the base, under spring pressure.

The balance diaphragm extends outward from the baffle plate to the outflow valve, which creates a sealed chamber between the fixed baffle and outflow valve. Cabin air enters the chamber via a set of holes in the side of the outflow valve. The air in this area exerts a force against the inner face, which opposes spring tension and causes the valve to open. The actuator diaphragm extends outward from the outflow valve to the cover assembly, creating a second chamber.

The second chamber lies between the unit's cover and the outflow valve. Air from the head and reference chamber section flows through holes in the cover, filling this chamber and exerting a force against the outflow valve that assists spring tension and attempts to keep the valve closed. Cabin pressure is controlled as the position of the outflow valve varies the rate at which cabin air is vented to the atmosphere. The action of components in the head and reference chamber section controls the movements of

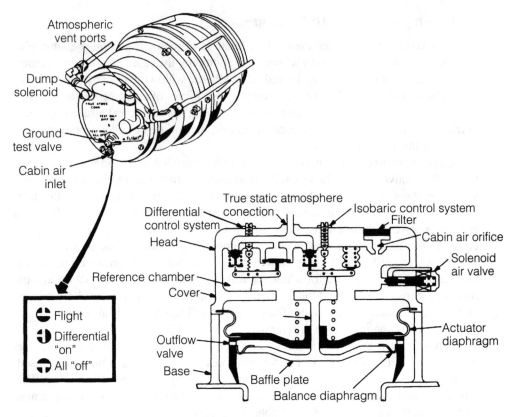

Fig. 8-9. *Cabin air pressure regulator.*

the outflow valve by varying the pressure of reference chamber air being exerted against the outer face of the valve.

The head and reference chamber section include an isobaric control system, a differential control system, a filter, a ground test valve, a true static atmosphere connection, and a solenoid air valve. The area inside the head is called the *reference chamber*.

The isobaric control system incorporates an evacuated bellows, a rocker arm, a follower spring, and an isobaric metering valve. One end of the rocker arm is connected to the head by the evacuated bellows. The other end of the arm positions the metering valve to a normally closed position against a passage in the head. A follower spring between the metering valve seat and a retainer on the valve causes the valve to move away from its seat as the rocker arm permits.

Whenever the reference chamber air pressure is great enough to compress the bellows, the rocker arm pivots about its fulcrum. This allows the metering valve to move from its seat an amount proportionate to the amount of compression in the bellows. When the metering valve is open, reference chamber air flows to atmosphere through the true static atmosphere connection.

The differential control system incorporates a diaphragm, a rocker arm, a metering valve, and a follower spring. One end of the rocker arm is attached to the head by the diaphragm. The diaphragm forms a pressure-sensitive face between the reference chamber and a small chamber in the head. This small chamber is opened to an atmosphere connection. Atmospheric pressure acts on one side of the diaphragm and reference chamber pressure acts on the other. The opposite end of the rocker arm positions the metering valve to a normally closed position against a passage in the head. A follower spring between the metering valve seat and a retainer on the valve causes the valve to move away from its seat as the rocker arm permits.

When reference chamber pressure exceeds atmospheric pressure sufficiently to move the diaphragm, the metering valve is allowed to move from its seat an amount proportionate to the movement of the diaphragm. When the metering valve is open, reference chamber air flows to atmosphere through the true static atmosphere connection.

By regulating reference chamber air pressure, the isobaric and differential control systems control the actions of the outflow valve to provide for three modes of operation called *unpressurized, isobaric*, and *differential*.

During unpressurized operation (FIG. 8-10), reference chamber pressure is sufficient to compress the isobaric bellows and open the metering valve. Cabin air entering the reference chamber through the cabin-air orifice flows to the atmosphere through the isobaric metering valve. Because the cabin-air orifice is smaller than the orifice formed by the metering valve, reference chamber pressure is maintained at a value slightly less than cabin pressure. As pressure increases in the cabin, the differential pressure between the outflow valve inner and outer face increases. This unseats the outflow valve and allows cabin air to flow to the atmosphere.

As the isobaric range is approached, reference chamber pressure, which has been decreasing at the same rate as atmospheric pressure, will have decreased

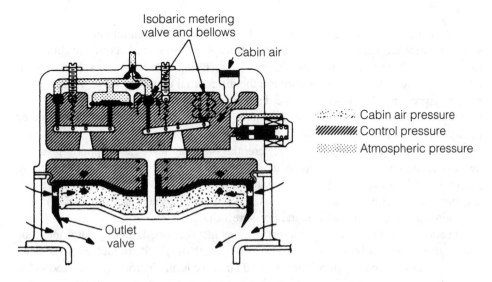

Fig. 8-10. *Cabin air pressure regulator in the unpressurized mode.*

enough to allow the isobaric bellows to expand and move the metering valve toward its seat (FIG. 8-11). As a result, the flow of reference chamber air through the metering valve is reduced, preventing further decrease in reference pressure. In response to slight changes in reference chamber pressure, the isobaric control system modulates to maintain a substantially constant reference pressure in the chamber throughout the isobaric range of operation. Responding to the differential between the constant reference chamber pressure and the variable cabin pressure, the outflow valve opens or closes, metering air from the cabin to maintain a constant cabin pressure.

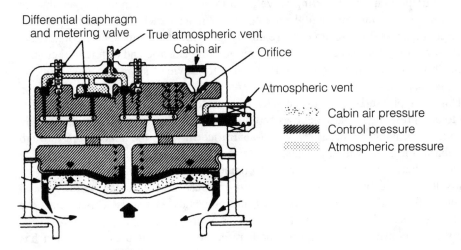

Fig. 8-11. *Cabin air pressure regulator in the isobaric range.*

As the differential range is approached, the pressure differential between the constant reference pressure and the decreasing atmospheric pressure becomes sufficient to move the diaphragm and open the differential metering valve. As a result, reference chamber air flows to atmosphere through the differential metering valve, reducing the reference pressure. Responding to the decreased reference pressure, the isobaric bellows expands and closes the isobaric metering valve completely. Reference chamber pressure is now controlled, through the differential metering valve, by atmospheric pressure being reflected against the differential diaphragm. As atmospheric pressure decreases, the metering valve opens more and allows reference pressure to decrease proportionately. Responding to the pressure differential between cabin and reference pressures, the outflow valve opens or closes as required to meter air from the cabin and maintain a predetermined differential pressure value.

In addition to the automatic control features just described, the regulator incorporates a ground test valve and a solenoid air valve, both of which are located in the head and reference chamber section. The solenoid air valve is an electrically activated valve spring-loaded to a normally closed position against a passage through the head that opens the reference chamber to atmosphere. When the flight deck pressure switch is positioned to "ram," the regulator solenoid opens, causing the regulator to dump cabin air to the atmosphere.

The ground test valve, as shown in the left illustration of FIG. 8-9, is a three-position, manually operated control that allows for performance checks of the regulator and cabin pressurization system. In the TEST ONLY—ALL OFF position, the valve renders the regulator completely inoperative. In the TEST ONLY—DIFFERENTIAL ON position, the valve renders the isobaric control system inoperative so that the operation of the differential control system can be checked. In the FLIGHT position, the valve allows the regulator to function normally. The ground test valve should always be lockwired in the FLIGHT position unless being tested.

Cabin-air pressure safety valve. The cabin-air pressure safety valve combines several important functions into one unit: over-pressure relief, vacuum relief, and dump valve functions (FIG. 8-12). The primary purpose of the pressure relief valve is to prevent cabin pressure from exceeding the manufacturer's maximum differential pressure limitation. The purpose of the vacuum relief function is to prevent the ambient pressure from exceeding the cabin pressure. The valve does so by allowing ambient air to enter the cabin whenever ambient air pressure exceeds the cabin pressure. The dump valve in the illustration is controlled by a flight-deck switch. When the switch is placed in the RAM position, a solenoid valve opens, which causes the valve to allow the cabin pressure to equalize with the ambient air pressure.

The safety valve consists of an outflow valve section and a control chamber that are separated by a flexible, sealed diaphragm. The diaphragm is exposed to cabin pressure on the outflow valve side and control chamber pressure on the opposite side. Movement of the diaphragm causes the outflow valve to open or close. A filtered opening in the outflow valve allows cabin air to enter the reference chamber. The outflow valve pilot extends into this opening to limit the flow of air into the chamber. Air pressure inside the reference chamber exerts a force against the inner face of the outflow valve to aid spring tension in holding the valve closed. The pressure of cabin air against the outer face of the outflow valve provides a force opposing spring tension to open the valve. Under normal conditions, the combined forces within the reference chamber are able to hold the outflow valve in the closed position. The movement of the outflow valve from closed to open allows cabin air to escape to atmosphere.

The head incorporates an inner chamber, called the *pressure relief control chamber*, which has two pressure relief diaphragms, the calibration spring, the calibration screw, and the spring-loaded metering valve. The action of these components within the chamber controls the movement of the outflow valve during normal operation.

The two diaphragms form three pneumatic compartments within the control chamber. The inner compartment is open to cabin pressure through a passage in the outflow valve pilot. The middle compartment is open to the reference chamber and is vented to the outer compartment through a bleed hole in the metering valve. The flow of reference chamber air from the middle compartment to the outer compartment is controlled by the position of the metering valve, which is spring-loaded to a normally

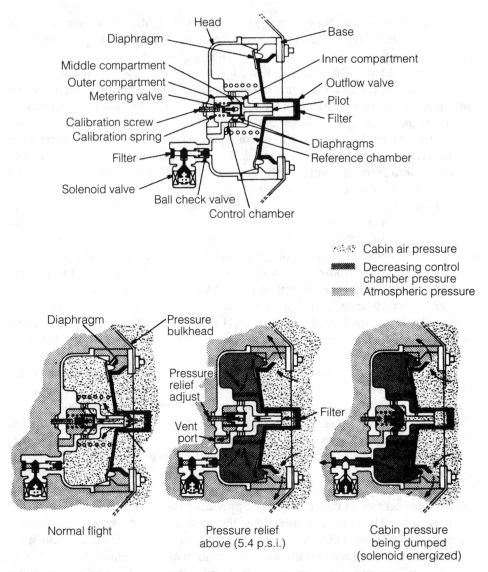

Fig. 8-12. *Cabin air pressure safety valve.*

closed position. The outer compartment, in which the calibration spring and screw are located, is opened to atmosphere through a passage in the head. Atmospheric pressure, reflected against the diaphragms, aids the calibration spring in keeping the metering valve closed. Cabin pressure, acting on the diaphragms through the inner compartment, tries to open the metering valve by moving the metering valve back against the calibration screw. Under normal conditions, the combined forces of atmospheric pressure and the calibration spring hold the metering valve away from the calibration screw, keeping it closed.

Pressure relief occurs when the cabin pressure exceeds atmospheric pressure by a predetermined value. At this point, cabin pressure overcomes the combined forces of atmospheric pressure and spring tension in the control chamber, moving the metering valve back against the calibration screw, opening the metering valve. With the valve open, reference chamber air can escape through the outer compartment to the atmosphere. As the reference chamber air pressure is reduced, the force of cabin pressure against the outflow valve overcomes spring tension and opens the valve, allowing cabin air to flow to atmosphere. The rate-of-flow of cabin air to atmosphere is determined by the amount the cabin-to-atmosphere pressure differential exceeds the calibration point. As cabin pressure is reduced, the forces opening the valve will be proportionately reduced, allowing the valve to return to the normally closed position as the forces become balanced.

AIR DISTRIBUTION

Research has indicated that people feel more comfortable in an environment where their feet are slightly warmer than their head. In addition, it has been shown that rates of air movement fewer than 10 feet per minute produce a feeling of stagnation and are insufficient to remove odors and airborne contaminants. On the other hand, rates in excess of 60 feet per minute result in uncomfortable cooling drafts and increased ozone levels at altitude that further result in reduced relative humidity. While these are hardly earthshaking news items, they do illustrate that something as apparently trivial as air distribution can be more complicated than would appear at face value.

The following discussion is representative of air distribution in a corporate-class turboprop aircraft. Cabin-air distribution systems common to that type of aircraft will generally have the following types of components:

- Air ducts
- Filters
- Heat exchangers
- Silencers
- Check valves
- Humidifier
- Mass flow control sensors
- Mass flow meters

The system in FIG. 8-13 is representative of a cabin-class turboprop air distribution system. Air enters the cabin supercharger via the left-engine oil cooler air scoop that is mounted behind a screen that reduces the potential for foreign object damage. If the screen ices over while flying, a spring-loaded door beside the screen will automatically open and allow the airflow to bypass the screen. Pressurized air that leaves the cabin

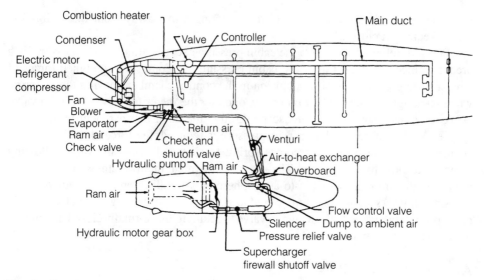

Fig. 8-13. *Typical air distribution system.*

supercharger passes through a firewall shutoff valve, pressure relief valve, and finally a silencer that will help reduce the loud whining and pulsating sounds of the super-charger. Finally, the air goes through the flow-control valve that regulates how much air, in pounds per minute, flows into the cabin.

Air ducts. Circular and rectangular cross section cabin air ducts are preferred among numerous sizes and shapes. Distribution outlets fall into three basic categories: cabin zones, individual passenger outlets, and window demisters. A filter or filters are also generally included in the duct-work to remove dust particles, oil mist, and other impurities.

Air supply ducts are made out of stainless steel, aluminum alloys, or plastic. Those ducts that channel air with temperatures in excess of 200°C are typically made from stainless steel. Rigid and flexible plastic tubing are used as outlet distribution ducts.

One of the primary considerations in the selection of the proper ducting material is heat expansion. Ducting expands when hot air is flowing and contracts when cool air is flowing. This expansion and contraction phenomenon must not interfere with the duct's pressure-tight integrity. To reduce that chance, expansion bellows and supports are occasionally incorporated along the ducting to reduce the effect of expansion and contraction (FIG. 8-14).

AIR-CONDITIONING SYSTEMS

The primary purpose of the air-conditioning system is to maintain a comfortable cabin temperature throughout all conditions of flight. It must be capable of adjusting cabin temperature to be able to sustain a range from approximately 70°–80°F under normally anticipated outside air temperatures. To do so, the system must have the ability to heat

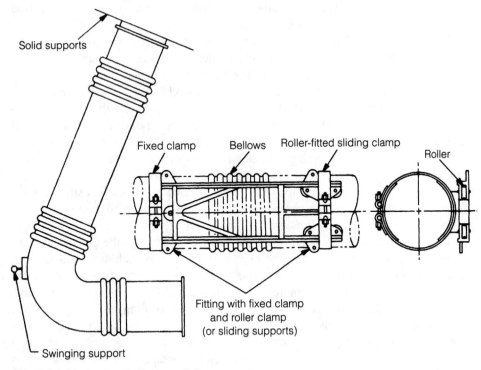

Fig. 8-14. *Expansion bellows and duct supports.*

and cool the flow of air. Finally, the system must also reasonably control cabin humidity to assure passenger comfort and prevent window fogging. Research has shown that long exposure to low levels of humidity results in an acceleration of the aging effect on skin, dehydration, and possible kidney and liver diseases. It is worth noting that diuretics such as coffee exacerbate these problems.

Conditioned air is distributed throughout the flight deck and cabin via ductwork that attempts to reduce, if not eliminate, temperature stratification (hot and cold layers) in the cabin. The duct outlets should be positioned to maintain the temperature of the cabin wallpanels and floors.

Air-cycle cooling systems

An *air-cycle machine* (ACM) consists of two main units, the expansion turbine, and the air-to-air heat exchanger. The ACM essentially does the opposite of a turbocompressor; the turbocompressor causes the air to compress and heat; the ACM causes air to expand and cool. A portion of the hot, compressed air that was destined for the cabin is rerouted to the ACM. It first passes through the air-to-air *heat exchanger*, which is simply a radiator, to extract as much of the heat of compression as possible by using ambient air as a cooling medium. The ambient air enters the aircraft as ram air via a scoop, where it is channeled around the coils that carry the hot, compressed cabin air.

The system works much more efficiently at altitude, where the temperature is cold, and less efficiently on the ground or at low altitudes. Certain heat exchangers incorporate an impeller to force the ambient air through the heat exchanger at a higher rate of speed to compensate for the lack of ram air during ground and low-airspeed operations.

Compressed air that leaves the heat exchanger is channeled to the expansion turbine, causing the turbine to turn. Two things happen at this point. The turbine is connected to the same shaft as the impeller and causes the impeller to turn at the same rate. One use for the impeller is forcing cold ambient air over the heat exchanger coils at a faster rate. Certain manufacturers use the impeller to further compress the turbocompressor output air to hit the turbine at a faster speed. Secondly, heat is extracted from compressed air when the compressed air turns the turbine, which causes the compressed air's temperature and pressure to drop, sufficiently producing cold cabin air.

The system is primarily controlled by a valve that regulates the amount of compressed air that is allowed through the expansion turbine. For additional cooling air, the valve is opened wider, causing more air to flow from the turbocompressor to the ACM turbine. The only power required to operate the ACM is the compressed air turning the turbine; therefore, the greater the demand for cooling air, the greater the amount of compressed air required. As cooling increases, so too does the demand on the turbocompressor, which must supply pressurization air and cooling air. Few systems are able to provide maximum pressurization and cooling simultaneously, and as a result, a compromise must often be made by reducing one or the other.

It is for that reason that many aircraft require the ACM to be deactivated immediately prior to takeoff and not be turned on until after the aircraft is well established in the climb. The purpose for this lies in the relationship of bleed air to engine power output. Recall that the bleed-air source used to operate the cabin pressurization and cooling systems is also the same bleed-air source that supplies compressed air to the engine's burners and ultimately becomes thrust that propels the aircraft. More bleed air diverted to other purposes translates into less thrust produced by the engine. Shutting off the air-conditioning system prior to takeoff reduces the bleed air load imposed upon the engine. In the event of engine failure, maximum performance out of the other engine(s) is more desirable than cooling the cabin air. Figure 8-15 depicts a typical combined air-conditioning and pressurization system.

Vapor-cycle (Freon) systems

The vapor-cycle system is used on a number of aircraft because it has a greater cooling capacity than most air-cycle systems and can be used even when the engines are not operating. This system is typically powered by the aircraft electrical system that might be serviced by a GPU or APU. Vapor-cycle is very similar to a home air-conditioner system.

The vapor-cycle system works because a liquid can be vaporized at any temperature by changing the pressure acting upon it. For instance, at a standard sea level pres-

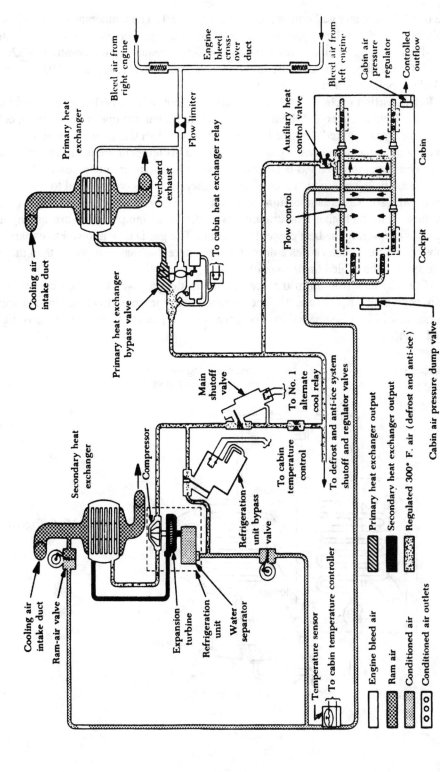

Engine bleed air □

Ram air ▨

Conditioned air ░

Conditioned air outlets ⊙⊙⊙

Primary heat exchanger output ▨

Secondary heat exchanger output ■

Regulated 300° F. air (defrost and anti-ice) ░

Fig. 8-15. *Cabin air conditioning and pressurization system flow schematic.*

sure of 14.7 psi, water will boil at approximately 212°F. The temperature at which any liquid boils is dependent upon the surrounding pressure; therefore, the same water placed in a closed container with a pressure of 90 psi requires a temperature of 320°F to boil. Conversely, if the pressure in the closed container is reduced to 0.95 psi, then the boiling point of water becomes 100°F.

Refrigeration cycle. One of the fundamental laws of thermodynamics states that heat flows from hot to cold. If heat is to be made to flow in the opposite direction, cold to hot, energy is required. The method used to accomplish this in an air-conditioner is based upon the fact that when a gas is compressed, its temperature is raised, and, similarly, when a compressed gas is allowed to expand, its temperature is lowered.

To achieve the required "reverse" flow of heat, a gas is compressed to a pressure high enough so that its temperature is raised above that of the outside air. Heat will now flow from the higher temperature gas to the lower temperature surrounding air (*heat sink*), lowering the heat content of the gas. The gas is now allowed to expand to a lower pressure, which causes a drop in temperature that makes it cooler than the air in the space to be cooled (*heat source*).

Heat will now flow from the heat source to the gas, which is then compressed again, beginning a new cycle. The mechanical energy required to cause this apparent reverse flow of heat is supplied by a compressor. A typical refrigeration cycle is illustrated in FIG. 8-16.

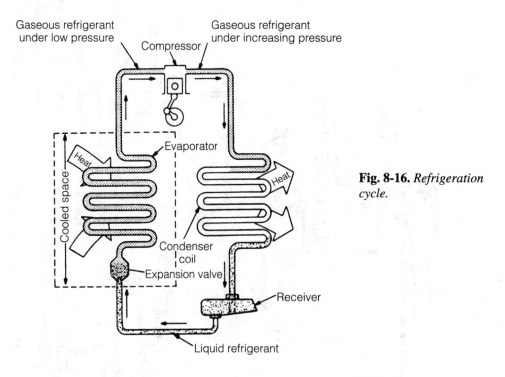

Fig. 8-16. *Refrigeration cycle.*

This refrigeration cycle is based upon the principle that the boiling point of a liquid is raised when the pressure of the vapor around the liquid is raised. The cycle operates as follows: A liquid refrigerant confined in the receiver at a high pressure is allowed to flow through the expansion valve into the evaporator. The pressure in the evaporator is low enough so that the boiling point of the liquid refrigerant is below the temperature of the air to be cooled, or heat source. Heat flows from the space to be cooled to the liquid refrigerant, causing it to boil (to be converted from liquid to a vapor). Cold vapor from the evaporator enters the compressor, where its pressure is raised, thereby raising the boiling point. The refrigerant at the high pressure and high temperature flows into the condenser. Here heat flows from the refrigerant to the outside air, condensing the vapor into a liquid. The cycle is repeated to maintain the cooled space at the selected temperature.

It is easy to understand that liquids that will boil at low temperatures would be preferred for the operation of a vapor-cycle system. This is because comparatively large quantities of heat are absorbed when liquids are changed to a vapor. For this reason, liquid Freon is used in most vapor-cycle refrigeration units whether used in aircraft or home air conditioners.

The approximate sea level boiling point of Freon is 39°F and can be raised to as high as 150°F under a pressure of 96 psi. As with other fluids, Freon has the characteristic of absorbing heat when it changes from a liquid to a vapor. Conversely, the fluid releases heat when it changes from a vapor to a liquid. In the Freon cooling system, the change from liquid to vapor (evaporation or boiling) takes place at a location where heat can be absorbed from the cabin air, and the change from vapor to liquid (condensation) takes place at a point where the released heat can be ejected to outside the aircraft. The pressure of the vapor is raised prior to the condensation process so that the condensation temperature is relatively high; therefore, the Freon, condensing at approximately 150°F, will lose heat to the outside air that might be as hot as 100°F.

Refrigeration effect refers to the amount of heat each pound of refrigerant liquid absorbs while flowing through the evaporator. A given pound of refrigerant will only absorb the amount of heat needed to cause it to vaporize, if no *superheating* (raising the temperature of a gas above that of the boiling point of its liquid state) takes place. If the liquid approaching the expansion valve were at exactly the temperature at which it was vaporizing in the evaporator, the quantity of heat that the refrigerant could absorb would be equal to its *latent heat*. That is the amount of heat required to change the state of a liquid, at the boiling point, to a gas at the same temperature.

When liquid refrigerant is admitted to the evaporator, it is completely vaporized before reaching the outlet. Because the liquid is vaporized at a low temperature, the vapor is still cold after the liquid has completely evaporated. Cold vapor that flows through the remainder of the evaporator continues to absorb heat and become superheated.

The vapor absorbs *sensible heat* (heat that causes a temperature change when added to, or removed from, matter) in the evaporator as it becomes superheated. This, in effect, increases the refrigerating effect of each pound of refrigerant. This means that each pound of refrigerant absorbs not only the heat required to vaporize the refrigerant, but also an additional amount of sensible heat that superheats the refrigerant.

Major components of a typical Freon system include:

- Evaporator
- Compressor
- Condenser
- Expansion valve

The compressor is powered by either an electric motor or by an air-turbine drive mechanism. The compressor increases the pressure of the Freon in vapor form. The result is that the Freon's high pressure raises the condensation temperature and results in the force necessary to circulate the Freon through the system.

The condenser receives the Freon in a gaseous state where it passes through the heat exchanger that uses ambient air to remove heat. When heat is removed from the high-pressure gas, Freon is converted to a liquid. This condensation releases the Freon's heat that was acquired from the cabin air.

The receiver takes this liquid Freon and serves as a reservoir; the fluid level varies with system demand. During heavy demand, the reservoir might be near empty and during periods of reduced cooling demand the reservoir might be full. The main purpose for the receiver is to ensure that the thermostatic expansion valve is not starved for refrigerant under heavy cooling load conditions.

The subcooler is used on select vapor-cycle systems to reduce the liquid temperature after it leaves the receiver. Cooling the refrigerant at that point reduces the possibility of *premature vaporization (flash-off)*. Because the greatest amount of cooling occurs as the refrigerant changes from a liquid to a gaseous state, and the evaporator is designed to accomplish that task in the most effective manner possible, flash-off reduces the overall cooling efficiency of the system.

The filter/drier is simply a sheet-metal housing with inlet and outlet connections. The housing contains an *alumina desiccant*, a filter screen, and a filter pad. The desiccant absorbs moisture; the screen and filter pad trap any contaminants. This step is very important because the expansion valve has very close tolerances. Absolutely dry air must enter the valve to preclude the possibility of moisture freezing in the valve and blocking the passage. Similarly, because the passage is so narrow, even small, particulate contaminants might block them.

A sight glass is included in the system to help the maintenance technician visually determine if sufficient refrigerant is present. It simply permits a view of the fluid passing through the line. The technician activates the system on the ground, then looks through the sight glass. In normal operation, a steady flow of refrigerant moves past the sight glass. If the unit is low on refrigerant, bubbles will be present in the fluid passing the sight glass.

The Freon flows from the condenser to the expansion valve as a high-pressure liquid. The expansion valve lowers Freon pressure and, thus, Freon temperature. As a result, the cooler liquid Freon is able to extract heat from the cabin air as it passes through the evaporator. The expansion valve meters the flow of refrigerant into the evaporator.

Metering, or regulation, of the amount of Freon that is sprayed into the evaporator is necessary because the heat load does not remain constant. If, for instance, the amount of heat to be extracted never varied, it would be a simple matter of determining precisely how much refrigerant was required to carry away that specific amount of heat; however, in reality, the heat load varies and the amount of refrigerant sprayed into the evaporator must also vary. The regulating system consists of a thermostatic expansion valve that senses evaporator conditions and appropriately meters the refrigerant.

The final unit of the system is the evaporator. All other system components exist to support the operation of this one component. The evaporator is simply a heat exchanger with passages through which warm cabin air flows around passages that contain Freon refrigerant. Freon changes in the evaporator from a liquid to a vapor; essentially, the Freon boils in the evaporator at a temperature lower than the cabin temperature, which causes the passages that contain the Freon to be at a temperature lower than the cabin air flowing around them. The result is that the Freon extracts heat from the cabin air.

HEATING SYSTEMS

In most situations, heating cabin air is automatically accomplished by the normal temperature rise associated with compressing the air in the turbocompressor. In fact, the primary concern is how to cool the air down prior to entering the cabin. Heated air is used for cabin heating, deicing, and anti-icing of aircraft components; however, there are operating conditions where the outside air temperature is cold enough that additional heat is required.

Radiant panels. An effective form of radiant heat can be obtained by embedding electric wires in wall and floor panels. Wires and panels become hot when electric power is applied to the wires. This system operates off the normal aircraft electrical system.

Electric heaters. An electric heater in an air duct is a series of high-resistance wire coils that become hot when electric power is applied. A fan forces air through the duct and around the hot coils. This transfers heat to the air that is then blown into the cabin. The system is very effective, but does require a significant amount of electrical power. The system is ideal when being operated by a ground power unit for a cold morning cabin warm-up.

Compressed air heating. This system reroutes the hot air output of the cabin compressor back into its own intake. This double compression and heating is sufficiently hot enough to eliminate the need for additional heating.

SUPPLEMENTAL OXYGEN SYSTEMS

Corporate turbine aircraft routinely operate at altitudes that require cabin pressurization. Recall that pressurization systems maintain cabin altitudes in the 8000–15,000-

foot range, regardless of the actual altitude of the aircraft. Under normal operating conditions, that is sufficient to provide a comfortable cabin environment, but what happens under abnormal or emergency conditions? If, for instance, an aircraft operating at FL 310 (31,000 feet) suddenly experiences a loss of cabin pressurization, the results could be fatal for the occupants; therefore, supplemental oxygen systems provide a measure of safety against such situations. Before discussing the supplemental oxygen options, let's take a look at sudden loss of pressurization.

Rapid cabin depressurization. Hollywood has fostered an image of rapid cabin depressurization that has come to be known as explosive decompression. Someone on the ground shoots a hole into the side of an aircraft and it results in total loss of cabin pressure with paper, food trays, and baggage flying everywhere. In another movie, a popular spy fires a gun that shoots out a small passenger window resulting in an explosive decompression that sucks the villain out the window. Reality simply does not work that way. A bullet hole in a cabin wall would have no perceived effect on cabin pressure; assuming it did not hit a component of the pressurization unit which would result in a system failure. A bullet hole is far smaller than the opening of the outflow valve. In fact, such a hole would account for less air leakage than what is normally lost around door and window seals. The outflow valve would simply adjust accordingly and the cabin pressure would remain unchanged.

If a window totally blew out of the airplane, it certainly would have a significant effect on cabin pressure. If the aircraft were operating at a very high cabin pressure differential, perhaps a passenger sitting directly next to the window, who did not have a seat belt on, might be drawn outward. More than likely, though, the passenger remains inside unless the window was large.

In the event of an actual explosive decompression, such as a structural failure causing a door to blow open or a portion of the aircraft outer skin to separate, there are several serious consequences in addition to the obvious loss of pressurization. First, the cabin temperature would drop from approximately 70°F to as low as –60°F. This rapid, instantaneous change in temperature would cause the cabin air to be chilled far below its dew point; a cloud would form inside the cabin making visibility virtually zero. The crew, if they had the presence of mind to be trying to fly the aircraft, might not be able to see the instruments, let alone outside the window.

Second, with temperatures that low, it is only a matter of seconds before hypothermia sets in and everyone begins to freeze to death. Meanwhile, everyone is feeling around in the freezing fog for supplemental oxygen because the air has been pulled out of their lungs. It's not a particularly pleasant thought, but fortunately it is also an extremely rare occurrence. In the end, if the crew can keep warm and get the oxygen masks on in time, the cloud in the cabin will rapidly dissipate and they can concentrate on getting the aircraft down to a lower altitude and a safe landing.

Continuous-flow oxygen

A simple schematic of a basic continuous-flow oxygen system is illustrated in FIG. 8-17. When the line valve is placed in the ON position, oxygen flows from the cylinder to the pressure-reducing valve. The valve reduces the compressed oxygen to the lower pressure required for use by individual oxygen masks. A calibrated orifice in the outlets actually controls the amount of oxygen that flows into individual masks.

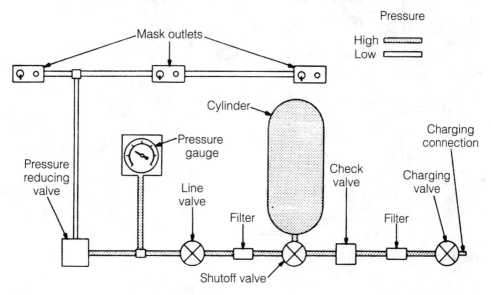

Fig. 8-17. *Continuous-flow oxygen system.*

Passenger oxygen systems come in two forms. In one type, a series of plug-in supply sockets are located in the cabin walls next to the passenger seats. If supplemental oxygen is required, the passenger takes the oxygen mask and plugs it into the closest supply socket. In the second type of system, the masks are installed in overhead containers that open automatically in the event of an excessively low cabin pressure. The masks drop out of the containers and dangle in front of each passenger seat. In each case, the flow of oxygen is automatically turned on by an automatic barometric control valve. An override function for both systems allows the crew to make oxygen available to the passengers regardless of the cabin pressure.

Pressure-demand oxygen

Pressure-demand provides individual pressure-demand regulators for each crewmember (FIG. 8-18). This allows each individual to adjust the airflow to their own requirements. Otherwise, both systems are very similar.

Diluter-demand regulators. The name indicates the manner in which it functions by delivering oxygen to the user's lungs in response to the suction of the user's breath. Whenever the unit is operated below 34,000 feet, it automatically dilutes the oxygen in the regulator with appropriate amounts of atmospheric air.

Figure 8-19 depicts a diluter-demand regulator. It is a diaphragm-operated valve that opens automatically when a slight suction is applied as the user breathes in. When

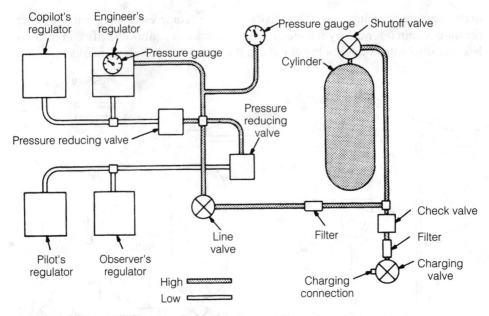

Fig. 8-18. *Typical pressure-demand oxygen system.*

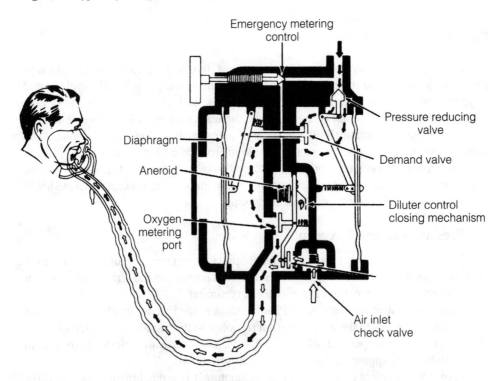

Fig. 8-19. *Schematic of a diluter-demand regulator.*

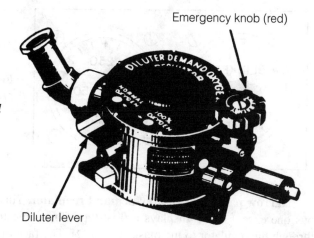

Emergency knob (red)

Fig. 8-20. *Diluter-demand regulator control.*

Diluter lever

the user exhales, the valve automatically closes, preventing unwanted oxygen from flowing. As altitude increases, the air inlet is gradually closed by the bellows, which results in an increasingly high concentration of oxygen until, at approximately 34,000 feet, the air inlet is shut and 100 percent oxygen is supplied.

The diluter control (FIG. 8-20), can be set by turning the lever to give 100 percent oxygen at any altitude; however, the obvious drawback is the oxygen supply will not last as long. Unless there is a reason for operating at 100 percent oxygen, such as smoke in the cabin, the control usually remains in the normal oxygen mode. Finally, a red emergency knob is located on the regulator control that, when activated, provides a steady stream of pure oxygen to the mask. This pure oxygen stream will occur regardless of altitude and will not require the suction of a breath to draw it into the mask.

One problem associated with supplemental oxygen is that it is not always easy to know if the unit is actually providing oxygen. The following procedure illustrates a typical method for checking the operation of a diluter-demand regulator.

First, check the oxygen system pressure gauge, which should indicate between 425 and 450 psi, then check out the system using the following steps:

1. Connect an oxygen mask to each diluter-demand regulator.
2. Turn the auto-mix lever on the diluter-demand regulator to the "100 percent oxygen" position and listen carefully to make certain that no oxygen is escaping.
3. Breathe oxygen normally from the mask. The oxygen flowmeter should blink once for each breath (FIG. 8-21).
4. With the auto-mix lever in the "100 percent oxygen" position, place the open end of the mask-to-regulator hose against the mouth and blow gently into the hose. Do not blow hard because the relief valve in the regulator will vent. Resistance should be positive and continuous; if not, the diaphragm or some part of the air-metering system might be leaking.
5. Return the auto-mix lever to the "normal oxygen" position.

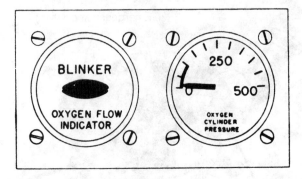

Fig. 8-21. *Flow indicator and pressure gauge.*

Narrow panel type diluter-demand regulator. This unit is similar to the previous one except that it displays a float-type flow indicator that signals oxygen flow through the regulator to the mask (FIG. 8-22). The face of the regulator also has three manual control levers. The supply lever opens and closes the oxygen supply valve to the unit. The emergency lever will initiate pure oxygen under pressure directly to the mask. The third lever, an oxygen selector, allows the crewmember to choose either an air/oxygen mixture or pure oxygen only.

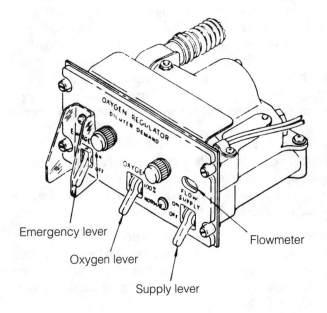

Fig. 8-22. *Typical narrow panel oxygen regulator.*

Emergency lever

Oxygen lever

Supply lever

Flowmeter

Dangers associated with use of oxygen

It is critical to keep in mind that oxygen is one of the three elements necessary for fire: oxygen, fuel, and heat. Most things will burn under the proper conditions but oils, grease, and nonmetallic materials are particularly susceptible to fire when exposed to pure oxygen under pressure. As a result, all oxygen equipment must be kept absolutely free of any type of oil or petroleum-based product.

Fire potential is directly related to the proportional mix of all three elements. Oxygen alone is not flammable, but oxygen supports and intensifies a fire with any combustible material. Never permit smoking or any open flame in an environment in which oxygen is being used.

MANUFACTURER DOCUMENTATION

(The following information is extracted from Beech Aircraft Corporation pilot operating handbooks. The information is to be used for educational purposes only. This information is not to be used for the operation or maintenance of any aircraft.)

Beechjet 400A

Environmental system. The pressurization and air-conditioning systems utilize bleed air to pressurize and air-condition the cabin, and defog the cockpit windows. During normal operation, most functions are automatic. The only manual adjustments required are for individual comfort, such as cabin rate of climb and temperature. Ram air for cabin ventilation is available when the pressurization system is not in use.

Air-conditioning. Engine bleed air is used to heat, cool, and pressurize the cabin. Hot compressed air is tapped off the gas generator case of each engine (FIG. 8-23). Bleed air coming from the engine will not normally exceed 600°F (315°C) and 150 psi at takeoff-rated power. The bleed air is routed into the refrigeration unit located in the aft fuselage. Prior to reaching the refrigeration unit, the bleed air passes through a pressure regulator/shutoff valve, a venturi, and a check valve. The pressure regulator/shutoff valve reduces bleed pressure to 30 psi and serves as a system shutoff. The inlet venturi is installed to restrict the volume of air that can be extracted from the engine. At low power settings (below 60 percent N_1 during ground operations) the volume of air flowing through the venturi is inadequate; therefore, a flow-increasing valve is used to increase the airflow. The flow-increasing valve is automatically controlled by switches located on the thrust lever quadrant and the squat switch.

Refrigeration unit. Hot bleed-air passes through the primary heat exchanger, which decreases the bleed-air temperature. The heat exchangers are cooled by ram air supplied through the flush scoop on both sides of the dorsal fin. The ram air is augmented by a bleed-air ejector. The ejector shutoff valve is an electrical solenoid valve that is normally open (powered closed). It is open during all ground operations. Air flowing through the primary heat exchanger is partially cooled and then directed to the compressor where its pressure and temperature are increased. A thermal switch installed in the compressor outlet duct is used to sense abnormally high temperatures that might occur during low-speed flight with low ram pressure. If the thermal switch senses a temperature of 350°F (176°C) or above, the ejector shutoff valve opens to al-

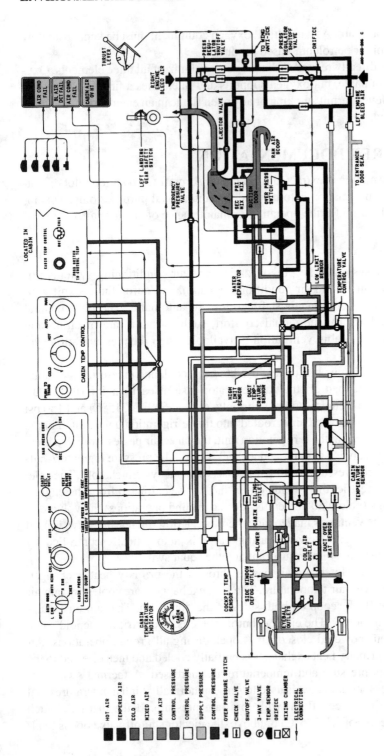

Fig. 8-23. *Environmental system.* Beech Aircraft Corporation

low greater cooling airflow. The valve will close five minutes later, provided the temperature has returned to normal.

The air then enters the secondary heat exchanger where it is partially cooled again. After the second cooling, the air is expanded through the cooling turbine where its pressure and temperature are reduced. Energy from the expanding air is converted to shaft power to drive the compressor. At the cooling turbine outlet, hot engine bleed-air is mixed with the cold turbine air to maintain a constant 39°F (4°C) at the water separator discharge. The air temperature control prevents ice buildup at the cooling turbine outlet and prevents icing at the water separator. Some of the cold air is routed to the eyeball, cockpit, and cabin ceiling outlets. The rest is routed to a mixing chamber where it is combined with hot bleed-air to achieve the temperature selected by the temperature control.

An overpressure switch located in the inlet of the compressor and an overtemperature switch in the outlet duct of the compressor are provided to protect the cooling package from the effects of a system failure. Temperature above 350°F (177°C) will open the ejector shutoff valve, allowing the ejector to introduce additional cooling air for the heat exchanger. Pressure above 53 psi or temperature above 400°F (204°C) will close both pressure regulator/shutoff valves, the emergency pressure valve will open and the AIR COND FAIL annunciator will illuminate.

Cabin-air distribution. Cabin-air distribution lines are composed of cold air lines and conditioned air lines. Cold air is delivered to the cockpit and cabin overhead eyeball outlets and cabin ceiling outlet. For maximum cooling in the cabin, the cabin ceiling outlet can be opened by a CABIN CEIL OUTLET switch on the instrument panel. NOTE: The cabin ceiling outlet is usually used on the ground on a hot day. Using the outlet in flight results in excessive cooling.

Conditioned air is fed through check valves at the aft pressure bulkhead to the cabin floor outlets on both sides and to the cockpit. These bulkhead check valves are used to prevent cabin depressurization in the event of a duct rupture upstream. The cockpit system includes floor outlets and a windshield and side window defogger. DEFOG SELECT levers are provided for the pilot and copilot to select defog, floor outlets, or a combination of both. Thermal switches, installed in the cockpit side-window defog ducts, illuminate the DEFOG AIR OV HT annunciator if duct temperatures above 200°F (93°C) are encountered. This is to prevent the side windows from being damaged by heat. Manually controlled air outlets are provided in the upper area of the cockpit to supply cold air to the pilots.

Temperature control. The temperature control system is a pneumatically operated system that manually or automatically regulates the temperature of air delivered to the cabin from the refrigeration unit. The Beechjet 400A is equipped with a two-zone temperature control system that permits independent control of cabin and cockpit temperatures.

Automatic operation—cockpit. When the mode-select switch on the instrument panel is set to AUTO, the cockpit temperature is automatically controlled between 60°F and 90°F (16°C to 32°C) by rotating the cockpit temperature control switch located on the instrument panel.

Manual operation—cockpit. When the mode select switch on the instrument panel is set to MAN, the cockpit temperature is manually controlled by rotating the cockpit-temperature control switch. In this case, the cockpit-temperature control switch directly controls the temperature control valve.

Automatic operation—cabin. When the mode-select switch on the copilot's side panel is set to AUTO, the cabin temperature is automatically maintained between 60°F to 90°F (16°C to 32°C) after rotating the cabin-temperature control switch adjacent to the mode-select switch.

Manual operation—cabin. When the mode select switch on the copilot's side panel is set to MAN, the cabin temperature is manually controlled by rotating the cabin temperature control switch. In this case, the cabin temperature control switch directly controls the temperature control valve.

Temperature control transfer switches. Cabin temperature control can be transferred to the cabin by pushing the illuminated push-button switch adjacent to the cabin temperature control switch. The transfer switch controls a three-way valve that transfers temperature control between cockpit and cabin.

Ventilation blower. The ventilation blower introduces outside fresh air or additional conditioned air into the cockpit depending upon whether environmental control system OFF or ON operation is selected. An illumination push-button switch, located adjacent to the cockpit mode-select switch, controls the ventilation blower operation.

Pressurization. The cabin is pressurized by the flow of air from the cockpit and cabin air outlets. Cabin pressurization control is accomplished by modulating discharge air from the cabin. The system's major components are the outflow safety valves, cabin air regulator, air filters, solenoid valves, and a quick dump valve (FIG. 8-24). This system uses a variable isobaric controller to drive two outflow safety valves through a pneumatic relay. Both outflow safety valves modulate the flow of air discharged from the cabin during normal operation. Either or both valves open automatically, as required, to provide positive or negative pressure relief protection. Both valves are connected to cabin altitude pressure regulators that automatically override a valve failure and prevent the cabin altitude from exceeding 12,500 ±1500 feet.
Cabin pressurization control system

Cabin pressure switch. The CABIN PRESS switch determines the source of air for cabin pressurization. It is a six position rotary type switch with positions placarded OFF, L ENG, BOTH NORM, BOTH HIGH, R ENG, and EMER:

OFF   No bleed air is supplied. Ram air is supplied to the cabin in flight. No temperature control is provided.

L ENG   Bleed air is supplied by the left engine. The right bleed air is shut off. Normal temperature control is provided.

BOTH NORM   Bleed air is supplied by both engines. Normal temperature control is provided.

BOTH HIGH   Bleed air is supplied by both engines at a greater rate than in the BOTH NORM position by opening the flow increasing valve. Normal temperature control is provided.

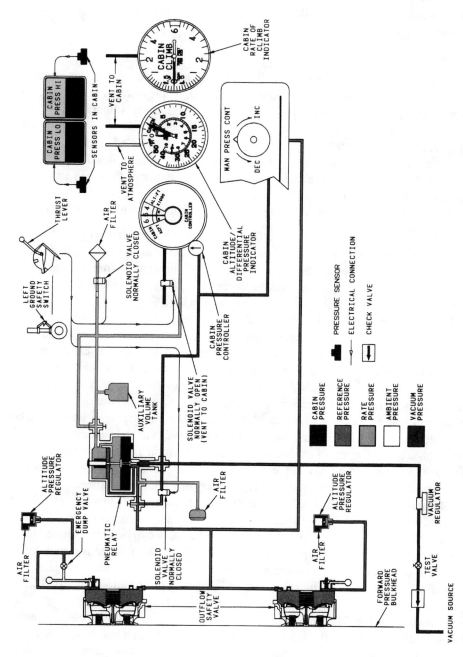

Fig. 8-24. *Cabin pressurization control system.* Beech Aircraft Corporation

R ENG   Bleed air is supplied by the right engine. The left bleed air is shut off. Normal temperature control is provided.

EMER   Bleed air is supplied to the cabin for emergency pressurization. No temperature control is provided.

Caution annunciators.

AIR COND FAIL   Illuminates automatically for bleed air overtemperatures or overpressure conditions and results in actuation of emergency pressurization. Placing the CABIN PRESS switch to the EMER position will also illuminate this light.

DEFOG AIR OV HT   Illuminates if the air supply to the cockpit side-window defog system exceeds 200°F (93°C).

BL AIR DCT FAIL   Illuminates if any four of the sensor switches installed in the aft fuselage detects excessively high temperature due to a rupture of the bleed air duct or leakage of wing anti-ice lines in the cabin.

CABIN AIR OV HT   Illuminates if the air supply to the cabin exceeds 250°F (121°C).

Outflow safety valves. The two outflow safety valves are installed on the forward pressure bulkhead. The outflow safety valves regulate cabin air flow through the pressure bulkhead. The valves are controlled by an air-pressure signal from the cabin air-pressure controller through the pneumatic relay. The valve contains a positive differential pressure relief (9.1 psig) and a negative relief.

Cabin air pressure controller. The cabin air pressure controller is mounted on the right-hand instrument panel and controls the cabin altitude and cabin rate of climb. This unit is connected to the vacuum source, the pneumatic relay, and the cabin ambient pressure. The controller face has a cabin-altitude selector knob, a cabin-rate-control selector knob, and a dial showing cabin-altitude setting and the altitude at which the airplane will reach maximum differential pressure for the selected cabin-altitude pressure. Desired cabin altitude and maximum airplane altitude at which the cabin altitude can be maintained is selected by rotating the cabin-altitude select knob. The knob, to the left of the dial, controls the rate of change of cabin altitude within a range of approximately 50 to 2000 feet per minute. When the arrow on the knob is halfway between MIN and MAX (straight up), a rate of approximately 500 feet per minute is obtained.

Manual control valve. The MAN PRESS CONT valve mounted on the instrument panel is used to manually depressurize the cabin. This valve is connected to a vacuum source and the outflow safety valve control line. The cabin can be depressurized in an emergency by turning the knob counterclockwise. This action proportionately vents the control chambers of both outflow safety valves to vacuum, resulting in opening of the outflow safety valves. The cabin can be depressurized to an altitude of 12,500 ±1500 feet.

Pneumatic relay. The pneumatic relay is mounted in the cabin-pressure control module that is installed on the cabin side of the forward pressure bulkhead. The pneumatic relay is a high-gain device that operates by a signal from the cabin-air pressure controller and from cabin pressure to regulate the outflow safety valve control pressure to the desired level.

Altitude pressure regulator. The altitude pressure regulator directly senses cabin pressure. When the cabin altitude pressure reaches 12,500 ±1500 feet, the poppet valve in the regulator opens to direct cabin pressure to both outflow safety valve control chambers. The outflow safety valves will close to a modulating position, thus preventing excessive exhaust of cabin air.

Quick dump valve. The quick dump valve is mounted on the instrument panel and provides rapid cabin depressurization in case of emergency. This valve is opened by turning the knob counterclockwise. This results in complete and rapid depressurization of the cabin.

Indicators and annunciators. The cabin altitude/differential pressure indicator and cabin rate of climb indicator are installed on the instrument panel. The CABIN PRESS LO annunciator will illuminate to warn the pilot when the cabin altitude exceeds 10,000 feet. The CABIN PRESS HI annunciator will illuminate to warn the pilot any time cabin differential pressure exceeds 9.1 psi.

Pressurization control during takeoff. When the thrust levers are advanced to the T.O. position on the ground, the cabin is pressurized to a positive 80 feet differential. This prevents a pressure bump at liftoff.

Pressurization control during climb. The cabin altitude and rate controls enable the pilot to select the desired cabin altitude and the desired cabin rate of climb. The selected values can be maintained until pressure differential between the cabin and the atmosphere reaches 8.9–9.1 psig.

Pressurization control during descent. When preparing to descend, the crew should select the landing field elevation plus 500 feet on the controller. When the cabin reaches the selected landing altitude, the system maintains the cabin at 500 feet above field elevation until the airplane descends below this level. The outflow safety valves are opened while descending through the 500 feet level assuring an unpressurized cabin during landing.

Oxygen system. The system consists of an oxygen cylinder-regulator assembly mounted in the right electronics bay compartment. The 77-cubic-foot cylinder is of composite construction and stores oxygen at a maximum pressure of 2000 psig. Normal oxygen pressure is 1850 psig. The oxygen duration chart (found in the emergency procedures section of the FAA-approved airplane flight manual) is based on a flow rate of 3.7 LPM per passenger mask and on an altitude schedule for the diluter-demand crew masks. The supply pressure regulator, installed on the cylinder, is capable of delivering up to 300 LPM-NTPD regulated oxygen at 70 ±10 psi with cylinder pressure from 200–2000 psig. The regulator incorporates an ON—OFF valve that vents low-side pressure when in the OFF position. It is actuated by a push-pull SYS READY control located adjacent to the lower right corner of the copilot's instrument panel. A high-pressure rupture fitting relieves high side cylinder pressure and dumps oxygen overboard (if it exceeds) 2775 psig. When this occurs, the vent line receives pressure in excess of 60 psi and ruptures a green indicator disk located on the right side nose skin, indicating that an overpressure condition has occurred and oxygen was routed overboard.

System operation. The oxygen system should be armed prior to takeoff by pulling out the SYS READY control. This opens the oxygen regulator at the cylinder, charges the lines to the crew masks and provides oxygen immediately to the crew upon donning the masks (FIG. 8-25). Oxygen supply to the passenger masks is controlled by the cabin oxygen shutoff valve. Normally closed, it may be opened either electrically or manually. Electrically, the shutoff valve is opened by the cabin barometric switch when the cabin altitude exceeds 12,500 ±500 feet. Manually, it is opened from the cockpit at any cabin altitude by pulling out on the PASS OXYGEN control knob located adjacent to the lower right corner of the copilot's instrument panel.

Filler valve. Access to the filler valve is gained by opening the right electronics-bay compartment door. When the oxygen supply line is connected to the filler valve and supplying oxygen, the poppet is unseated to allow oxygen to flow to the storage cylinder. Loss of oxygen is prevented when the supply line is removed due to the re-seating of the poppet.

Pressure gauge. The oxygen pressure gauge is located adjacent to the lower right corner of the copilot's instrument panel. The gauge is illuminated and is a direct pressure reading instrument. The range markings are yellow arc (0 to 200 psi), green arc (1600 to 1850 psi), and redline (2000 psi).

Outlet receptacle. Two identical outlet receptacles are provided for the crew oxygen system. When the oxygen mask supply tube plug is inserted into the outlet, the poppet unseats allowing oxygen flow to the mask. When the plug is removed, the poppet reseats and shuts off the oxygen flow.

Overboard discharge indicator. The overboard discharge indicator is located on the right side of the nose and is mounted flush with the airplane skin. A low-pressure 60 ±20 psig disc is installed at the overboard discharge port to prevent dust and contamination from entering the oxygen system. The indicator line is connected to the high-pressure rupture fitting of the pressure regulator. When the green indicator disc is ruptured or missing from the indicator, an oxygen cylinder overpressure condition has occurred and oxygen has vented overboard. CAUTION: (If) an overpressure condition (occurs) (2700 to 3000 psig), the cylinder must be replaced.

Oxygen system annunciation. A pressure switch located downstream of the furthest-most aft passenger mask senses when oxygen is being supplied to the passenger masks and illuminates a green PASS OXYGEN ON annunciator located on the shroud indicator panel.

Crew oxygen system. The crew is provided with . . . automatic pressure-breathing diluter-demand, quick-donning oxygen masks with integral microphones. To don the mask, remove the mask from its stowage cup, inflate the mask harness by squeezing the red lever on the left side of the regulator, then don the mask and release the

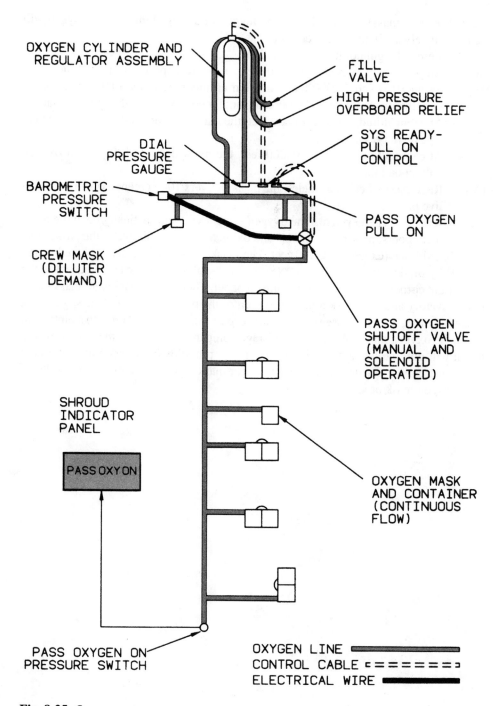

Fig. 8-25. *Oxygen system.* Beech Aircraft Corporation

lever. The crew masks should always be plugged in during flight so that oxygen will be immediately available when required.

The demand regulator has three positions:

NORM   Automatically supplies an air-oxygen mixture appropriate for altitudes between 8000 and 30,000 feet. At cabin altitudes between 30,000 and 35,000 feet, the mask delivers 100 percent (undiluted) oxygen only upon inhalation. At cabin altitudes above 35,000 feet, the mask delivers 100 percent oxygen at a positive pressure.

100%   At cabin altitudes below 35,000 feet, the mask delivers 100 percent oxygen only upon inhalation.

EMER   Regardless of cabin altitudes, the mask delivers 100 percent oxygen at a positive pressure.

Passenger oxygen system. The oxygen system will automatically open the mask compartment doors, present the masks, and oxygen will be available to the passengers (if) cabin altitude (exceeds) 12,500 feet. The masks can be manually deployed at any altitude by pulling out on the PASS OXYGEN push-pull control located adjacent to the lower right corner of the copilot's instrument panel. There are 11 passenger masks; two in the lavatory and one mask adjacent to each of the passenger seats. One spare mask is centrally located. The passenger masks incorporate a lanyard attached to a pintle pin. When the oxygen mask falls out of its storage compartment, pulling the mask down to don it will pull the lanyard and in turn the pintle pin allowing oxygen to flow to the mask. Instructions for the use of the passenger masks are located on the inside lid of each oxygen mask box.

9
Landing gear systems

AIRPLANE LANDING GEAR SYSTEMS CONSIST OF MAIN AND AUXILIARY UNITS
The main landing gear provides the principal support for the aircraft on the ground while the auxiliary (nose) gear provides supplemental support and steering capability. The earliest airplanes had tail skids and eventually tailwheels, but cabin-class aircraft are fitted with a nose gear. Three primary reasons for the change to nose gear were:

- More forceful application of the brakes for higher landing speeds without nosing over.
- Better forward visibility for the crew during landing and taxiing.
- Prevent aircraft ground-looping (a quick 180° pivot that might occur after landing) by moving the aircraft center of gravity (c.g.) ahead of the main wheels. (Forces acting on the c.g. tend to keep the aircraft moving forward on a straight line rather than ground-looping.)

Landing gear have many variables. The number and location of wheels on the main gear might vary. Some aircraft have a single wheel per strut; other aircraft have multiple wheels per strut (FIG. 9-1). More wheels per strut create a wider safety margin. The aircraft's weight is spread over a larger area, and the degradation in control associated with tire failure will be reduced.

Fig. 9-1. *Dual main landing gear wheel arrangement.*

Very heavy aircraft will often have four or more wheels per strut. When more than two wheels are located with a single strut, the attaching mechanism is called a *bogie* (FIG. 9-2). The manufacturer determines the number of wheels included in the bogie based upon the aircraft's maximum certificated gross weight and the surfaces on which it is intended to land. Regardless of the number of wheels, the landing gear is made up of many subassemblies including: air/oil shock struts, main-gear alignment units, support units, retraction and safety devices, auxiliary protective devices, nosewheel steering system, aircraft wheels, tires, tubes, and aircraft brake systems.

SHOCK STRUTS

These units take the brunt of the shock imparted on the landing gear by taxiway and runway bumps and holes, and hard landings. Each strut is a self-contained hydraulic shock absorber unit that shares the weight of the aircraft on the ground with the other shock struts.

The *pneumatic/hydraulic* shock strut is typical on aircraft (FIG. 9-3). The strut uses compressed air combined with hydraulic fluid to absorb and dissipate shock loads. This type of unit is also referred to as an *air/oil* or *oleo* strut.

A shock strut is simply two telescoping cylinders with sealed ends. Known as cylinder and piston, when the two pieces are put together they form an upper and lower chamber for movement of the fluid. The lower chamber is filled with the

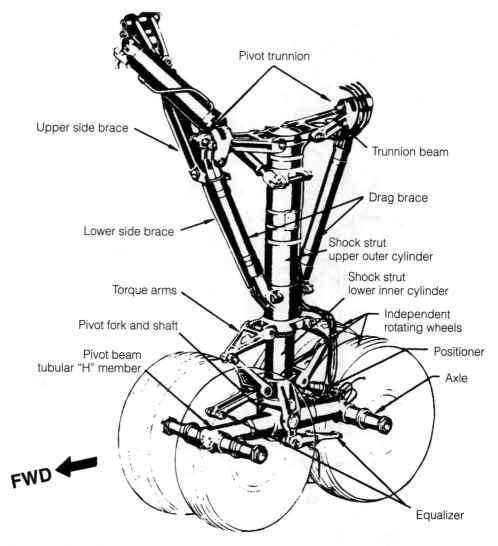

Fig. 9-2. *"Bogie" truck main landing gear assembly.*

fluid; the upper chamber contains the compressed air. The two chambers are separated by an orifice that allows the fluid to flow into the upper chamber when the two cylinders compress together, and return to the lower chamber when they are pulled apart.

Shock absorbers also include a *metering pin* that controls the rate that the fluid flows from the lower to the upper chamber during compression. As a result, during the compression stroke, the fluid flow rate varies according to the variable shape of the metering pin as the fluid passes through the orifice.

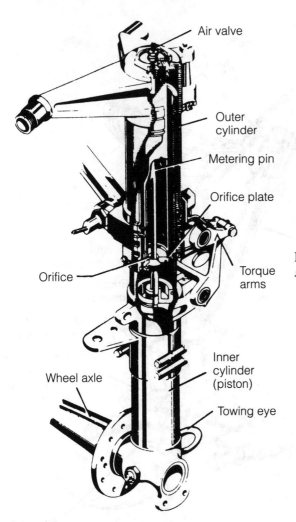

Air valve

Outer
cylinder

Metering pin

Orifice plate

Orifice

Torque
arms

Fig. 9-3. *Landing gear shock
strut of the metering pin type.*

Inner
cylinder
(piston)

Wheel axle

Towing eye

MAIN LANDING GEAR

The main landing gear is made up of a number of components that vary among the different types of systems. Typical components include torque links, trunnion and bracket arrangements, drag strut linkages, electrical and hydraulic gear-retraction devices, and gear indicators.

Alignment. Figure 9-4 illustrates landing gear torque links that keep the landing gear pointed straight ahead. Two torque links are connected in the center with a hinge that permits up/down movement, but no side-to-side movement. One link is affixed to the shock strut cylinder and the other to the piston.

Support. A major consideration is securing the main gear to the aircraft structure. This is typically accomplished through the use of a trunnion and bracket arrangement as depicted in FIG. 9-5. This method allows the strut to pivot left and right when the

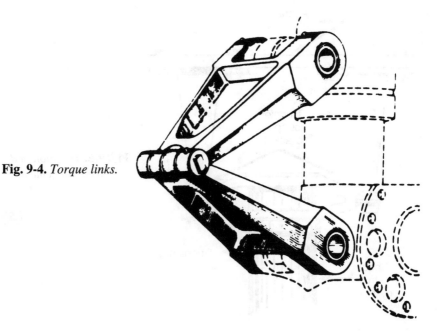

Fig. 9-4. *Torque links.*

gear is being steered, and swing fore and aft when it is being retracted. Different link-ages are employed to control the pivot action during ground movement. One linkage is the *drag strut* (FIG. 9-6). The upper portion of the strut connects to the aircraft structure; the lower portion is connected to the shock strut. The drag strut is also hinged to allow for landing gear retraction.

Electrical retraction system

The system depicted in FIG. 9-7 is typical of an electric-driven landing gear retraction system. Note the following features:

- A motor for converting electrical energy into rotary motion.
- A gear reduction system for decreasing the speed and increasing the force of rotation.
- Other gears for changing rotary motion into push-pull movement.
- Linkage for connecting the push-pull movement to the landing gear shock struts.

The key to the system is an electrically-driven screw jack that raises and lowers the landing gear. When the crew selects the GEAR UP position on the flight deck, an electric motor activates, and through a series of gears, a torque tube, and an actuator screw, the force is transmitted to the drag strut linkages. This causes the gear to retract and lock in the upright position. If the crew moves the switch into the GEAR DOWN po-sition, the electric motor reverses its direction of rotation and the gear moves down and locks into position. The sequence of operation of the respective gear doors is similar to a hydraulic retraction system.

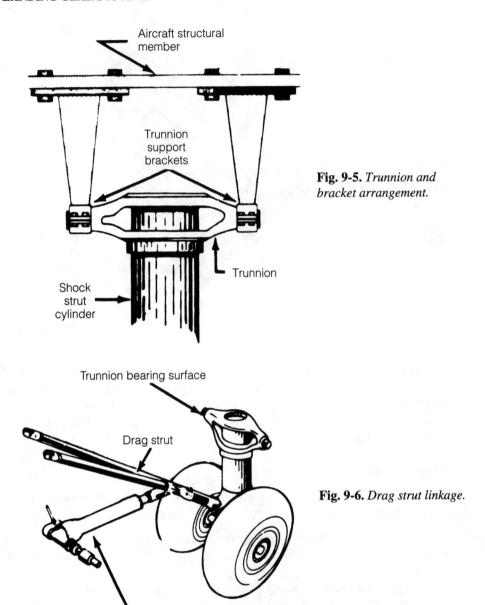

Fig. 9-5. *Trunnion and bracket arrangement.*

Fig. 9-6. *Drag strut linkage.*

Hydraulic retraction system

Devices used in a typical hydraulic gear retraction system include actuating cylinders, selector valves, uplocks, downlocks, sequence valves, tubing, and other standard hydraulic components. The main gear and nose gear and the various gear doors are interconnected to permit proper sequencing of all units and operate all components off a

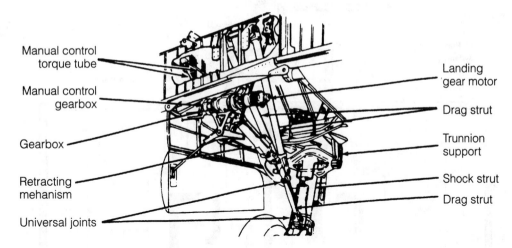

Manual control
torque tube

Manual control
gearbox

Gearbox

Retracting
mehanism

Universal joints

Landing
gear motor

Drag strut

Trunnion
support

Shock strut

Drag strut

Fig. 9-7. *Electrical retraction system.*

single hydraulic system. Figure 9-8 illustrates a typical hydraulic landing gear retraction system schematic.

Because the hydraulic landing gear system is so common in aircraft, it is worth looking more closely at how it works. Consider what happens when the landing gear is retracted. When the crew moves the landing gear selector switch to the GEAR UP position, the selector valve is placed in the "up" position. This causes pressurized hydraulic fluid to enter the gear-up line and flow to eight different units:

1. C valve
2. D valve
3. Left landing gear down lock
4. Right landing gear down lock
5. Nose gear down lock
6. Nose gear cylinder
7. Left main gear actuating cylinder
8. Right main gear actuating cylinder

Note that because sequence valves C and D are closed at the time, fluid cannot flow to the door cylinders yet. The lines are pressurized and waiting for the sequence valves to open; therefore, the doors cannot close. The fluid that has been routed to the three downlock cylinders is not delayed, resulting in the gear becoming unlocked. Simultaneously, pressurized fluid also enters the up side of the gear-actuating cylinders causing the gear to retract. Due to the small size of its actuating cylinder, the nose gear retracts and engages its uplock first. In addition, the nose gear door is operated only by linkage from the nose gear, so this door closes as soon as the nose gear has been retracted.

At the same time, the main landing gear is still en route, which forces hydraulic fluid to leave the downside of each main gear cylinder. The fluid that is exiting flows

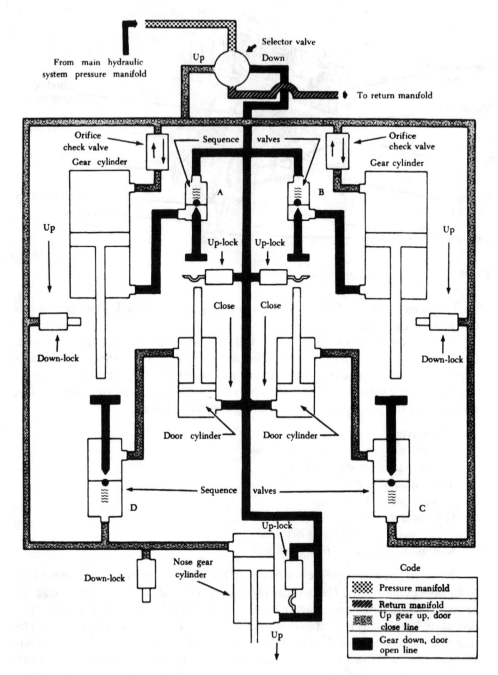

Fig. 9-8. *Hydraulic landing gear retraction system schematic.*

through an orifice check valve that opens the sequence check valve A or B, as appropriate. From there it flows through the landing-gear selector valve and into the system return line. When the main gear is fully retracted a spring-loaded uplock is engaged by the main gear that causes the gear linkage to strike the plungers of sequence valves C and D. That action causes the sequence check valves to open and allow pressurized fluid to flow into the door cylinders that close the gear doors.

Wing landing gear operation

Retracted main landing gear can be stored in the wing or fuselage. By far, the most common location for main gear attachment and storage is the wing. Figure 9-9 illustrates a common wing landing gear operating sequence. Hydraulic pressure causes the gear to raise and lower by exerting pressure on the up or down side of the gear actuator as appropriate. The actuator operates in conjunction with a walking beam to apply force to the wing-gear shock strut. This causes the strut to swing inboard and forward into the wheel well. Both the actuator and the walking beam are connected to lugs on the landing gear trunnion. The outboard ends of the actuator and walking beam pivot on a beam hanger that is attached to the aircraft structure. A wing landing gear locking mechanism located on the outboard side of the wheel well locks the gear in the up position. Locking of the gear in the down position is accomplished by a downlock bungee that positions an upper and lower jury strut so that the upper and lower side struts will not fold.

EMERGENCY EXTENSION AND SAFETY SYSTEMS

Numerous methods are employed by manufacturers to extend landing gear in an emergency. Some systems are as simple as pulling an emergency release handle on the flight deck that disconnects the gear uplocks and allows the gear to literally fall, under their own weight, to the downlock position. Other manufacturers prefer to use compressed air to release the uplock release cylinders. Some aircraft are so designed that gravity alone will not extend the landing gear. Manufacturers of those aircraft must install a more positive method of gear extension using hydraulic fluid or compressed air to provide the required pressure. Probably the most positive method employed utilizes a mechanical handcrank system or a backup hydraulic system with either an auxiliary hand pump or an electrically powered backup hydraulic pump.

Landing gear safety devices

The overriding concern is inadvertent gear retraction while on the ground. Manufacturers have taken great pains to design systems that are essentially foolproof so that landing gear is not retracted while the aircraft is on the ground. Preventive devices most commonly used include mechanical downlocks, safety switches, and ground locks.

Mechanical downlocks are built-in parts of a gear-retraction system and are operated automatically by the gear-retraction system. To prevent accidental operation of downlocks, electrically operated safety switches are installed.

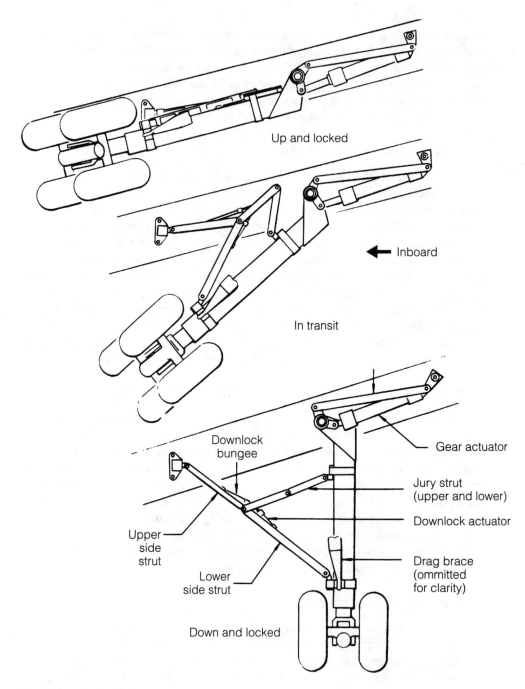

Up and locked

← Inboard

In transit

Downlock
bungee

Gear actuator

Jury strut
(upper and lower)

Downlock actuator

Upper
side
strut

Lower
side strut

Drag brace
(ommitted
for clarity)

Down and locked

Fig. 9-9. *Wing landing gear operating sequence.*

Safety switches. A landing gear safety switch similar to that pictured in FIG. 9-10 is often mounted in a bracket on one of the main gear shock struts. The switch is activated by a linkage to the landing gear torque links. When the gear are on the ground, the torque links are pressed close together, which causes the safety switch to open. During takeoff, as the weight of the aircraft leaves the struts and the links extend, the safety switch closes, which completes a ground; the solenoid energizes and unlocks the selector valve so that the gear handle can be moved into the GEAR UP position.

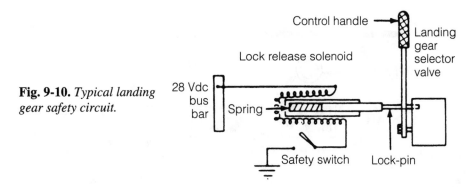

Fig. 9-10. *Typical landing gear safety circuit.*

Ground locks. While the safety switch goes a long way toward preventing the crew from inadvertently retracting the gear, it is still possible for the gear to collapse. To reduce this possibility, the crew installs an additional safety device, a ground lock, in the main gear when the aircraft is secured.

Commonly, a metal-pin ground lock is inserted through aligned holes drilled into two or more units of the landing gear support structure. A somewhat similar system is the spring-loaded clip that is designed to fit around and hold two or more units of the support structure together. Both types must be removed during preflight because either will prevent the gear from being retracted.

Gear indicators. Most aircraft system failures provide the crew with enough clues in sufficient time for them to take corrective action. Due to the way landing gear are situated on most aircraft, the failure of a gear could go undetected until the aircraft hits the runway. To prevent this, the manufacturer provides landing gear indicators on the flight deck.

Several methods are used by manufacturers to alert the crew of an unsafe landing gear configuration. Most aircraft have aural and visual systems. For instance, a horn will sound and a light illuminate if one or more throttles are retarded and the gear is in any position other than down and locked.

Gear position indicators come in several varieties (FIG. 9-11). One of the simplest systems uses a single light to indicate that one or more landing gear are in transit and three green lights to indicate that each gear is down and locked. Another system displays miniature, movable landing gear that are slaved to the actual gear to show the appropriate position. Yet another type of gear indicator utilizes a tab indicator with the

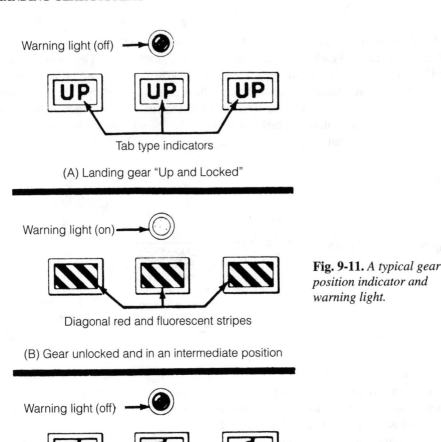

Warning light (off) →

UP   UP   UP

Tab type indicators

(A) Landing gear "Up and Locked"

Warning light (on) →

Diagonal red and fluorescent stripes

Fig. 9-11. *A typical gear position indicator and warning light.*

(B) Gear unlocked and in an intermediate position

Warning light (off) →

Silhouette of wheels

(C) Landing gear "Down and Locked"

word UP displayed in a small window for each gear that is in the up and locked position. When the gear is in transit, those same windows display a red-and-white striped flag. Finally, when the gear is down and locked, the windows display a silhouette of a landing gear.

Nosewheel centering. The nosewheel is the only landing gear that has the ability to rotate left or right (for steering). This might cause problems during retraction if the gear is not centered. To prevent that from happening, there are several types of centering devices such as *internal centering cams* (FIG. 9-12). The cam system automatically centers the nose gear during retraction as it moves into the wheel well.

During nose gear retraction, the strut hangs below the aircraft extended by the air pressure within the strut and by gravity acting upon its own weight. As the strut ex-

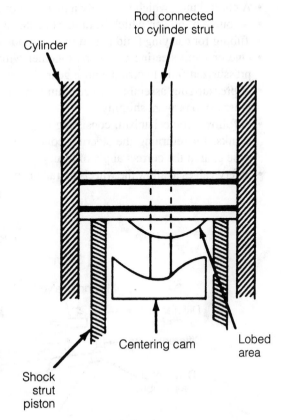

Fig. 9-12. *Cutaway view of a nose gear internal centering arm.*

Cylinder

Rod connected to cylinder strut

Centering cam

Lobed area

Shock strut piston

tends, the raised area of the piston strut contacts the sloping area of the fixed centering cam and slides along it. This action causes it to align with the centering cam that forces the nose gear piston to face forward.

NOSEWHEELS

Complexity of a nosewheel steering system varies, depending upon the size and weight of the aircraft. Light aircraft utilize a very simple system of mechanical linkage hooked to rudder pedals. Such systems typically use push-pull rods that connect the pedals to horns located on the pivotal portion of the nosewheel strut. Aircraft that have a larger mass, which requires more positive control of steering, use a power steering system. Among several different methods employed to accomplish the same objective, all systems have some common features that are displayed in FIG. 9-13:

- A flight deck control, such as a wheel, handle, lever, or switch to allow starting, stopping, and to control the action of the system.
- Mechanical, electrical, or hydraulic connections for transmitting flight deck control movements to a steering control unit.

- A control unit, which is usually a metering or control valve.
- A source of power, which is, in most instances, the aircraft hydraulic system.
- Tubing for carrying fluid to and from various parts of the system.
- One or more steering cylinders, together with the required linkages, for using pressurized fluid to turn the nose gear.
- A pressurizing assembly to keep fluid in each steering cylinder always under pressure to prevent shimmy.
- A follow-up mechanism, consisting of gears, cables, rods, drums, and/or bell-cranks, for returning the steering control unit to neutral and thus holding the nose gear at the correct angle of turn.
- Safety valves to allow the wheels to trail or swivel in the event of hydraulic failure.

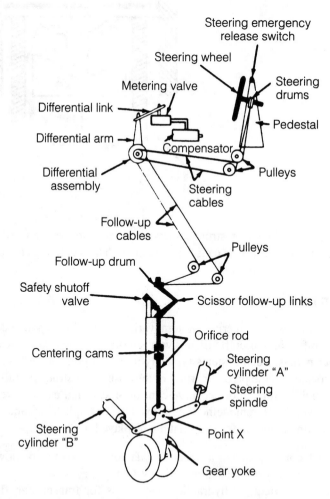

Fig. 9-13. *Nosewheel system mechanical and hydraulic units.*

Nosewheel steering operation

A common hydraulic nosewheel steering system is depicted in FIG. 9-14. The nose-wheel steering wheel is connected to a steering drum via a steel shaft. The drum is contained inside the pilot's control pedestal. As the drum turns, its motion is passed on to the differential assembly drum through a series of cables and pulleys. Movement of that assembly is further transmitted by a differential link to the metering valve assembly, where it moves the selector valve to the appropriate position. That allows hydraulic pressure to turn the nose gear.

As illustrated by FIG. 9-14, hydraulic fluid under pressure goes through the safety shutoff valve to the metering valve. From there, the fluid goes out port A, through the right-turn alternating line, and into steering cylinder A. Because this is a one-port cylinder, fluid pressure results in piston extension. The rod of the piston, which is connected to the nose steering spindle, causes it to pivot at point X. Piston extension causes the nose gear to rotate to the right, while contraction causes the nose gear to rotate to the left. As the nose gear turns right, the fluid is forced out of cylinder B, through the left-turn alternating line, and into port B of the metering valve. The metering valve sends this return fluid into a compensator, which routes the fluid into the aircraft system return manifold.

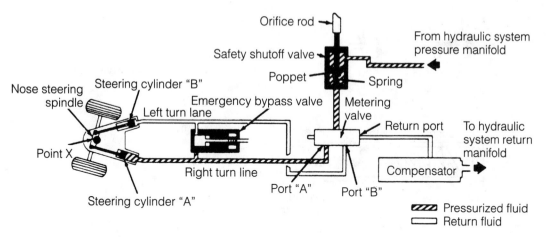

Fig. 9-14. *Nosewheel steering hydraulic flow diagram.*

Because hydraulic pressure does activate nose gear rotation (steering), the crew must be careful to avoid turning too far during operation. To help avoid this situation, the nose gear steering system contains devices to stop and hold the gear at the selected angle of turn.

Follow-up linkage. Recall that the nose gear is turned by the steering spindle as the piston of cylinder A extends. But the rear of the spindle has gear teeth that mesh with a gear on the bottom of the orifice rod; thus, as the nose gear and its spindle turn, the orifice rod also turns, although in the opposite direction. This rotation is transmitted by the two sections of the orifice rod to the scissor follow-up links (FIG. 9-13) located at the top of the nose gear strut. As the follow-up links return, they rotate the connected follow-up drum, which transmits the movement by cables and pulleys to the differential assembly. Operation of the differential assembly causes the differential arm and links to move the metering valve back toward the neutral position.

The compensator unit (FIG. 9-15) that is a part of the nosewheel system keeps fluid in the steering cylinders pressurized at all times. This hydraulic unit consists of a three-port housing, which encloses a spring-loaded piston and poppet. The left port is an air vent, which prevents trapped air at the rear of the piston from interfering with movements of the piston. The second port, located at the top of the compensator, connects through a line to the metering valve return port. The third port is located at the right side of the compensator. This port, which is connected to the hydraulic return manifold, routes the steering system return fluid into the manifold when the poppet valve is open.

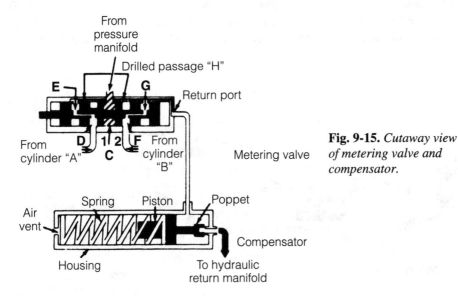

Fig. 9-15. *Cutaway view of metering valve and compensator.*

The compensator poppet opens when pressure acting upon the piston becomes high enough to compress the spring. Because pressure in a confined fluid is transmitted equally and undiminished in all directions, there is equal pressure in metering valve passage H and in chambers E, D, G, and F. This same pressure is also applied in the right- and left-turn alternating lines, as well as in the steering cylinders.

Shimmy dampers

The purpose of a shimmy damper is to control nosewheel shimmy (vibration) through hydraulic damping while taxiing, landing, or taking off. The damping device is attached, or built, into the nose gear. Three primary types of shimmy dampers are piston, vane, and power steering.

Piston shimmy damper. This damper is used primarily on lighter aircraft and is illustrated in FIG. 9-16. It is made up of two primary components: cam assembly and damper assembly. The shimmy damper is attached to the nose gear shock strut outer cylinder at the lower end. The cam assembly is attached to the inner cylinder of the shock strut and rotates with the nosewheel. The damper assembly consists of a spring-loaded reservoir piston that maintains the confined fluid under pressure and an operating cylinder and piston. A ball check permits the flow of fluid from the reservoir to the operating cylinder to make up for any fluid loss in the operating cylinder. Because of the rod on the operating piston, its stroke away from the filler end of the piston displaces more fluid than its stroke toward the filler end. This difference is taken care of by the reservoir orifice, which permits a small flow both ways between the reservoir and operating cylinder.

As the nosewheel fork rotates in either direction, the shimmy damper cam displaces the cam follower rollers, causing the operating piston to move in its chamber. This movement forces fluid through the orifice in the piston. Because the orifice is very small, rapid movements of the piston, which commonly occur during landing and takeoff, are restricted, and nosewheel shimmy is eliminated.

Vane shimmy damper. This damper is located on the nosewheel shock strut immediately above the nosewheel fork and can be mounted either internally or externally. When this unit is mounted internally, its housing is fitted and secured inside the shock strut, and the shaft is splined to the nosewheel fork. If the unit is mounted externally, the housing of the unit is bolted to the side of the shock strut, and the shaft is connected by mechanical linkage to the nosewheel fork.

The shimmy damper housing is divided into three main parts (FIG. 9-17): replenishing chamber, working chamber, and lower shaft packing chamber.

The replenishing chamber, located at the top of the housing, is a reservoir for hydraulic fluid. Pressure is applied to the fluid by the spring-loaded replenishing piston and piston shaft that extends through the upper housing and serves as a fluid-level indicator. The area above the piston contains the piston spring and is open to atmosphere to prevent hydraulic lock. Fluid is prevented from leaking past the piston by O-ring packings. A grease fitting provides the means for filling the replenishing chamber with fluid.

The working chamber is separated from the replenishing chamber by the abutment and valve assembly. The working chamber contains two one-way ball check valves, which will allow fluid to flow from the replenishing chamber to the working chamber only. This chamber is divided into four sections by two stationary vanes called abutment flanges, which are keyed to the inner wall of the housing, and two rotating vanes, which are an integral part of the wing shaft. The shaft contains the valve orifice that fluid passes through when going from one chamber to another.

Turning the nosewheel in either direction causes the rotating vanes to move in the housing. This results in two sections of the working chamber growing smaller, while

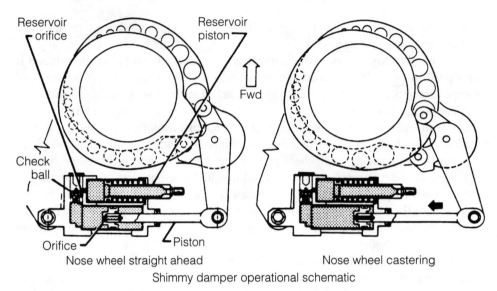

Fig. 9-16. *Typical piston-type shimmy damper.*

the opposite two chambers grow larger. The rotating vane can move only as fast as the fluid can be displaced from one chamber to the other. All of the fluid being displaced must pass through the valve orifice in the shaft. Resistance to the flow of fluid through the orifice is proportional to the velocity of flow. This means that the shimmy damper offers little resistance to slow motion, such as that encountered during normal steering of the nose gear or ground handling, but offers high resistance to shimmy on landing, takeoff, and high-speed taxiing.

280

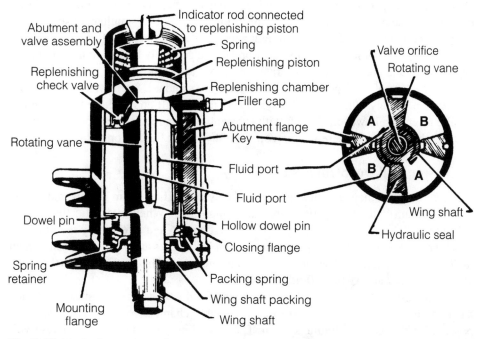

Fig. 9-17. *Typical vane-type shimmy damper.*

An automatic orifice adjustment compensates for temperature changes. A bimetallic thermostat in the shaft opens and closes the orifice as the temperature and viscosity change. This results in a constant resistance over a wide temperature range. In case an exceptionally high pressure is suddenly built up in the working chamber by a severe twisting force on the nosewheel, the closing flange moves down, compressing the lower shaft packing spring, allowing fluid to pass around the lower ends of the vanes, preventing structure damage.

Steer dampers. A steer damper is another hydraulically operated dampening device that performs the dual role of steering and eliminating shimmy.

BRAKE SYSTEMS

It would be difficult to overstate the case for properly maintaining the aircraft brake system. Brakes are in near-constant use whenever the aircraft is on the ground. They help control taxi speed, assist in steering the aircraft, secure it for parking, must hold against engine thrust during runups, assure that the aircraft does not run off the end of the runway after landing, and brakes are critical to safety during an aborted takeoff.

The common configuration for brake systems is that each main landing gear has a brake that can be activated simultaneously with, or independent of, the other main gear. This is done by incorporating toe brakes with the rudder pedals. When the pilot moves the rudder pedals fore and aft, the action turns the nosewheel and rudder appropriately.

Depressing the upper portion of each rudder pedal activates and linearly increases brake pressure on the corresponding brake. This allows the pilot to simultaneously activate the brakes on both sides for even, straight ahead braking, or apply differential braking, if necessary, to assist in a turn. Nose gear brakes are generally limited to transport category aircraft and not found on corporate turboprops and turbojets.

In light aircraft, independent brake systems are standard. Such systems are totally independent of any other hydraulic system on the aircraft and include their own hydraulic reservoir; however, aircraft with substantial mass require very powerful brake systems that, in turn, require a large volume of hydraulic fluid to operate.

Power brake control systems

Figure 9-18 depicts a typical power brake system. The system is powered by the aircraft's main hydraulic system through a single line. When hydraulic fluid under pressure is diverted to the brake system, the first component affected by the fluid is a check valve that protects the brake system from losing pressure in the event of a main hydraulic system failure. The check valve traps pressurized hydraulic fluid in the brake system and only allows fresh fluid to enter.

After passing through the check valve, the fluid goes to the accumulator, which has two purposes. One purpose is to act as a surge chamber for excessive loads imposed upon the brake hydraulic system. The second purpose is to serve as a reservoir

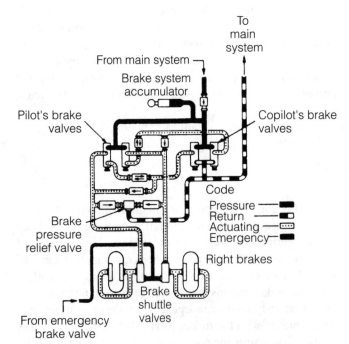

Fig. 9-18. *Typical power brake control valve system.*

for pressurized fluid. Pressure drops in the accumulator when the pilot presses down on the brakes. The result is a demand for additional hydraulic fluid from the main system.

From the accumulator, the fluid travels to the pilot's and first officer's control valves. The control valve controls the volume and pressure of the fluid used to actuate the brakes. Four check valves and two orifice check valves are associated with these control valves. The check valves permit the hydraulic fluid to flow in only one direction. The orifice check valves allow unrestricted flow of fluid in one direction from the pilot's brake control valve; flow in the opposite direction is restricted by an orifice in the poppet. Orifice check valves help prevent chatter.

The hydraulic fluid eventually encounters the pressure relief valve. Under normal conditions, the fluid merely bypasses the valve, but unseats under overpressure conditions. The exact pressure settings vary from system to system, but the valve's purpose is to protect the system from excess pressure. For instance, the valve might be preset to open up if the system pressure reached 825 psi. The valve would reroute pressurized fluid back to the main hydraulic system until brake system pressure dropped to fewer than 760 psi, at which time the valve would close and allow the system to continue undisturbed.

Each brake actuating line incorporates a shuttle valve for the purpose of isolating the emergency brake system from the normal brake system. When brake actuating pressure enters the shuttle valve, the shuttle is automatically moved to the opposite end of the valve. This closes off the hydraulic brake system actuating line. Fluid returning from the brakes travels back into the system to which the shuttle was last open.

Power boost brake systems

We have seen that independent brake systems are used in light aircraft and power brake systems in large aircraft. There is a middle ground between light and large: aircraft that land fast enough to build up more energy than an independent system can handle, but are still light enough in weight that they do not require the complex power brake system. Certain manufacturers simply install the more sophisticated power brake system; another solution is the power boost system.

The power boost system uses aircraft hydraulic system fluid to assist the pedals through a power boost master cylinder. Actual hydraulic system fluid never enters the brake system. A typical power boost brake system is depicted in FIG. 9-19. The system is made up of a reservoir, two power boost master cylinders, two shuttle valves, and the brake assembly in each main gear. The system depicted also utilizes a bottle of compressed air as an emergency backup. Main hydraulic system pressure is routed from the pressure manifold to the power master cylinders. When the brake pedals are depressed, fluid for actuating the brakes is routed from the power boost master cylinders through the shuttle valves to the brakes.

When the brake pedals are released, the main system pressure port in the master cylinder is closed. Fluid that was moved into the brake assembly is forced out the return port by a piston in the brake assembly, through the return line to the brake reser-

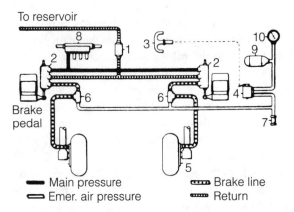

1. Brake reservoir
2. Power boost master cylinder
3. Emergency brake control
4. Air release valve
5. Wheel brake
6. Shuttle valve
7. Air vent
8. Main system pressure manifold
9. Emergency air bottle
10. Emergency air gauge

━━ Main pressure
⬤━ Emer. air pressure
▭▭▭ Brake line
▭▭▭ Return

Fig. 9-19. *Power boost master cylinder brake system.*

voir. The brake reservoir is connected to the main hydraulic system reservoir to assure an adequate supply of fluid to operate the brakes.

Multiple-disk brakes

High speed, heavy aircraft develop substantial kinetic energy that must be dissipated upon landing in order to slow it down. The purpose of the brake is to convert that kinetic energy into heat through the use of constructive friction. Two major considerations are related to accomplishing this task. The first is that the brakes must be capable of creating sufficient friction to slow down and stop the aircraft under all normally anticipated loads and speeds. The other consideration is that they be able to efficiently and effectively deal with the tremendous amount of heat that is generated by the friction. The standard single- and even dual-disk brake systems associated with light aircraft are typically going to be inadequate for turboprop and turbojet aircraft.

Multiple-disk brakes are designed to be used with either power brake control valves or power boost master cylinders. An exploded view of a multiple-disk brake assembly is shown in FIG. 9-20. The brake consists of a bearing carrier, four rotating disks called *rotors*, three stationary disks called *stators*, a circular actuating cylinder, an automatic adjuster, and various support components.

Regulated hydraulic pressure is applied through the automatic adjuster to a chamber in the bearing carrier. The bearing carrier is bolted to the shock strut axle flange and serves as a housing for the annular actuating piston. Hydraulic pressure forces the annular piston to move outward, compressing the rotating disks, which are keyed to the landing wheel and compressing the stationary disks, which are keyed to the bearing carrier. The resulting friction causes a braking action on the wheel and tire assembly.

When the hydraulic pressure is relieved, the retracting springs force the actuating piston to retract into the housing chamber in the bearing carrier. The hydraulic fluid in the chamber is forced out by the returning of the annular actuating piston and is bled through the automatic adjuster to the return line. The automatic adjuster traps a prede-

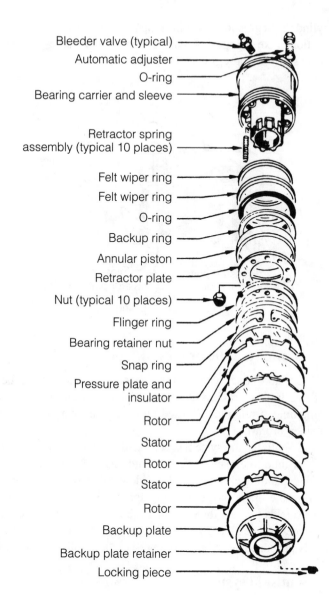

Bleeder valve (typical)
Automatic adjuster
O-ring
Bearing carrier and sleeve

Retractor spring
assembly (typical 10 places)

Felt wiper ring
Felt wiper ring
O-ring
Backup ring
Annular piston
Retractor plate
Nut (typical 10 places)
Flinger ring
Bearing retainer nut
Snap ring
Pressure plate and
insulator
Rotor
Stator
Rotor
Stator
Rotor
Backup plate
Backup plate retainer
Locking piece

Fig. 9-20. *Multiple-disk brake.*

termined amount of fluid in the brake, an amount just sufficient to give correct clearances between the rotating disks and stationary disks.

Segmented rotor brakes

Segmented rotors are another heavy-duty brake system that is particularly adaptable to heavier aircraft with high-pressure hydraulic systems. A segmented rotor system is functionally very similar to the multiple-disk system. The segmented rotor brake is also usable with either power brake control valves or power boost master cylinders. Segmented rotor braking is accomplished by several sets of stationary, high-friction brake linings that make contact with rotating (rotor) segments. Figure 9-21 is a cross section of a segmented rotor brake.

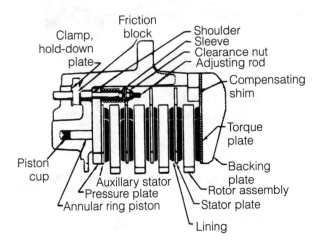

Fig. 9-21. *Cross section of a segmented rotor brake.*

Expander tube brakes

One additional type of brake system is the expander tube. This low-pressure brake utilizes 360° of braking surface making it very efficient for use on large aircraft. At the same time, a relatively limited number of moving parts makes it attractive for some smaller aircraft.

Essentially, the system is activated when pressurized hydraulic fluid causes the expander tube to expand (FIG. 9-22 and FIG. 9-23). Because the frame prevents the tube from expanding either inward or sideways, the expander tube moves outward. This results in the brake blocks moving against the brake drum, which creates friction. The tube shields prevent the expander tube from extruding between the blocks; the torque bars prevent the blocks from rotating with the drum. Friction created by the brake is directly proportional to brakeline pressure as applied to the rudder pedals by the pilot.

Antiskid systems

Because braking is supposed to stop the aircraft, the brake system must function properly and effectively. An area beyond the crew's control that can significantly decrease brake effectiveness is the condition of the surface. Patches of snow, ice, standing water, and other slick surfaces can have a very negative effect on braking action. If an aircraft is on a landing roll with the brakes being applied, a slick patch will greatly reduce the tire-to-surface friction, causing the wheel to stop rotating because of brake pressure. As soon as the nonrotating wheel has slipped off the slick patch, the tire will recontact a surface area of greater friction with resulting damage to the tire and a sig-

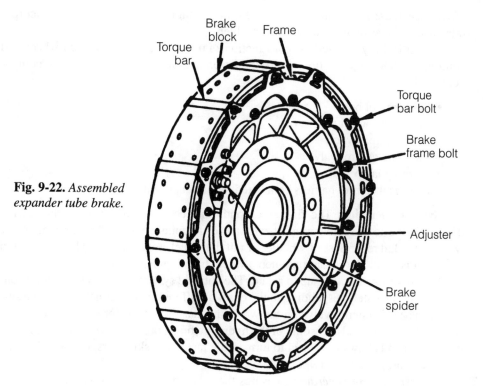

Fig. 9-22. *Assembled expander tube brake.*

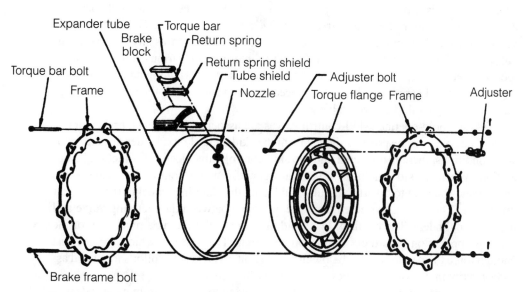

Fig. 9-23. *Exploded view of expander tube brakes.*

nificant decrease in braking effectiveness. Select high-performance aircraft incorporate an antiskid system to prevent this problem.

The antiskid system serves four functions: normal skid control, locked-wheel skid control, touchdown protection, and fail-safe protection. The primary system components include:

- Skid control box
- Two skid control generators
- Skid control switch
- Two skid control valves
- Warning lamp
- Electrical control harness connected to the squat switch

System operation. The first skid condition occurs when the wheel is slowing down relative to the remaining wheels; the differential rotational speed among the wheels is a skid regardless of whether or not the wheel stops rotating. The second skid condition is a full skid situation when a wheel has completely stopped rotating while the remaining wheels continue to rotate. The antiskid system prevents the second condition by immediately responding to the onset of differential rotational speed. The system opens a skid control valve on the wheel that has the slower (or slowest) rotational speed. The valve bleeds off some of the hydraulic brake pressure, which reduces braking friction and allows the wheel to turn a little faster. If a skid worsens, more braking pressure is proportionally removed.

The *skid control generator* measures the rotational speed of a wheel. The small, independent electrical generator is mounted in the axle of each wheel. The armature of each generator is driven by the respective wheel. As the wheel's rotational speed increases, the generator produces a greater voltage and current signal. The signal travels via wires to the skid control box that interprets the signal and senses any change in signal strength. The box is sophisticated enough to interpret the various signals as skids, locked wheels, brake applications, and brake releases. Analyzing the data from all the wheels, the box determines which wheel needs attention and sends the appropriate signal to that wheel's skid control valve.

Two solenoid-operated skid-control valves are associated with each brake control valve. The solenoids respond to signals from the skid control box. Under normal, non-skid conditions, there is no signal and the skid control valve does not affect brake operation. When a skid condition arises, a signal from the control box causes the appropriate solenoid to divert hydraulic fluid from the skidding wheel to the reservoir return line. The fluid is extracted from the brakeline between the metering valve and the brake cylinders. The result of less fluid pressure in the line is a relaxation of the braking application regardless of the pilot's input. The system, which applies just enough force to operate slightly below the skid point, overrides the pilot's input and assures maximum braking efficiency, given the surface conditions.

In the event of a system failure, a fail-safe protection circuit automatically deactivates the system, returns full control to the pilot, and illuminates the warning light. In addition, the crew may elect at any time to deactivate the system and resume total control over the braking. Whenever the system is deactivated, a warning lamp illuminates.

Another function of the antiskid system is *locked-wheel* skid control. This system, which automatically activates whenever aircraft ground speed exceeds approximately 20 mph, causes full brake release if a wheel locks up. If, for instance, a wheel rolls over a patch of ice, the dramatic reduction in surface friction might cause the wheel to completely stop rotating. When the tire leaves the patch and encounters normal pavement again, there is a serious threat that the tire will blow because the stationary tread would be scraped off by the pavement. The system prevents that from occurring by bleeding off brake system pressure longer than if it were a minor difference in rotational speeds. This gives the stopped wheel time to regain speed.

One final aspect of the system is *touchdown protection*. This circuit prevents brake application at any time during flight. The protective system assures that the wheels have sufficient time to rotate before they are forced to carry the full weight of the aircraft, which essentially eliminates pilot-induced braking on touchdown in the event the aircraft lands with the pilot's feet depressing the brakes. Brake pedal inputs are ineffective until two conditions are met:

- The squat switch must signal that the weight of the aircraft is on the wheels.
- The wheel generators must sense a wheel rotation speed in excess of approximately 20 mph.

If both conditions are not met, the skid control box continues to activate the solenoids that prevent fluid pressure from activating the brakes.

WHEELS AND TIRES

Aircraft landing wheels

Pilots sometimes give little thought to the aircraft wheel, but all of the landing shock, turn skidding, and other actions related to taxi, takeoff, and landing are transmitted from the tire to the wheel. It is a testimony to the manufacturer that few, if any, problems are the result of the wheel. A typical aluminum or magnesium wheel is strong, lightweight, and relatively maintenance-free. Figure 9-24 shows a typical forged aluminum split wheel for a heavy aircraft, which is the most popular wheel in aviation.

Aircraft tires

Aircraft tires have the toughest job on the airplane. They are used, abused, and cursed, but day after day they compensate for a myriad of static and dynamic stresses under a very wide range of conditions. While all these stresses certainly take their toll on the tire, it is the buildup of heat that has the most profound and devastating effect on the life of the tire. Figure 9-25 illustrates the tough construction of a common aircraft tire.

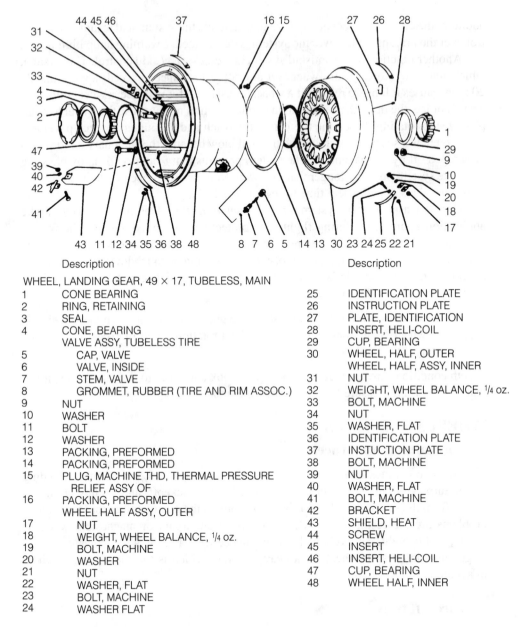

Description

WHEEL, LANDING GEAR, 49 × 17, TUBELESS, MAIN

1	CONE BEARING
2	RING, RETAINING
3	SEAL
4	CONE, BEARING
	VALVE ASSY, TUBELESS TIRE
5	CAP, VALVE
6	VALVE, INSIDE
7	STEM, VALVE
8	GROMMET, RUBBER (TIRE AND RIM ASSOC.)
9	NUT
10	WASHER
11	BOLT
12	WASHER
13	PACKING, PREFORMED
14	PACKING, PREFORMED
15	PLUG, MACHINE THD, THERMAL PRESSURE RELIEF, ASSY OF
16	PACKING, PREFORMED
	WHEEL HALF ASSY, OUTER
17	NUT
18	WEIGHT, WHEEL BALANCE, 1/4 oz.
19	BOLT, MACHINE
20	WASHER
21	NUT
22	WASHER, FLAT
23	BOLT, MACHINE
24	WASHER FLAT

Description

25	IDENTIFICATION PLATE
26	INSTRUCTION PLATE
27	PLATE, IDENTIFICATION
28	INSERT, HELI-COIL
29	CUP, BEARING
30	WHEEL, HALF, OUTER
	WHEEL, HALF, ASSY, INNER
31	NUT
32	WEIGHT, WHEEL BALANCE, 1/4 oz.
33	BOLT, MACHINE
34	NUT
35	WASHER, FLAT
36	IDENTIFICATION PLATE
37	INSTUCTION PLATE
38	BOLT, MACHINE
39	NUT
40	WASHER, FLAT
41	BOLT, MACHINE
42	BRACKET
43	SHIELD, HEAT
44	SCREW
45	INSERT
46	INSERT, HELI-COIL
47	CUP, BEARING
48	WHEEL HALF, INNER

Fig. 9-24. *Heavy aircraft split wheel.*

Tread. Aircraft tires are built from a rubber compound that is known for its toughness and durability. After years of study, manufacturers have determined that the circumferential ribbed pattern is the most effective for most passenger aircraft operations.

Tread reinforcement. Multiple layers of reinforced nylon cord fabric give unusually high strength to the tread. This is necessary because of the fast touchdown speeds

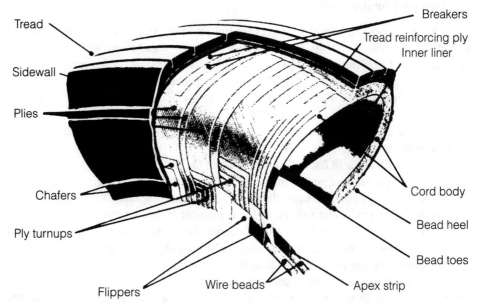

Fig. 9-25. *Aircraft tire construction.*

associated with certain aircraft. The tire must be capable of accelerating from zero to much faster than 100 knots in an instant. The tremendous stress imparted to the tire when this happens is displayed when an approaching aircraft, especially an airliner, first touches the runway and puffs of smoke curl up where the tire initially touched the surface.

Casing plies/cord body. Diagonal layers of rubber-coated nylon cord fabric run at opposite angles to each other. These plies provide the majority of the strength of the tire, totally encompassing the tire body, folding around the wire beads and back against the tire sidewalls.

Beads. The beads are steel wires embedded in the rubber and wrapped in fabric. The beads anchor the carcass plies and provide a firm mounting surface on the wheel.

Flippers. Fabric and rubber flippers provide insulation for the carcass from the bead wires, resulting in improved tire durability.

Chafers. Layers of fabric and rubber protect the carcass from damage during mounting and demounting. The chafers insulate the carcass from brake heat and provide a good seal against movement during dynamic operations.

Innerliner. On tubeless tires, an inner layer of less permeable rubber acts as a built-in tube that prevents air from seeping through casing plies. For tube tires, a thinner rubber liner prevents tube chafing against the inside ply.

Sidewall. Cord body sides are protected from abrasion and exposure by the sidewalls. One unusual sidewall is the *chine tire*, which is designed to be used on the nosewheel. The chine is a built-in deflector to divert runway water and slush to the side to reduce the possibility of splashing up and aft into tail-mounted engine intakes.

Tire and brake care

It is very important to realize that every time the pilot steps on the brakes it is like lighting a cigar with a $10 bill. Brakes are very expensive and every application shortens their useful life; the same applies to tires because braking causes tread abrasion. Avoiding unnecessary braking should be a major consideration for all flight crews. Unless the conditions truly warrant, avoid using brakes in turns or turning at an excessively fast speed. For the same reasons, avoid fast taxi speeds, too. Turboprop aircraft have the edge here because it is possible to use differential prop blade angles in beta mode to assist in turns.

Another problem area for brakes exists when the pilot inadvertently depresses the brake portion of the rudder pedal while taxiing. This continuous dragging of the brakes eats up brake linings and overheats the brakes at an alarming rate. Whenever feasible, land on the long runway and roll out as far as possible before applying brakes.

Aircraft tire care is similar to auto tire care. The same rules of safe driving and careful inspection apply on the runway and on the highway. Areas of similarity include speed control, braking considerations, and proper cornering, plus inspecting the tires for proper inflation, cuts, bruises, and signs of tread wear. Most pilots are unaware that the major concern about aircraft tires is not the impact of hard landings but rather the rapid heat buildup during lengthy ground operations.

Excessive temperatures are a result of the way aircraft tires are manufactured. Tires are designed to have a significant amount of flex that is twice as much as automotive tires. The advantage is that aircraft tires are far better prepared to meet the tremendous demands put upon them, but the problem is that flexing causes internal stress and friction that increases the temperature build-up. Excessive temperatures lead to long-term tire body damage.

Tire inflation is also an area of potential trouble. Aircraft tires should always be properly inflated to the value specified by the tire dealer or airframe manufacturer. Only a tire gauge can accurately indicate the state of a tire's inflation, but a quick visual inspection of the tread can reveal if air pressure has been consistently high or low. Excessive wear in the shoulder area of the tire is an indication of under inflation. Excessive wear in the center of the tire suggests over inflation.

Tires should also be routinely checked for cuts and bruises. The best method to avoid damage is to taxi slowly and watch runway and taxi surface conditions carefully. Potholes, heaved surfaces, and foreign objects all cause cuts and bruises. Be particularly mindful of making tight turns on loose gravel or other foreign objects. Unlike cars, which have four wheels, it is possible to pivot the aircraft on a single tire. If a rock or other object with a point lodges underneath the pivoting tire, the object might puncture the tread.

When changing a tire or tube, the pressure should be checked daily for several days. During installation, pockets of air might become trapped between the tire and the tube, which will result in a false reading. As time goes on, the trapped air slowly finds its way out by seeping underneath the beads and around the valve hole. The result might be a significant under-inflation situation. Nylon tires will stretch after the first 24-hours of installation, causing a change in tire pressure. It is particularly important on a dual-tire installation to assure that both tires are always within 5 psi of each other, otherwise one tire will carry more load than the other, leading to differential wear.

Tire preflight inspection

Look for a number of tire-related items when preflighting. A brief list includes:

- Inspect visible tread area for abnormalities such as cuts or other abrasions. Remove anything caught or stuck in the treads such as glass or pebbles. Any metal objects should be examined by an A&P immediately because a nail, screw, or heavy wire staple has probably punctured the tire and might be acting as a seal if the tire is not already deflated or noticeably low.
- Check sidewalls for weather or ozone checking and cracking, radial cracks, cuts, snags, gouges, and the like. No cord should be exposed.
- Check tire surface for skid burns or other flat spots. Any tire with a flat spot that exposes carcass cord should be changed. Tires should not show more than approximately 80 percent tread wear.
- Check the tire for spotty or uneven wear patterns that would indicate faulty brakes resulting in differential braking effect.

Tire tips

Advisory Circular 65-15A Airframe and Powerplant Mechanics Airframe Handbook details specific operating instructions to increase the useful life of tires. Flight crews should incorporate the following recommendations into their normal operating procedures.

Taxiing. Needless tire damage or excessive wear can be prevented by proper handling of the aircraft during taxiing. Most of the gross weight of any aircraft is on the main landing gear wheels on two, four, eight, or more tires. The tires are designed and inflated to absorb shock of landing and will deflect (bulge at the sidewall) about two and one-half times more than a passenger car or truck tire. The greater deflection causes more working of the tread, produces a scuffing action along the outer edges of the tread, and results in rapid wear.

Also, if an aircraft tire strikes a chuck hole, a stone, or foreign objects lying on the runway, taxi strip, or ramp, there is more possibility of its being cut, snagged or bruised because of the percentage of deflection. Or, one of the main landing gear wheels, when making a turn, might drop off the edge of the paved surface causing severe sidewall or shoulder damage. The same type of damage might also occur when the wheel rolls back over the edge of the paved surface.

With dual main landing gear wheels, one tire might be forced to take a damaging impact (which two could withstand without damage) simply because all the weight on one side of the plane is concentrated on one tire instead of being divided between two.

As airports grow in size and taxi runs become longer, chances for tire damage and wear increase. Taxi runs should be no longer than absolutely necessary and should be made at speeds no greater than 25 mph, particularly for aircraft not equipped with nosewheel steering.

For less damage in taxiing, all personnel should see that ramps, parking areas, taxi strips, runways and other paved areas are regularly cleaned and cleared of all objects that might cause tire damage.

Braking and pivoting. Increasing airport traffic, longer taxi runs, and longer runs on takeoff and landing are subjecting tires to more abrasion resulting from braking, turning, and pivoting.

Severe use of brakes can wear flat spots on tires and cause them to be out of balance, making premature recapping or replacement necessary. Severe or prolonged application of the brakes can be avoided when ground speed is reduced.

Careful pivoting of aircraft also helps to prolong tire tread life. If an aircraft turned as an automobile or truck does—in a rather wide radius—the wear on the tire tread would be materially reduced; however, when an aircraft is turned by locking one wheel (or wheels), the tire on the locked wheel is twisted with great force against the pavement. A small piece of rock or stone that would ordinarily cause no damage, can, in such a case, be literally screwed into the tire. This scuffing or grinding action takes off tread rubber and places a very severe strain on the sidewalls and beads of the tire at the same time.

To keep this action at a minimum, it is recommended that whenever a turn is made, the inside wheel (wheels) be allowed to roll on a radius of 20 to 25 feet and up to 40 feet for aircraft with bogies (two or more tires per strut).

Takeoff and landings. Aircraft tire assemblies are always under severe strain on takeoff or landing. But under normal conditions, with proper control and maintenance of tires, they are able to withstand many such stresses without damage.

Tire damage on takeoff, up to the point of being airborne, is generally the result of running over some foreign object. Flat spots or cuts incurred in pivoting can also be a cause of damage during takeoff or landing.

Tire damage at the time of landing can be traced to errors in judgment or unforeseen circumstances. Smooth landings result in longer tread wear and eliminate much of the excessive strain on tires at the moment of impact.

Landings with brakes locked, while almost a thing of the past, can result in flat spotting. Removal of the tire for recapping or replacement is almost invariably indicated. Brakes-on landings also cause very severe heat at the point of contact on the tire tread and might even melt the tread rubber (skid burn). Heat has a tendency to weaken the cord body and places severe strain on the beads. In addition, heat buildup in the brakes might literally devulcanize the tire in the bead area. Under these circumstances, blowouts are not uncommon because air under compression must expand when heated.

Sometimes an aircraft will be brought in so fast that full advantage will not be taken of runway length and brakes are applied so severely that flat spots are produced on the tires. Or, if brakes are applied when the plane is still traveling at a fast speed and still has considerable lift, tires might skid on the runway and become damaged beyond further use or reconditioning.

The same thing might occur during a rough landing if brakes are applied after the first bounce. For maximum tire service, delay brake application until the plane is definitely settled into its final roll.

More tires fail on takeoff than on landing and such failures on takeoff can be extremely dangerous. For that reason, emphasis must be placed on proper preflight inspection of tires and wheels.

Condition of landing field. Regardless of the preventive maintenance and extreme care taken by the pilot and ground crew when handling the aircraft, tire damage is almost sure to result if runways, taxi strips, ramps, and other paved field areas are in a bad condition or poorly maintained.

Chuck holes, pavement cracks, or step-offs from these areas to the ground all can cause tire damage. In cold climates, especially during the winter, all pavement breaks should be repaired immediately.

Another hazardous condition often overlooked is accumulated loose material on paved areas and hangar floors. Stones and other foreign materials should also be swept off all the paved areas. In addition, tools, bolts, rivets and other repair materials are sometimes left lying on the aircraft, and when the aircraft is moved, these materials drop off. The objects picked up by the tires of another aircraft can cause punctures, cuts, or complete failure of the tire, tube, and even the wheel. With jet aircraft, it is even more important that foreign material be kept off areas used by aircraft.

Hydroplaning. A wave of water can build up in front of spinning tires on a wet runway or taxiway, and when the wave is overrun, tires will no longer make contact with the runway. This results in the complete loss of steering capability and braking action. Hydroplaning can also be caused by a thin film of water on the runway mixing with the contaminants present: oil, rubber, and more.

Most major airport runway surfaces have crosscutting that has greatly reduced the danger of hydroplaning; however, the ridges of concrete created by this crosscutting can cause a chevron cutting of tread ribs, particularly with the high-pressure tires on jet aircraft. The cuts are at right angles to the ribs and rarely penetrate to the fabric tread reinforcing strip. Such damage would not be considered cause for removal unless fabric was exposed due to a piece of tread rib tearing out.

MANUFACTURER DOCUMENTATION

(The following information is extracted from Beech Aircraft Corporation pilot operating handbooks and Falcon Jet Corporation aircraft and interior technical descriptions. The information is to be used for educational purposes only. This information is not to be used for the operation or maintenance of any aircraft.)

King Air C90

Landing gear construction. The tricycle gear, when fully extended, is a braced semicantilevered construction. The system utilizes folding braces called *drag legs* that lock in place when the gear is fully extended. Oleo struts form the semicantilevered beams. The oleo landing gear struts are attached to the airplane structure, in pinned joints. Knee braces are employed to prevent rotation between oleo piston and cylinder. The forward oleo strut is fitted with a mechanism for nosewheel steering.

Hydraulic extension and retraction system. The nose and main landing gear assemblies are extended and retracted by a hydraulic power pack in conjunction with hydraulic cylinders. The hydraulic power pack is located forward of the center section main spar. One hydraulic actuator is located at each landing gear. The power pack consists of a hydraulic pump, a 28-Vdc motor, a two-section fluid reservoir, filter screens, a gear selector valve and solenoid, a fluid level sensor, and a gear-up pressure switch (FIG. 9-26). For manual extension, the system has a hand-lever-operated pump.

The pump handle is located on the floor, to the left of the pedestal, in the pilot's compartment. Three hydraulic lines are routed to the nose and main gear actuators: One for normal extension and one for retraction are routed from the power pack, and one for emergency extension is routed from the hand pump. The normal extension lines and the manual extension lines are connected to the upper end of each hydraulic actuator; hydraulic lines for retraction are fitted to the lower ends of the actuators.

An internal mechanical lock in the nose gear actuator and the over-center action of the nose gear drag leg assembly lock the nose gear in the down position. Notched hook, lock link, and lock link guide attachments fitted to each main gear upper drag leg, provide positive downlock action for the main gear.

Electrical overload to the system is prevented through the use of a 60-ampere circuit breaker located under the cabin floor in the wing center section.

The landing gear hydraulic power pack motor is controlled by the use of the landing gear switch handle. LDG GEAR CONT—UP—DN appears adjacent to the switch handle on the pilot's subpanel. The switch handle must be pulled out of a detent before the handle can be moved from either the UP or the DN position.

Safety switches, called squat switches, on the main gear torque knees open the control circuit when the oleo strut is compressed. The squat switches must close to actuate a solenoid that moves the down-lock hook on the LDG GEAR CONT switch to the released position. This mechanism prevents the LDG GEAR CONT switch handle from being placed in the UP position when the airplane is on the ground. The down-lock hook disengages when the airplane leaves the ground because the squat switches close and a circuit is completed through the solenoid that moves the hook. In the event of a malfunction of the solenoid or the squat switch circuit, the down-lock hook can be overridden by pressing downward on the red DOWN LOCK REL button just left of the LDG GEAR CONT switch handle.

In flight, with the LND GEAR CONT switch handle in the DN position, as the landing gear moves to the full down position, the down-lock switches are actuated and they

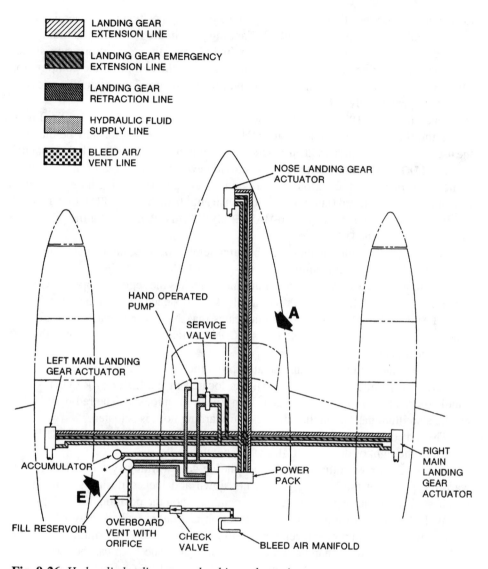

LANDING GEAR
EXTENSION LINE

LANDING GEAR EMERGENCY
EXTENSION LINE

LANDING GEAR
RETRACTION LINE

HYDRAULIC FLUID
SUPPLY LINE

BLEED AIR/
VENT LINE

NOSE LANDING GEAR
ACTUATOR

HAND OPERATED
PUMP

SERVICE
VALVE

A

LEFT MAIN LANDING
GEAR ACTUATOR

ACCUMULATOR

POWER
PACK

RIGHT
MAIN
LANDING
GEAR
ACTUATOR

E

FILL RESERVOIR

OVERBOARD
VENT WITH
ORIFICE

CHECK
VALVE

BLEED AIR MANIFOLD

Fig. 9-26. *Hydraulic landing gear plumbing schematic.* Beech Aircraft Corporation

cause the landing gear relay to interrupt current to the pump motor. When the red in-transit lights in the LDG GEAR CONT switch handle extinguish and the green NOSE L R annunciators illuminate, the landing gear is in the fully extended position.

A gear select solenoid located on the valve body of the pump is energized when the LDG GEAR CONT switch handle is in the UP position and actuates the gear select valve, allowing system fluid to flow to the up side of the system. The gear select valve is spring loaded in the down position and will only move to the up position when energized.

Hydraulic system pressure holds the landing gear in the retracted position. When the hydraulic pressure reaches approximately 1850 psi, the gear-up pressure switch will cause the landing gear relay to open and interrupt the current to the pump motor. The same pressure switch will cause the pump to activate if the hydraulic pressure drops to approximately 1600 psi.

An annunciator, HYD FLUID LOW, will illuminate whenever the hydraulic fluid is low in the fluid reservoir. Functional check of the fluid level sensor may be made by the use of the HYD FLUID LEVEL SENSOR—TEST switch located on the pilot's subpanel.

The LDG GEAR CONT switch handle should never be moved out of the DN detent while the airplane is on the ground. If it is, the landing gear warning horn will sound intermittently and the red (landing gear in-transit) lights in the LDG GEAR CONT switch handle will illuminate (provided the MASTER SWITCH is ON), warning the pilot to return the handle to the DN position.

Landing gear position is indicated by an assembly of three annunciators in a single unit that has a light transmitting cap.

One light in each segment, when illuminated, makes the segment appear green and indicates that particular gear is down and locked. Absence of illumination with the LDG GEAR CONT switch handle down indicates that the landing gear is not safe. Absence of illumination with the LDG GEAR CONT switch handle up indicates that the landing gear is up.

To check the landing gear annunciator, press the face of the annunciator.

Two red, parallel-wired indicator lights located in the LDG GEAR CONT switch handle illuminate to show that the gear is in transit or unlocked. The red lights in the handle also illuminate when the landing gear warning horn is actuated.

The red control handle lights may be checked by pressing the HDL LT TEST button located to the right of the LDG GEAR CONT switch handle.

Landing gear warning system. The landing gear warning system is provided to warn the pilot that the landing gear is not down and locked during specific flight regimes. Various warning modes result, depending upon the position of the flaps.

With the flaps in UP or APPROACH position and either or both power levers retarded below a certain power level, the warning horn will sound intermittently and the LDG GEAR CONT switch handle lights will illuminate. The horn can be silenced by pressing the GEAR WARN SILENCE button adjacent to the LDG GEAR CONT switch handle; the lights in the LDG GEAR CONT switch handle cannot be canceled. The landing gear warning system will be rearmed if the power lever(s) are advanced sufficiently.

With the flaps beyond APPROACH position, the warning horn and landing gear switch handle lights will be activated regardless of the power settings, and neither can be canceled.

Brake system. The dual hydraulic brakes are operated by depressing the toe portion of either the pilot's or copilot's rudder pedals. The series system plumbing enables braking by either the pilot or copilot.

Dual parking brake valves are installed adjacent to the rudder pedals between the master cylinders of the pilot's rudder pedals and the wheel brakes. A control for the valves, placarded PARKING BRAKE—PULL ON is located on the pilot's left subpanel. After the pilot's brake pedals have been depressed to build up pressure in the brake lines, both valves can be closed simultaneously by pulling on the parking brake handle. This retains the hydraulic pressure in the brake lines.

The parking brake is released by depressing the brake pedals briefly to equalize the hydraulic pressure on both sides of the valves, then pushing in on the parking brake handle to open the valve, releasing the hydraulic oil pressure. **Caution:** The parking brake should be left off and wheel chocks installed if the airplane is to be left unattended. Changes in temperature can cause the brakes to release or to exert excessive pressure.

Tires. The airplane is normally equipped with 8.50 x 10, 8-ply rated, tubeless, rim-inflated tires on each main gear. For increased service life, 10-ply rated tires of the same size may be installed.

Falcon 900B

Landing gear. The retractable tricycle landing gear features dual wheels on each vertical oleostrut (FIG. 9-27). The main wheels are fitted with 29 in.¥ 7.7 in. 210-mph radial tires. Main wheel tire pressure is 190 psi. The main gear retract by swinging laterally inward. The wheel wells are closed by doors with hydraulic actuators. The nosewheels are fitted with 17.5 in.¥ 5.75 in. 210 mph radial tires. The nose gear retracts by swinging forward. Nosewheel tire pressure is 130 psi. The wheel well is closed after gear retraction by mechanically operated doors. The nosewheel steering is centered automatically upon retraction.

Operation. The No. 1 hydraulic system is used for normal operation and locking of the landing gear and doors. Proximity switches are used for landing gear and door operation and for position indication. An alternate extension system utilizes No. 1 hydraulic system pressure but bypasses normal electrical sequencing. A manual extension system is provided for emergency free-fall extension of the gear. No electrical power or hydraulic pressure is required. Release handles are located in the flight deck.

Brakes and nosewheel steering. The carbon disc brakes are normally powered by the No. 1 hydraulic system and controlled by the pilot's or copilot's brake pedals. An antiskid system is provided. An independent standby brake system powered by the No. 2 hydraulic system is also operated by the pilot's and copilot's brake pedals. No antiskid protection is provided.

The parking brake system, powered by an independent No. 2 hydraulic system accumulator, can be used for emergency braking. It is able to stop the aircraft even with both hydraulic systems inoperative. It is operated by extending the parking handle. Pressure is fully modulated. The braking is not differential. The capacity of the accu-

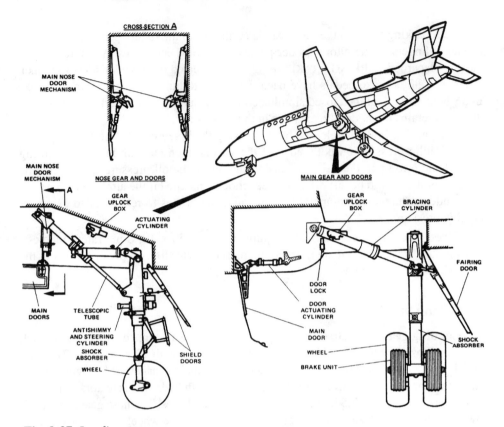

Fig. 9-27. *Landing gear.* <small>Falcon Jet Corporation</small>

mulator will permit at least five emergency applications. A warning light indicates low pressure in the accumulator.

The No. 2 hydraulic system is used for operation of the nosewheel steering system. Nosewheel deflection is obtained through a servo-valve actuated by a control wheel on the pilot side console. In the event of failure of the nosewheel steering system, a shimmy damper will center the nosewheel assembly and the aircraft can be steered with differential braking.

10
Fire protection systems

IT IS HARD TO IMAGINE ANY GREATER THREAT, OR A MORE FRIGHTENING situation, than an in-flight fire. Potential fire zones are protected by a fixed (permanent) fire protection system. A *fire zone* is an area or region of an aircraft designated by the manufacturer to require fire detection and/or fire extinguishing equipment and a proper level of inherent fire resistance.

FIRES

The National Fire Protection Association has classified fires according to three basic types:

- Class A, defined as fires in ordinary combustible materials such as wood, cloth, paper, upholstery materials, and the like.
- Class B, defined as fires in flammable petroleum products or other flammable or combustible liquids, greases, solvents, paints, and the like.
- Class C, defined as fires involving energized electrical equipment where the electrical nonconductivity of the extinguishing media is of importance. In most cases where electrical equipment is de-energized, extinguishers suitable for use on Class A or B fires can be employed effectively.

The definitions help identify the nature of fires and the types of fire extinguishers that can be used to combat them, but it is critical to realize that a given fire might encompass more than one type of fire; therefore, it is important that detection systems, extinguishing systems, and extinguishing agents be carefully considered before installation in the aircraft. For instance, fire extinguishing agents that are appropriate for Class A fires cannot be safely used on Class B or Class C fires. Agents effective on Class B or C fires will have some effect on Class A fires, but are not as effective as a Class A extinguisher.

PORTABLE EXTINGUISHERS

Water extinguishers. Water is used on Class A fires only; water must never be used on Class C (electrical) fires because water is electrically conductive. Use of a water extinguisher on an electrical fire can result in the electrocution of the extinguisher operator. A water extinguisher on Class B fires can also be counter-productive depending upon the exact type of flammable material involved. The consideration here is that water is heavier than most petroleum-based products, such as aviation fuel or jet fuel. If a water extinguisher is used on an aviation or jet fuel fire, the liquid fuel will float on top of the water and potentially spread to an even greater area. Though rarely used in aircraft, the water extinguisher can be used to extinguish smoldering fabric, cigarettes, or trash containers. The common method of deployment of a water-based fire extinguisher is through a hand-held unit.

Carbon dioxide extinguishers. This agent is specifically used to combat electrical fires. The unit consists of a long, hinged tube with a nonmetallic megaphone-shaped nozzle that permits discharge of the CO_2 gas close to the fire source to smother the fire. One significant problem associated with the CO_2 extinguisher is that it robs the area in which it is sprayed of oxygen. Excessive use in a closed pressure vessel (the cabin) can affect passengers.

Dry chemical fire extinguisher. Dry chemical can be used on Class A, B, and C fires. The agent is highly effective, but precautions must be observed. The extinguisher discharges a dense cloud of fine, nonconductive particulate matter. Discharge of this agent on the flight deck can lead to temporary, severe visibility restriction. In addition, because the agent is nonconductive, it is possible that the agent might interfere with the electrical contacts of surrounding equipment.

Halogenated hydrocarbon fire extinguishers. *Halon* is a common reference to this agent that has low toxicity, leaves no residue, and does not deprive passengers of oxygen. Portable aircraft fire extinguishers are almost exclusively halon.

ZONE CLASSIFICATIONS

Sections of an aircraft are classified by zones based upon the amount of air that flows through them.

Class A zones. A zone having large quantities of air flowing past regular arrangements of similarly shaped obstructions. The power section of a reciprocating engine is typically a Class A zone.

Class B zone. This is a zone that has large quantities of air flowing past aerodynamically clean obstructions. Heat exchanger ducts and exhaust manifold shrouds are usually of this type. Also, Class B zones are the inside of nacelles or other closures that are smooth, free of pockets, and adequately drained so leaking flammables cannot puddle. Turbine engine compartments can be considered in this class if engine surfaces are aerodynamically clean and all airframe structural formers are covered by a fireproof liner to produce an aerodynamically clean enclosure surface.

Class C zone. Zones having relatively small amounts of air flow, for instance an engine accessory compartment separated from the power section.

Class D zone. Zones having very little or no air flow, for instance wing compartments and wheel wells where little ventilation is provided.

Class X zone. Zones having large quantities of air flowing through them and that are of unusual construction making uniform distribution of the extinguishing agent very difficult. Zones containing deeply recessed spaces and pockets between large structural formers are of this type. Tests indicate agent requirements to be double those for Class A zones.

DETECTION SYSTEMS

Fire detection systems that could be used in turbine-powered aircraft include:

- Observation of crew and passengers
- Rate-of-temperature-rise detectors
- Radiation sensing detectors
- Smoke detectors
- Overheat detectors
- Carbon monoxide detectors
- Combustible mixture detectors
- Fiber-optic detectors

An underlying principle of fire detection is total automation; therefore, human observation is not relied upon as a primary method of detection. Of the eight possible methods listed, the most commonly used are: rate-of-rise, radiation sensing, and overheat detectors.

Detection requirements

Attributes of an ideal fire detection system would be:

- A system that does not cause false warnings during any flight or ground operating conditions.
- Rapid indication and accurate location of a fire.
- Accurate indication that a fire has been extinguished.
- Indication that a fire has reignited.
- Continuous indication for the duration of a fire.

- Means for electrically testing the detector system from the aircraft flight deck.
- Detectors that are lightweight and easily adaptable to any mounting position.
- Detectors that resist exposure to oil, water, vibration, extreme temperatures, and handling by maintenance personnel.
- Detector circuitry that operates directly from the aircraft power system without inverters.
- Minimum electrical current requirements when not indicating a fire.
- Each detection system should actuate a flight deck light indicating the location of the fire and an audible alarm system.
- A separate detection system for each engine.

DETECTORS

The purpose of a fire detection system is to alert the crew to the presence and location of a fire. Most systems incorporate two warning components: an audible warning bell or horn and a visual warning light. Every area of the aircraft does not contain fire detection equipment, but the detectors are located in those areas that are most susceptible to fire. Detector systems are thermal switch, thermocouple, continuous-loop detector, and continuous element.

Thermal switch detector

A thermal detector senses heat that is in excess of the preset value assigned to each individual switch. The heat-sensitive thermal switch acts as an ON/OFF switch for a flight-deck warning light. The thermal switch is normally in an open position. When the temperature surrounding the switch exceeds the programmed value, the switch closes and allows system current to pass through. Thermal switches are connected in parallel with each other but in series with the indicator lights (FIG. 10-1). If the critical switch temperature is exceeded in any section of the circuit, the thermal switch closes, which completes the circuit and illuminates the light indicating a fire or overheat situation.

A given circuit might contain any number of thermal switches. The precise number per circuit is determined by the manufacturer based upon the size of the area to be covered. Any given circuit, however, is localized to a specific site so that when the light illuminates, the crew can immediately identify the exact location of the fire; therefore, select circuits might have a number of thermal switches connected to a single light, other circuits might have a single thermal switch connected to a light.

Typically a warning light will include a push-to-test function that allows the crew to test the integrity of the circuit; depressing the warning light activates an auxiliary test circuit. Figure 10-1 also depicts a test relay. With the relay contact in the position shown, there are two possible paths for current flow from the switches to the light. Energizing the test relay completes a series circuit and checks all system wiring and the light bulb. It is important for the crew to be thoroughly familiar with the actual system operation as detailed in the pilot operating handbook because specific test functions

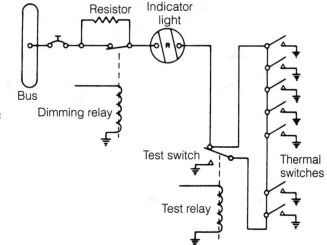

Fig. 10-1. *Thermal switch fire circuit.*

and features vary from aircraft to aircraft. The thermal switch system employs a bimetallic thermostat switch that is also known as a *spot detector*. Figure 10-2 illustrates a Fenwal spot detector.

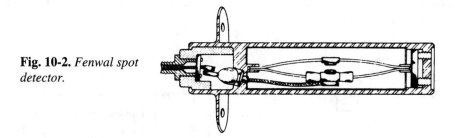

Fig. 10-2. *Fenwal spot detector.*

Fenwal spot detector. The Fenwal spot detector is wired in parallel between two complete loops of wiring as depicted in FIG. 10-3. This configuration gives the system the ability to withstand one fault, either an electrical open circuit or a short circuit, without activating a false fire alarm. In the event of an actual fire or overheat situation, the spot detector switch closes, which completes the circuit and causes alarm activation.

Thermocouple detector

So, the thermal switch operates whenever a preset temperature is exceeded; a thermocouple detector is activated when the rate of temperature rise exceeds a preset value. An important distinction is, unlike the thermal switch, the thermocouple system will not provide a warning when an engine slowly overheats or a short circuit develops.

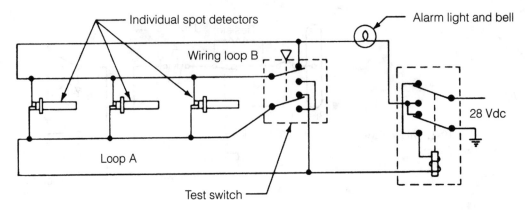

Fig. 10-3. *Fenwal spot-detector circuit.*

The thermocouple system is composed of a relay box, warning lights, and thermocouples. A wiring diagram for a typical thermocouple system is illustrated in FIG. 10-4, which depicts the detector, alarm, and test circuits. The system might have from one to eight identical circuits, depending upon the number of potential fire zones. The thermocouples control the relays that, in turn, activate respective warning lights. Any given circuit might consist of several thermocouples, each in series with the other and with the sensitive relay. The number of thermocouples employed in a given circuit depends upon the size of the fire zone and a technical limitation based upon a maximum total circuit resistance.

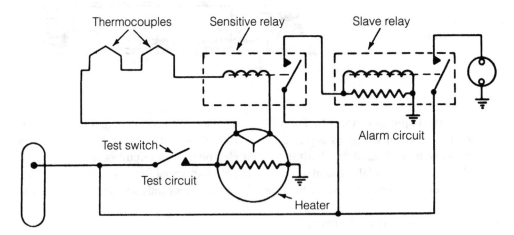

Fig. 10-4. *Thermocouple fire warning circuit.*

A thermocouple is built of two dissimilar metals that are joined together. The point where these metals are joined and will be exposed to the heat of a fire is called a *hot junction*. A *reference junction* is also enclosed in dead airspace between two insulation blocks. A metal cage surrounds the thermocouple to give mechanical protection without hindering the free movement of air to the hot junction.

If the temperature rises rapidly, the thermocouple produces a voltage because of the temperature difference between the reference junction and the hot junction. If both junctions are heated at the same rate, no voltage will result and no warning signal is given.

In case of fire, however, the hot junction will heat more rapidly than the reference junction. The ensuing voltage causes a current to flow within the detector circuit. Sufficient current will close the sensitive relay, completing a circuit from the aircraft power system to the coil of the slave relay. This, in turn, causes the relay to close and completes the circuit to the fire-warning light.

Continuous-loop detector

The continuous-loop system is a version of the thermal switch system that provides a significantly higher degree of coverage than any spot-type detector. Continuous-loop systems are heat-sensitive units that complete a circuit whenever a preset temperature is exceeded. The two most common types of continuous-loop systems are the Kidde and the Fenwal.

The *Kidde continuous-loop* system uses two wires embedded in a special ceramic core within an Inconel tube (FIG. 10-5). One of the two wires in the Kidde sensing system is welded to the case at each end of the wire and acts as an internal ground. The second wire is a hot lead with above ground potential that provides a current signal when the ceramic core material changes its resistance with a change in temperature.

The *Fenwal continuous-loop* system in FIG. 10-6 utilizes a single wire surrounded by a continuous string of ceramic beads contained in an Inconel tube. The beads are

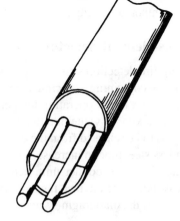

Fig. 10-5. *Kidde sensing element.*

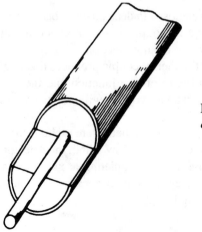

Fig. 10-6. *Fenwal sensing element.*

wetted with an eutectic salt and have a characteristic that causes them to lower their electrical resistance when the sensing element reaches its preset temperature value.

The core material in both elements prevents electrical current from flowing at normal temperatures. When the elements are exposed to increasing temperatures, the elements' core resistance drops and current flows between the signal wire and ground, which energizes the alarm system. Both systems will continuously monitor temperatures in the areas where they are installed, and both systems will automatically reset after the fire or overheat condition ends; however, there is a fundamental difference in how each system works.

In the Kidde system, the sensing elements are connected to a relay control unit. The purpose of the unit is to continuously measure the total resistance of the full sensing loop. The Kidde system is capable of sensing the average temperature around the loop as well as any single hot spot.

The Fenwal system uses a magnetic amplifier control unit. The system is non-averaging, but will activate the fire alarm if any portion of the sensing element reaches the preset alarm temperature value.

Continuous-element detector

The Lindberg fire detection system is a continuous-element detector consisting of a stainless steel tube containing a discrete element (FIG. 10-7). The discrete element absorbs gas in proportion to the operating temperature set point. When the temperature rises to the preset maximum value due to a fire or overheat condition, the heat generated causes the gas to be released from the element. Release of the gas causes the pressure in the stainless steel tube to increase. This pressure rise mechanically actuates the diaphragm switch in the responder unit, activating the warning lights and an alarm bell. A fire test switch is used to heat the sensors, expanding the trapped gas. The pressure generated closes the diaphragm switch, activating the warning system.

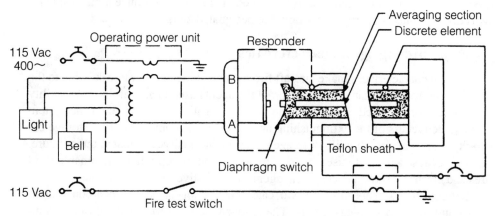

Fig. 10-7. *Lindberg fire detection system schematic.*

OVERHEAT WARNINGS

Select aircraft manufacturers use overheat warning systems to indicate excessively high temperatures in critical areas that might lead to a fire. Most common areas are the engine turbine, each engine nacelle, wheel wells, and the pneumatic manifold.

If an overheat condition occurs in an area using this detector system, the condition would be indicated on the flight deck as a flashing overheat warning light. Because the system uses a thermal switch for activation, the detectors automatically activate if the surrounding air temperature exceeds the system's preset maximum temperature. The switch contacts of the detector are on spring struts that close whenever the meter case is expanded by heat. One contact of each detector is grounded through the detector mounting bracket. The other contacts of all detectors connect in parallel to the closed loop of the warning light circuit; thus, the closed contacts of any one detector can cause the warning lights to illuminate.

When the detector contacts close, a ground is provided for the warning light circuit. Current then flows from an electrical bus through the warning lights and a flasher or keyer to ground. Because of the flasher in the circuit, the lights flash on and off to indicate an overheat condition.

EXTINGUISHER SYSTEMS

High-rate-of-discharge (HRD) extinguishing systems provide a fire extinguishing agent through short, highly pressurized feedlines using large discharge valves and outlets. Commonly, a halon extinguishing agent is employed, which is sometimes dispersed with a high-pressure, dry nitrogen (N_2) charge. An activated HRD extinguisher system discharges the agent and pressurizing gas into the designated zone in one second or less. The rapid discharge rate not only delivers the extinguishing agent, but also temporarily pressurizes the zone, which interrupts the normal flow of ventilating air, robbing the fire of its oxygen. Carefully placed agent outlets produce a high-velocity swirl that maximizes distribution within the designated zone.

Extinguishing agent characteristics

Several characteristics make aircraft fire extinguishing agents more suitable to the working environment than other agents. All the agents used in aircraft have the ability to be stored for extended periods of time without negative effect on either the system or the effectiveness of the extinguishing agent. Also in keeping with their storage capability, agents in aircraft will not freeze under any normally anticipated temperature. Finally, because agents are used in areas that could be susceptible to flammable fluid fires and electrical fires, all acceptable agents are effective in both situations.

Extinguishing agents fall into two categories: halogenated hydrocarbon (halon) agents and the inert cold-gas agents. The halogens used to form extinguishing compounds are fluorine, chlorine, and bromine. Iodine can be used but it is significantly more expensive with no advantage over the others. Halogenated agents put out fire by causing a chemical interference in the combustion process between the fuel and the oxidizer.

Inert cold-gas agents include carbon dioxide (CO_2) and nitrogen (N). Both agents are readily obtainable in either gaseous or liquid forms. The main difference is in the temperature and pressure required for storage in compact liquid form.

MANUFACTURER DOCUMENTATION

(The following information is extracted from Beech Aircraft Corporation pilot operating handbooks and Falcon Jet Corporation aircraft and interior technical descriptions. The information is to be used for educational purposes only. This information is not to be used for the operation or maintenance of any aircraft.)

Super King Air B200

Fire detection system. The fire detection system is designed to provide immediate warning in the event of fire in either engine compartment. The system consists of the following: three photoconductive cells for each engine; a control amplifier for each engine; two red warning lights on the warning annunciator panel, one placarded FIRE L ENG, the other FIRE R ENG; a test switch on the copilot's left subpanel; and a circuit breaker placarded FIRE DET on the right side panel (FIG. 10-8). The six photoconductive-cell flame detectors are sensitive to infrared radiation. They are positioned in each engine compartment so as to receive both direct and reflected rays, thus monitoring the entire compartment with only three photocells.

Conductivity through the photocell varies in direct proportion to the intensity of the infrared radiation striking the cell. As conductivity increases, the amount of current from the electrical system flowing through the flame detector increases proportionally. To prevent stray light rays from signaling a false alarm, a relay in the control amplifier closes only when the signal strength reaches a preset alarm level. When the relay

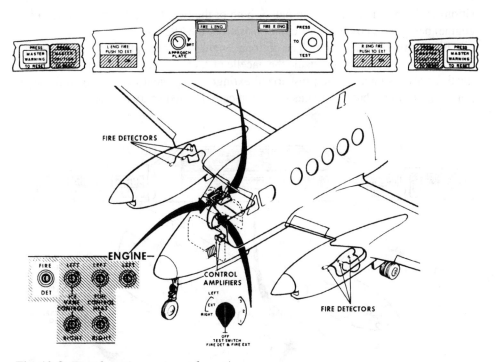

Fig. 10-8. *Fire detection system schematic.*   Beech Aircraft Corporation

closes, the appropriate left or right warning annunciators illuminate. When the fire has been extinguished, the cell output voltage drops below the alarm level and the relay in the control amplifier opens. No manual resetting is required to reactivate the fire detection system.

The test switch on the copilot's left subpanel, placarded TEST SWITCH—FIRE DET & FIRE EXT, has six positions: OFF—RIGHT EXT—LEFT EXT—3—2—1 (if the optional engine-fire-extinguisher system is not installed, the RIGHT EXT and LEFT EXT positions on the left side of the test switch will not be installed); the three test positions for the fire detector system are located on the right side of the switch (3—2—1).

When the switch is rotated from OFF (down) to any one of these three positions, the output voltage of a corresponding flame detector in each engine is increased to a level sufficient to signal the amplifier that a fire is present. The following should illuminate: the red pilot and copilot MASTER WARNING flashers, and, if the optional engine-fire-extinguisher system is installed, the red lenses placarded L ENG FIRE—PUSH TO EXT and R ENG FIRE—PUSH TO EXT on the fire-extinguisher activation switches.

The system may be tested anytime, either on the ground or in flight. The TEST SWITCH should be placed in all three positions, in order to verify that the circuitry for all six fire detectors is functional. Illumination failure of all the fire detection system annunciators when the TEST SWITCH is in any one of the three flame-detector-test positions indicates a malfunction in one or both of the two detector circuits (one in each engine) being tested by that particular position of the TEST SWITCH.

Fire extinguisher system. The optional engine-fire-extinguisher system incorporates a pyrotechnic cartridge inside the nacelle of each engine (FIG. 10-9). When the activation valve is opened, the pressurized extinguishing agent is discharged through a plumbing network that terminates in strategically located spray nozzles.

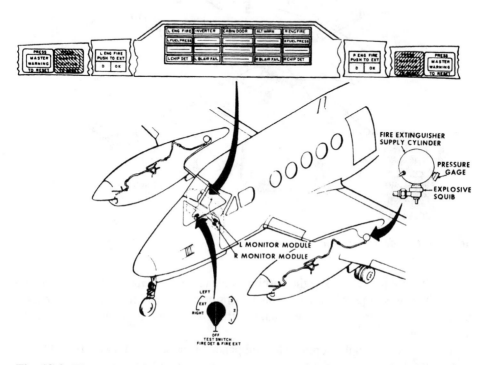

Fig. 10-9. *Fire extinguisher system schematic.*   Beech Aircraft Corporation

The fire extinguisher control switches used to activate the system are located on the glareshield at each end of the warning annunciator panel. Their power is derived from the hot battery bus. Each push-to-activate switch incorporates three indicator lenses. The red lens, placarded L (or) R ENG FIRE—PUSH TO TEST, warns of the presence of fire in the engine. The amber lens, placarded D, indicates that the system has been discharged and the supply cylinder is empty. The green lens, placarded OK, is provided only for the test function.

To discharge the cartridge, raise the safety-wired clear plastic cover and press the face of the lens. This is a one-shot system and will be completely expended upon activation. The amber D light will illuminate and remain illuminated, regardless of battery switch position, until the pyrotechnic cartridge has been replaced.

The fire-extinguisher-system test functions incorporated in the TEST SWITCH—

FIRE DET & FIRE EXT test the circuitry of the fire extinguisher pyrotechnic cartridges. During preflight, the pilot should rotate the TEST SWITCH to each of the two positions (RIGHT EXT and LEFT EXT) and verify the illumination of the amber D light and the green OK light on each fire-extinguisher-activation switch on the glareshield.

A gauge, calibrated in psi, is provided on each supply cylinder for determining the level of charge. The gauges should be checked during preflight.

Falcon 900B

Fire detection and extinguishing. A fire detection system is provided for the engines, APU, and main landing gear wheel wells (FIG. 10-10). Its continuous-loop sensors detect temperature increases, which provide both general and local overheat warnings.

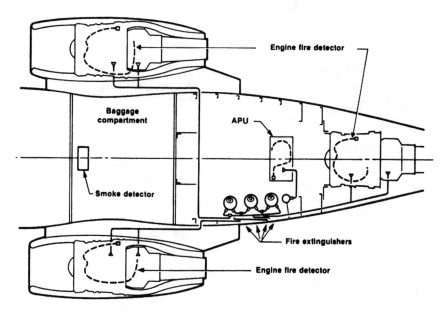

Fig. 10-10. *Fire protection.*   Falcon Jet Corporation

An optical smoke detector system is provided for the baggage compartment. A flight deck fire warning panel provides aural and visual warnings plus control of the extinguishing systems (FIG. 10-11). Four fixed dual-head extinguishers (Halon 1301) are provided for the engines. A fifth is installed for the APU and baggage compartment. Portable fire extinguishers are installed for the flight deck and passenger cabin.

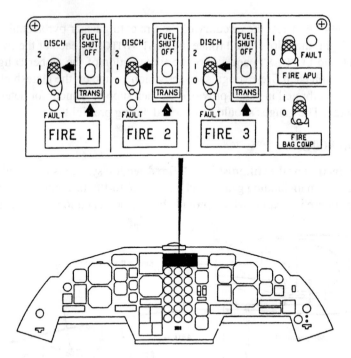

Fig. 10-11. *Fire panel.*   Falcon Jet Corporation

11
Aerodynamic control and protection systems

A PILOT TRANSITIONING FROM LIGHT, RECIPROCATING ENGINE AIRCRAFT to turboprop or turbojet aircraft will be introduced to some new concepts that accompany the basic controls found on all aircraft. It is true that aircraft have inherent stability, but that does not mean that they will always tend to fly straight and level. Stability is a function of total weight and load distribution; both are subject to the whims of the crew. In addition, aircraft speed affects the flight characteristics of the aircraft.

TABS

To offset the forces that tend to unbalance an aircraft in flight, ailerons, elevators, and rudders have auxiliary control surfaces known as *tabs*. These small, hinged controls are attached to the trailing edge of the primary control surface (FIG. 11-1). In light aircraft, where the aerodynamic forces are relatively light, tabs may be moved up or down via a flight deck crank. In faster and larger aircraft, the tabs are sometimes moved electrically. The tab balances the control forces, so the pilot does not have to fight the forces in order to maintain straight and level flight, or any other desired attitude.

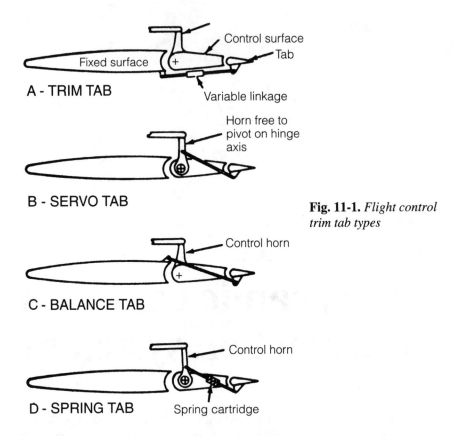

Fig. 11-1. *Flight control trim tab types*

Trim tabs. Trim tabs trim the aircraft in flight. *Trim* means correcting any tendency of the aircraft to move toward an undesirable flight attitude. Trim tabs control the balance of an aircraft so that it maintains straight and level flight without pressure on the control column, control wheel, or rudder pedals. A typical trim tab is illustrated in FIG. 11-1A. Note that the tab has variable linkage that is adjustable from the flight deck. Movement of the tab in one direction causes a deflection of the control surface in the opposite direction.

Servo tabs. As illustrated in FIG. 11-1B, servo tabs are very similar in operation and appearance to the trim tabs. Servo tabs, sometimes referred to as flight tabs, are used primarily on the large main control surfaces. They aid in moving the control surface and holding it in the desired position. Only the servo tab moves in response to movement of the flight deck control. The servo tab horn is free to pivot to the main control surface hinge axis. The force of the airflow on the servo tab then moves the primary control surface. With the use of a servo tab, less force is needed to move the main control surface.

Balance tabs. A balance tab is shown in FIG. 11-1C. The linkage is designed in such a way that when the main control surface is moved, the tab moves in the opposite direction; thus, aerodynamic forces acting upon the tab assist in moving the main control surface.

Spring tabs. The spring tab, which is illustrated in FIG. 11-1D, is similar in appearance to trim tabs, but serves an entirely different purpose. Spring tabs are used for the same purpose as hydraulic actuators—to help move a primary control surface. Various spring arrangements are used in the spring tab linkage.

On some aircraft, a spring tab is hinged to the trailing edge of each aileron and is actuated by a spring-loaded push-pull rod assembly that is also linked to the aileron control linkage. The linkage is connected in such a way that movement of the aileron in one direction causes the spring tab to be deflected in the opposite direction. This provides a balanced condition, reducing the amount of force required to move the ailerons.

The deflection of the spring tabs is directly proportional to the aerodynamic load imposed upon the aileron; therefore, at slow speeds the spring tab remains in a neutral position and the aileron is a direct, manually controlled surface. At high speeds, however, where the aerodynamic load is great, the tab helps move the primary control surface.

To reduce the force required to operate the control surfaces, the surfaces are usually balanced statically and aerodynamically. *Aerodynamic balance* is usually achieved by extending a portion of the control surface ahead of the hinge line. This utilizes the airflow about the aircraft to help move the surface. The various methods of achieving aerodynamic balance are shown in FIG. 11-2.

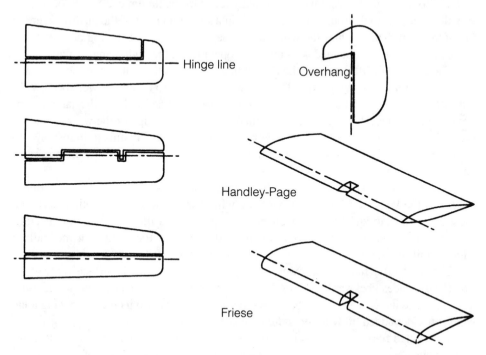

Fig. 11-2. *Three forms of aerodynamic balance.*

Static balance is accomplished by adding weight to the section in front of the hinge line until that front section weighs the same as the section behind the hinge line. When repairing a control surface, use care to prevent upsetting or disturbing the static balance. An unbalanced surface has a tendency to flutter when air passes over it.

HIGH-LIFT DEVICES

Turboprop and turbojet aircraft routinely operate at high altitudes and fast airspeeds. To achieve that type of performance, it was necessary for the engineer to design an aircraft with a relatively low-lift wing. The problem inherent in such an aircraft is that while they perform exceptionally well during cruise, the wing develops insufficient lift during takeoff and landing to sustain the weight of the aircraft; therefore, it is necessary to include a selectively deployable *high-lift device* to be used in combination with the airfoil during takeoff and landing. The high-lift device changes the lift characteristics of an airfoil during the takeoff and landing phase, and then is retracted into the wing to allow maximum performance of the airfoil during cruise operations.

Two high-lift devices that are commonly used on aircraft are illustrated in FIG. 11-3. They are known as the slot and the flap. The *slot* is a passageway cut through the leading edge of the wing. When the wing is at a high angle of attack, the air flows through the slot and smooths out the airflow over the top surface of the wing. The advantage is that at a point where the wing would normally reach its maximum coefficient of lift, the slot allows air to flow through, which results in a smoothing out of the airflow over the top of the wing. This enables the wing to pass beyond the normal stalling point without stalling. Greater lift is obtained with the wing operating at the higher angle of attack.

The other high-lift device is the *flap*, which increases the camber of the wing and therefore increases the wing's lift. This allows the aircraft to fly at a slower speed without stalling. Effects of the flap also permit a steeper glide angle in the landing approach. The flap is a hinged surface attached to the trailing edge of the wing. One flap is attached to the trailing edge of both wings. Both flaps move in unison in response to the proper flight deck control input. When the flaps are not in use, they are stored as part of the trailing edge of the wing.

Four flaps (FIG. 11-4) are *plain*, *split*, *Fowler*, and *slotted*. The plain flap is simply hinged to the wing and forms a part of the wing surface when operated. This is the most common flap and is used extensively in single- and light twin-engine aircraft. The split flap is named for the hinge at the bottom part of the wing near the trailing edge permitting the flap to be lowered from the fixed top surface.

The Fowler flap fits into the lower part of the wing, flush with the surface of the wing. The Fowler slides backward on tracks when positioned and tilts downward at the same time. All flaps increase wing camber; however, Fowler flaps also increase the wing area, which provides added lift without unduly increasing drag.

The slotted flap is similar to the Fowler flap in operation, but similar to the plain flap in appearance. This flap is equipped with tracks, rollers, or hinges of a special design. During operation, the flap moves downward and rearward away from the position

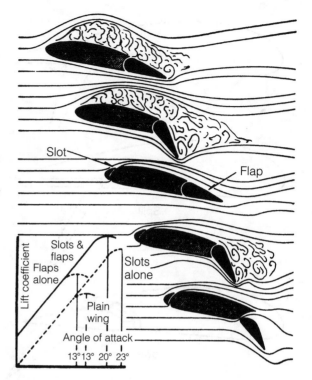

Fig. 11-3. *High-lift devices*

of the wing. The opened slot allows a flow of air over the upper surface of the flap to streamline the airflow and to improve the efficiency of the flap.

BOUNDARY-LAYER CONTROL DEVICES

Boundary layer control devices provide additional means of increasing the maximum lift coefficient of an airfoil section. Air encounters friction when moving over the surface of an airfoil. Airflow speed gradually decreases closer to the surface until the airflow is almost at a standstill directly next to the surface. The depiction of this airflow in FIG. 11-5 indicates a very thin boundary layer where the flow occurs in laminations of air sliding smoothly one over the other; hence, the name *laminar boundary layer*.

Due to the friction forces in the boundary layer, airstream energy dissipates as the airflow moves past the wing's surface. This constant decrease in the velocity of the airstream results in an increase in the thickness of the laminar boundary layer; therefore, there is a direct relationship between distance aft of the wing leading edge and the thickness of the boundary layer. At some point, the laminar flow no longer has sufficient energy to maintain a smooth flow and the flow begins a destabilizing oscillatory disturbance as depicted in Figure 11-5. The disturbance manifests itself as a waviness in the airflow that progressively worsens the farther back it gets from the leading edge. Ultimately, this results in the destruction of the smooth laminar flow;

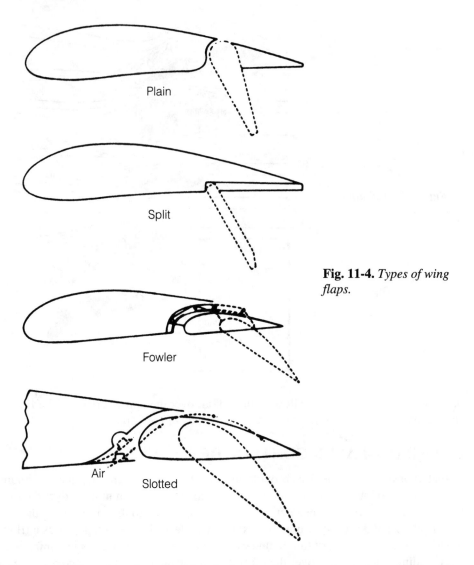

Fig. 11-4. *Types of wing flaps.*

thus, a transition takes place in which the laminar boundary layer decays into a turbulent boundary layer.

The onset of this condition, the point where the disturbance occurs over the wing, can be shortened by increasing the wing's angle of attack, which ultimately results in a *stall* condition and lift is no longer being created by a portion of the wing, if not the complete wing. The same sort of transition can be noticed in the smoke from a cigarette in still air. At first, the smoke ribbon is smooth and laminar, then develops a definite waviness, and decays into a random turbulent smoke pattern.

Boundary layer control devices, which control the airflow's kinetic energy, maintain a high velocity airflow in the boundary layer. This results in delayed separation of

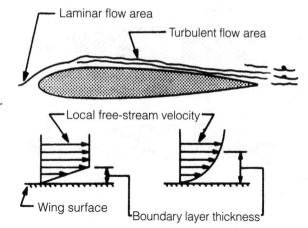

Fig. 11-5. *Boundary layer characteristics.*

the airflow. Some of the methods employed include the use of slots, slats, upper surface suction to draw off stagnant air, flap augmentation, and the re-energization of the boundary layer with high-velocity air.

Figure 11-6 illustrates several boundary layer control options. The *fixed slot* is merely an opening in the forward portion of the wing that permits higher energy air to flow through the wing to energize the boundary layer at high angles of attack. This method is simple, inexpensive, and effective.

An even more effective method is the movable control surface known as *slats* that attach to the leading edge of the wing. A closed slat forms the leading edge of the wing; when the slat is extended forward, a slot is created between the slat and the wing leading edge. As a result, high-energy air is directed to the boundary layer on the upper surface of the wing. This form of boundary layer control improves slow-airspeed handling characteristics, which allows the pilot greater lateral control of the aircraft at airspeeds otherwise slower than normal landing speed.

A third method of controlling the boundary layer to allow the wing to operate at higher angles of attack is *surface suction*. This method is an alternative to using a slot and has essentially the same effect. The suction draws the boundary layer down toward the surface of the wing and in the process helps maintain the energy of the airflow.

Finally, the boundary layer can also be controlled by injection of *high-pressure engine bleed air*. This method employs a narrow orifice immediately in front of the wing flap's leading edge (FIG. 11-6). Upon flap deployment, the orifice is exposed to emit high-energy airflow across the top of the flap, which re-energizes the boundary layer and helps maintain the laminar flow condition.

VORTEX GENERATORS

Many turbojet aircraft have vortex generators (FIG. 11-7). The small, vertical airfoils fulfill different purposes depending upon where they are located. Rows of vortex generators placed on the upper surface of a wing, just ahead of the ailerons, delay the on-

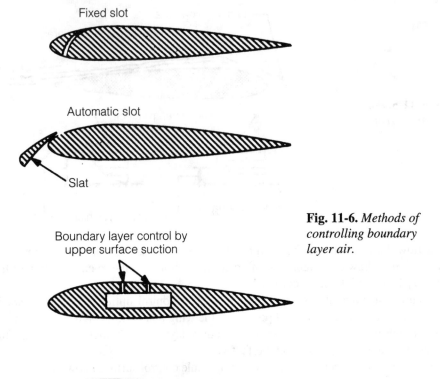

Fixed slot

Automatic slot

Slat

Fig. 11-6. *Methods of controlling boundary layer air.*

Boundary layer control by upper surface suction

Boundary layer control by flap augmentation

set of drag divergence at fast speeds. This has the effect of maintaining aileron effectiveness at fast operating speeds.

Vortex generators located on both sides of the vertical fin just upstream of the rudder prevent flow separation over the rudder during extreme angles of yaw that are attained only when rudder application is delayed after an engine loss at very slow speeds.

Finally, when rows of vortex generators are located underneath the horizontal stabilizer (occasionally on the upper surface) just upstream of the elevators, the generators are intended to prevent flow separation over the elevators at very slow speeds.

Synopsis: Vortex generators on wing surfaces improve fast-speed characteristics; vortex generators on tail surfaces improve slow-speed characteristics.

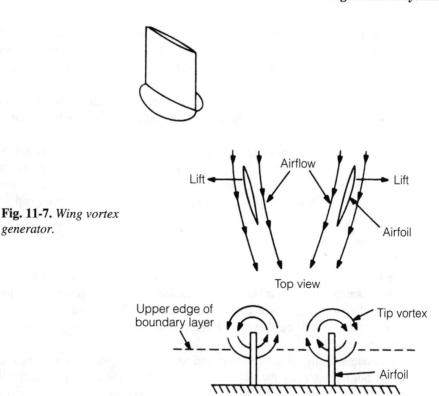

Fig. 11-7. *Wing vortex generator.*

FLIGHT CONTROL SYSTEMS

For the most part, the aerodynamic controls associated with turbine-powered aircraft are similar to those of reciprocating engine aircraft. The biggest difference is the autopilot and navigation equipment. Full examination of the differences is not within the scope of this book; however, a few observations are worth noting.

Hydraulically operated control systems. Aircraft with cruise speeds of approximately 300 knots and faster develop significant air loads on the control surfaces that are difficult for the pilot to overcome when operating the controls without mechanical advantage. As a result, aircraft in this category will typically employ hydraulically operated flight controls. Conventional cable or pushrod systems are installed in the aircraft as usual, but tied into a power transmission quadrant. When the system is activated, the pilot's control inputs do not go directly to individual control surfaces. Instead, the inputs open and close hydraulic valves that direct hydraulic fluid to individual actuators. The actuator moves the control surface to the requested position. To the pilot, there appears to be little, if any, difference between direct and indirect control.

There are two primary methods of providing for hydraulic control system failure. The first method connects the rod from the flight deck controls across the power transmission quadrant directly to the control surface. During normal hydraulic operation, the rod inputs to the power transmission quadrant, which boosts rod movement, which moves the control surfaces. In the event of hydraulic failure, manual reversion occurs and the pilot directly moves the control surfaces when manipulating the flight deck controls.

The second method of dealing with possible hydraulic failure is redundancy. Aircraft that do not have a manual reversion system will have a backup hydraulic system to take over in the event that the primary system fails. Select aircraft have three sources of hydraulic power: primary system, backup system, and auxiliary system.

PRECIP PROTECTION

Protection from the meteorological elements has been one of life's most fundamental challenges since the dawn of humanity. The history of transportation is littered with minor and major tragedies related to the elements. When the elements run amok with an aircraft, the results can severely impair, if not completely negate, its aerodynamic characteristics.

A solid understanding of meteorology and weather phenomenon is essential for all flight crews, but occasionally the best made plans deteriorate and the aircraft will be exposed to a hostile environment. How ice and rain protection systems work, and what they can, and cannot do for the pilot, is critical knowledge.

Icing

Of all the elements, ice is one of the most versatile destroyers of aircraft performance and handling characteristics. Two types of in-flight icing are *rime* and *glaze*. Rime ice forms a rough surface on the aircraft leading edges. The surface of rime ice is rough because the temperature is very cold and a droplet of water contacts the surface and freezes immediately, before spreading out. Conversely, glaze ice is smooth and forms a thick layer over the leading edges because the temperature is slightly below freezing, which allows the water droplet to hit the surface and spread out before freezing.

Icing conditions can be expected anywhere moisture is visible and the temperature is very near or below freezing. Even a small amount of structural icing can have a significant effect on the aerodynamic qualities of an airfoil. Ice accumulations can totally destroy the lift characteristics of the airfoil. The buildup of ice on an airframe increases drag and simultaneously decreases lift. The ice might cause vibrations as the air flows past, and when ice forms on antennae and other protrusions from the airframe, vibrations might be so severe than antennae subsequently break off the airframe. Additionally, radio antenna icing problems might interfere with communication and navigation equipment.

Control surfaces can freeze in one position or, if they have freedom of travel, might have a serious imbalance that causes destructive vibration. Fixed slots fill with ice; movable slots freeze in one position. Engine-air inlets can freeze, causing engine performance to degrade or causing the engine to flame-out. And, finally, ice on a windshield will severely reduce forward visibility making collision avoidance scanning techniques impossible and landing approaches very difficult, at best.

In-flight icing conditions can be dealt with three ways. First, avoidance: Even when flying aircraft certified for flight into known icing conditions, always critically analyze the situation. (Change course to avoid reported icing. Make pilot reports to assist other pilots flying through the same airspace and conversely rely upon other pilot reports about icing when you are flying through an area that might have icing conditions.) Second: Utilize anti-icing systems to prevent ice accumulation. Third: Utilize deicing systems to remove ice that has already formed.

Coping with ice

A number of methods are used to either prevent or control the formation of ice on aircraft: hot-air heating, electrical heating, mechanically breaking up ice formations, and alcohol spray. These methods can be employed three ways. A surface might be anti-iced by keeping the surface dry by heating the surface to a temperature that evaporates water upon impingement; or by heating the surface just enough to prevent freezing; or the surface might be deiced by allowing ice to form and then removing it.

Depending upon the surface, one or more methods can be employed.

Location of Ice	Method of Control
Leading edge of the wing	Pneumatic, thermal
Leading edges of vertical and horizontal stabilizers	Pneumatic, thermal
Windshields, windows, and radomes	Electrical, alcohol
Heater and engine air inlets	Electrical
Stall warning transmitters	Electrical
Pitot tubes	Electrical
Flight controls	Pneumatic, thermal
Propeller blade leading edges	Electrical, alcohol
Lavatory drains	Electrical

Pneumatic deicing systems

Deicing boots are standard equipment on all-weather, propeller-driven aircraft. These inflatable rubber deicers are attached to the leading edges of wings, and in some cases, horizontal stabilizers. The deicers are composed of a series of inflatable tubes that lie flat against the leading edge when not in use (FIG. 11-8). When activated, the tubes are alternately inflated with pressurized air, and deflated. This alternating inflation cycle causes the ice to crack and break off. The air stream carries the broken ice away.

Deicing system not
operating. Cells lie close
to airfoil section. Ice is
permitted to form.

Flexible hose

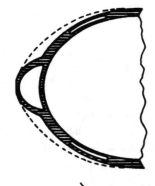

After deicer system has
been put into operation,
center cell inflates,
cracking ice.

Fig. 11-8. *Deicer
boot inflation cycle.*

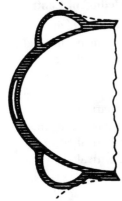

When center cell deflates,
outer cells inflate. This raises
cracked ice causing it to
be blown off by air stream.

In turbine-powered aircraft, pressurized air is provided by compressor bleed air.
The boot inflation sequence is controlled by a centrally-located distributor valve or,
depending upon the manufacturer, by a solenoid-operated valve located adjacent to the
deicer air inlet.

Rather than have a single deicer boot run the entire length of the leading edge of a
wing, they are installed in sections. Each section inflates and deflates separately from

the other sections to minimize the aerodynamic disturbance to the airflow that would result from completely changing the shape of the entire leading edge of the wing at once. Nevertheless, most operating handbooks still caution against using leading edge deice on approach to landing because approach airspeed is comparatively slow: a bad time to decrease the aerodynamic efficiency of the wing.

Deicer boot construction. Deicer boots are constructed with a number of considerations in mind. Built of a soft, pliable rubber or rubberized fabric, they contain tubular air cells. The outermost ply of the deicer is made of conductive neoprene for two reasons. First, excellent resistance to the elements and chemicals encountered in normal service. Second, neoprene is a conductive surface that will dissipate static electricity charges. (If electrical charges accumulated and discharged through the boot to the metal skin underneath, the discharge would cause static interference with the radio equipment.)

Early versions of deicer boots were secured to the leading edges of wing and tail surfaces with cement, fairing strips, and screws, or a combination of them. New versions use a special bonding agent (glue) that secures the boot to the surface. Top and bottom trailing edges of the boot are tapered to provide a smooth airfoil. Newer boots are also lighter in weight and aerodynamically cleaner when not in use (FIG. 11-9).

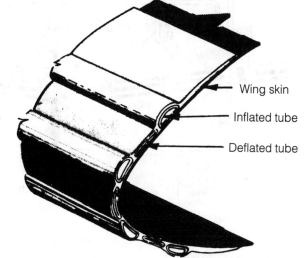

Fig. 11-9. *Deicer boot cross section.*

Wing skin

Inflated tube

Deflated tube

Deicer boot cells are connected to a source of pressure and vacuum by nonkinking, flexible hose. Major components of the deicer boot system are:

- Deicer boots
- Pressurized air source
- Oil separator
- Air pressure and suction relief valves

AERODYNAMIC CONTROL AND PROTECTION SYSTEMS

- Pressure regulator and shutoff valves
- Inflation timer
- Distributor valve or control valve

A typical deicer boot system schematic is depicted in FIG. 11-10. Air pressure for the system in FIG. 11-10 is supplied by bleed air tapped off the turbine engine's compressor. The pressurized air goes from the compressor to the pressure regulator that reduces the air pressure as necessary to operate the deicer system. The vacuum required to deflate the boots, and hold them flat when not in use, is provided by an ejector located downstream from the regulator.

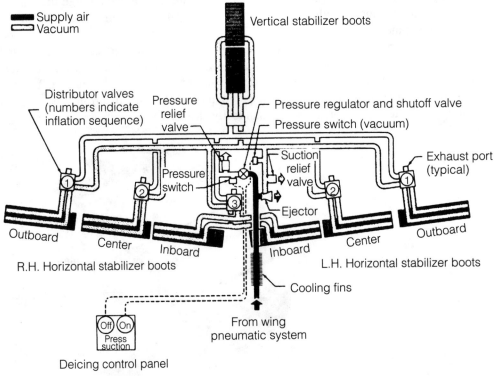

Fig. 11-10. *Schematic of a pneumatic deicing system.*

The air pressure and suction relief valves, and the regulators, maintain the pneumatic system pressure and suction at the desired settings. The timer is composed of a series of switch circuits actuated successfully by a solenoid-operated rotating step switch. The timer is energized when the deicing switch is placed in the ON position.

When the crew activates the system, the deicer port in the distributor valve is closed to vacuum. System operating pressure is then applied to the deicers that are connected to that port. At the end of the inflation cycle, the deicer pressure port is shut off, and the air in the deicer flows overboard through the exhaust port. When the air flowing from the deicers reaches a sufficiently low pressure, the exhaust port closes. Vacuum is then reapplied to exhaust the remaining air from the deicer. This cycling continues for as long as the system is activated. When the crew shuts off the system, the timer automatically returns to its starting position.

Deicer boot preventive maintenance. The life of the deicers can be greatly extended by observing these simple rules:

- Do not drag fuel hoses over the deicers.
- Keep deicers free of fuel, oil, grease, dirt and other deteriorating substances.
- Do not lay tools on deicers or lean equipment against them.
- Promptly have deicers repaired or resurfaced when abrasion or deterioration is discovered.

Thermal anti-icing systems

Two types of thermal systems are hot air and electrically heated elements. Thermal systems that are used to prevent the formation of ice—or to deice airfoil leading edges—are usually heated air ducts that run spanwise across the inside of the leading edge. Turbine-powered aircraft typically use bleed-air heat to provide the heated air necessary for operation. Some aircraft, however, use electrically heated elements for anti-icing and deicing airfoil leading edges.

When protection prevents the formation of ice, heated air is supplied continuously to the leading edges as long as the anti-icing system is operating. When protection is supposed to deice the leading edges, much hotter air is supplied for shorter periods on a cyclic system. Select aircraft even provide an automatic temperature control that mixes hot and cold air to maintain a predetermined temperature range.

Certain installations have a series of valves that permit the crew to control the hot-air flow and conserve bleed air in the event of an engine failure. Or the valve system might channel all the hot air to a specific location in the event of a critical buildup of ice. Those portions of the airfoil that must be protected from the formation of ice are typically provided with a closely spaced double skin (FIG. 11-11). This system carries heated air through ducting that allows the hot air to directly heat the leading edge from behind, which is sufficient to prevent ice formation. The hot air is then dumped overboard through a wing tip exhaust or at points where ice formation could be critical, for instance the leading edge of a control surface.

Propeller ice control systems

Ice formation on propeller blades causes the same problems as on any other airfoil with the compounding factor that the propeller is rotating at a high rate of speed. Be-

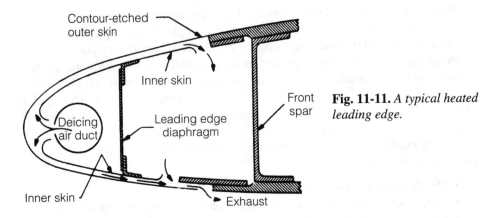

Fig. 11-11. *A typical heated leading edge.*

cause of this, ice generally forms unsymmetrically on a propeller blade, which results in prop imbalance. If severe enough, the resulting vibration can cause major structural damage to the propeller and engine. Two primary methods of controlling propeller ice are fluid and electrothermal.

A typical fluid system is illustrated in FIG. 11-12. The system includes an anti-icing fluid supply tank from which fluid travels through a filter, then to a fluid pump; the pump flow rate can be controlled by the crew from the flight deck to accommodate the severity of ice accretion. Fluid is transferred from a stationary nozzle on the engine nose case into a circular U-shaped channel, also called a *slinger ring*, that is mounted on the rear of the propeller assembly. Centrifugal force supplies the necessary pressure to cause the fluid to leave the nozzles and spray onto each blade shank. Anti-icing fluids include the very flammable isopropyl alcohol and a less flammable but more expensive phosphate compound.

The second method of dealing with propeller icing is the electrical deicing system. Figure 11-13 illustrates a typical electrical deicing system: an electrical energy source, a resistance heating element, system controls, and necessary wiring. The heating elements can be mounted on the propeller spinner and blades, either internally or externally, depending upon the manufacturer's preference. Electrical power from the aircraft system is transferred to the propeller hub through electrical leads, which terminate in slip rings and brushes. Flexible connectors are used to transfer power from the hub to the blade elements.

Icing control is accomplished by converting electrical energy to heat energy in the heating element. Balanced ice removal from all blades must be obtained as nearly as possible if excessive vibration is to be avoided. To obtain balanced ice removal, variation of heating current in the blade elements is controlled so that similar heating effects are obtained in opposite blades.

This type of electrical deicing system is usually designed for intermittent application of power to the heating elements, which removes ice after it forms, but prior to excessive accumulation. Proper control of heating intervals helps prevent runback because heat is applied just long enough to melt the ice face in contact with the blade.

330

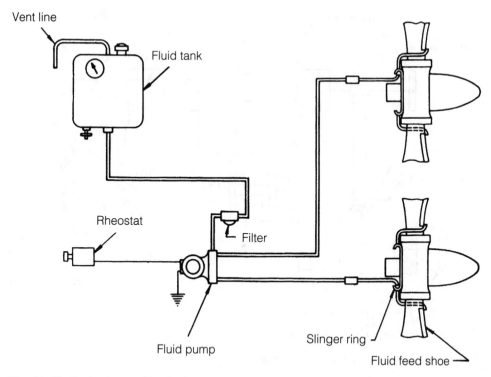

Vent line

Fluid tank

Rheostat

Filter

Fluid pump

Slinger ring

Fluid feed shoe

Fig. 11-12. *Typical propeller fluid anti-icing system.*

Runback is a condition that occurs when the heat applied to the surface is greater than the heat required to melt the inner ice face, but insufficient to evaporate all the water that is formed. The result is that the water will run back over the unheated surface and freeze. Runback of this nature causes ice formation on icing areas of the blade or surface that are not subject to ice control.

Cycling timers energize the heating element circuits for periods that are between 15 and 30 seconds. Normally, a complete cycle of all elements will take approximately two minutes.

Windshield icing control systems

Beyond leading edge wing surfaces and the propeller, the windshield is another important area to keep free of ice. Even a small layer of frost or ice can dramatically reduce, or eliminate, forward visibility. Scanning for other aircraft and obstructions, especially during approach and landing, becomes practically impossible without forward visibility.

Window anti-icing, deicing, defogging, and demisting systems keep window areas free of ice and frost. The systems vary according to the type of aircraft and its manufacturer. Select windshields are built with double panels that have a space between

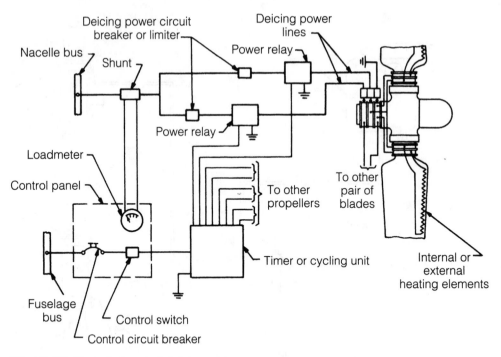

Fig. 11-13. *Typical electrical deicing system*

them, which permits the circulation of heated air between the surfaces to control icing and fogging. Other aircraft use windshield wipers and anti-icing fluid that is sprayed on.

One of the more common methods for controlling ice formation and fog on aircraft windows is the use of an electrical heating element built into the window. When this method is used on a pressurized aircraft, a layer of tempered glass gives strength to withstand pressurization. A layer of transparent conductive material is the heating element and a layer of transparent vinyl plastic adds a nonshattering quality to the window. The vinyl and glass plies are bonded by the application of pressure and heat (FIG. 11-14). The bond is achieved without cement because vinyl has a natural affinity for glass. The conductive coating dissipates static electricity from the windshield in addition to providing the heating element.

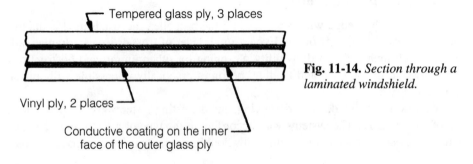

Fig. 11-14. *Section through a laminated windshield.*

On some aircraft, thermal electric switches automatically turn the system on when the air temperature is low enough for icing or frosting to occur. The system might stay on all the time during such temperatures, or on select aircraft the system might operate with a pulsating on-and-off pattern. Thermal overheat switches automatically turn the systems off in case of an overheating condition that could damage the transparent area.

A typical electrically heated windshield system schematic is illustrated in FIG. 11-15 and includes the following components:

- Windshield auto-transformers and heat control relays
- Heat control toggle switches
- Indicating lights
- Windshield control units
- Temperature-sensing elements (thermistors) laminated in the panel

The specific operation of this system is complicated and not of particular concern to the pilot, but several problems are associated with electrically heated windshields that should be understood. The primary problems include delamination, scratches, arcing, and discoloration.

Delamination. This is a separation of the windshield plies. Delamination is undesirable, but not a major problem. The two considerations are that the existing delamination falls within the manufacturer's established guidelines, and the separation does not block or restrict the crew's vision.

Arcing. When a windshield panel starts arcing, that usually indicates a breakdown in the conductive coating. Where chips or minute surface cracks are formed in the glass plies, simultaneous release of surface compression and internal tension stresses in the highly tempered glass can result in the edges of the crack and the conductive coating parting slightly. Arcing is produced where the current jumps this gap, particularly where these cracks are parallel to the window bus bars. Where arcing exists, there is invariably a certain amount of local overheating that, depending upon its severity and location, can cause further damage to the panel. Arcing in the vicinity of a temperature-sensing element is a particular problem because it could upset the heat control system.

Discoloration. Electrically heated windshields are transparent to directly transmitted light, but they have a distinctive color when viewed by reflected light. This color will vary from light blue, yellow tints, and light pink depending upon the manufacturer of the window panel. Normally, discoloration is not a problem unless it affects the optical qualities of the window.

Scratches. Windshield scratches are more prevalent on the outer glass ply where the windshield wipers are indirectly the cause of this problem. Any grit trapped by a wiper blade can convert the blade into an extremely effective glass cutter when the wiper is set in motion. The best solution for controlling windshield scratches is prevention; clean the windshield wiper blades as frequently as possible. Incidentally, windshield wipers should never be operated when dry because that increases the chances of damaging the surface.

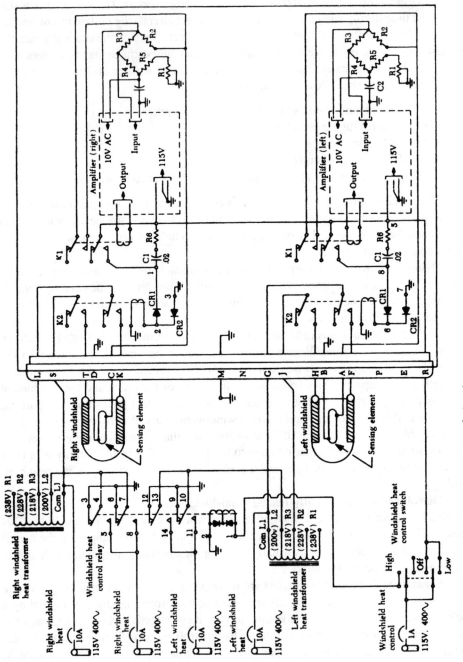

Fig. 11-15. *Windshield temperature control circuit.*

Rain elimination systems

Three ways to remove rain from a windshield in flight are wipe it off, blow it off, or use a chemical rain repellent.

Electrical windshield wiper systems. This system—very similar to the automotive windshield wiper—is powered by an electric motor. Larger aircraft might have dual systems, one for the captain and one for the first officer. Figure 11-16 depicts a typical electrical windshield wiper installation.

The system depicted in the illustration incorporates a single wiper on each of the two front windshields. Each wiper is driven by a motor-converter assembly that changes the rotary motion of the motor to the reciprocating motion required to operate the wiper. A wiper control switch is located in the flight deck that allows the crew to select the wiper speed most appropriate for the conditions. If the crew places the switch in the HIGH position, relays 1 and 2 are energized (FIG. 11-17). With both relays energized, field 1 and field 2 are energized in parallel. The circuit is completed and the motors operate at an approximate speed of 250 strokes per minute.

When the LOW position is selected, relay 1 is energized. This causes field 1 and field 2 to be energized in series. The motor then operates at approximately 160 strokes per minute. Setting the switch to the OFF position allows the relay contacts to return to their normal positions; however, the wiper motor will continue to run until the wiper arm reaches the PARK position. When both relays are open and the park switch is closed, the excitation to the motor is reversed, causing the wiper to move off the lower edge of the windshield, opening the cam-operated park switch. This de-energizes the motor and releases the brake solenoid. This ensures that the motor will not coast and reclose the park switch.

Hydraulic windshield wiper systems. This system is operated by the main hydraulic system. Figure 11-18 is a functional schematic of a typical hydraulic windshield wiper system. The speed control valve, which is a variable restrictor, controls the wiper's speed and its start and stop functions. Rotating the valve handle controls the flow of hydraulic fluid to the system and, as a result, the speed of the wipers. The control unit directs the flow of hydraulic fluid to the wiper actuators and returns fluid discharge from the actuators to the main hydraulic system. The control unit also alternates the direction of hydraulic fluid flow to each of the two wiper actuators. The wiper actuators convert hydraulic energy into reciprocating motion to drive the wiper arms back and forth.

Pneumatic rain removal systems. Two fundamental problems are associated with windshield wipers. The first deals with the slipstream aerodynamic forces that act on the wipers. The aerodynamic forces tend to reduce the wiper blade loading pressure on the window, which results in streaking. The second problem is achieving wiper oscillation that is fast enough to keep up with fast rain impingement rates during heavy rainfall. Most wiper systems fail to provide adequate vision in heavy rain.

Select turbine-powered aircraft use a pneumatic rain removal system that uses high pressure, high temperature compressor bleed air. The air is blown across the

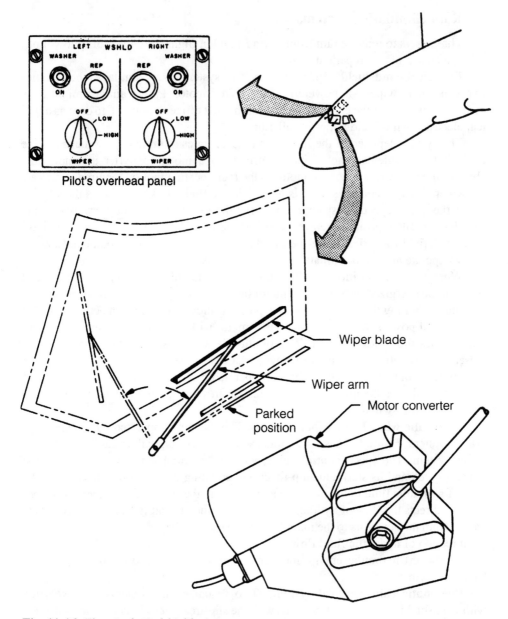

Pilot's overhead panel

Wiper blade

Wiper arm

Parked position

Motor converter

Fig. 11-16. *Electrical windshield wiper system.*

windshield forming a barrier that prevents raindrops from ever striking the windshield surface (FIG. 11-19).

Windshield rain repellent. Water poured onto clean glass spreads evenly across the glass. Even if the sheet of glass is held upward at an angle, it will remain wetted by

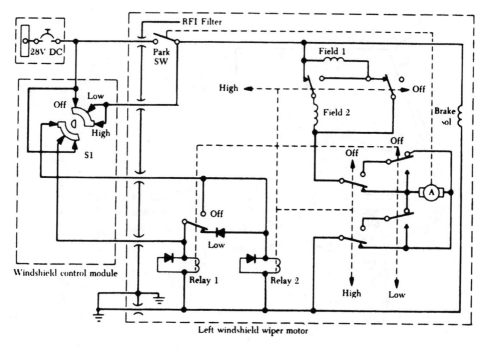

Fig. 11-17. *Windshield wiper circuit diagram.*

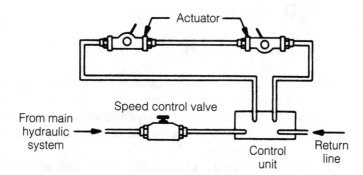

Fig. 11-18. *Hydraulic windshield wiper schematic.*

a thin film of water; however, certain chemicals do leave a uniform, transparent film on glass that chemically makes the water behave similarly to mercury. When the water contacts a surface that is coated with these chemicals, the water draws up into beads, leaving large areas of dry glass in between. The beaded water is easily removed from the glass by the natural airstream.

A rain repellent system permits application of the chemical repellent by a switch or push button on the flight deck. The proper amount of repellent is applied regardless of how long the switch is held. The repellent is marketed in pressurized disposable

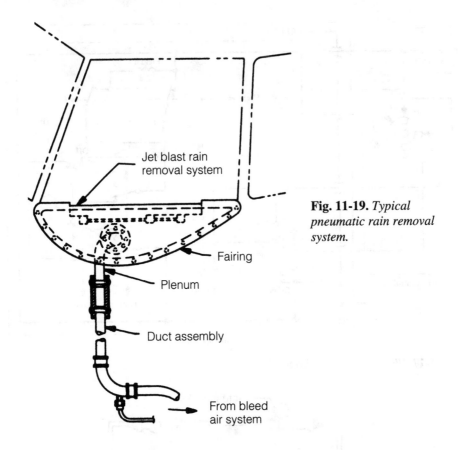

Fig. 11-19. *Typical pneumatic rain removal system.*

cans that screw into the aircraft system and provide the propelling force for application. The control switch opens an electrically-operated solenoid valve that allows repellent to flow to the discharge nozzles. The liquid repellent is squirted onto the exterior of the windshield and raindrops become the carrying agent to distribute the chemicals over the windshield surface.

The rain repellent system should not be operated on dry windows because heavy undiluted repellent will restrict window visibility. If the system is operated inadvertently, do not operate the windshield wipers or rain clearing system because that increases smearing. Also, the rain repellent residues caused by application in dry weather or very light rain cause staining or minor corrosion of the aircraft skin. To prevent this, any concentrated repellent or residue should be removed by a thorough freshwater rinse at the earliest opportunity.

After application, the repellent film slowly deteriorates with continuing rain impingement. This makes periodic reapplication necessary. The length of time between applications depends upon rain intensity, the type of repellent used, and windshield wiper usage.

MANUFACTURER DOCUMENTATION

(The following information is extracted from Beech Aircraft Corporation pilot operating handbooks and Falcon Jet Corporation aircraft and interior technical descriptions. The information is to be used for educational purposes only. This information is not to be used for the operation or maintenance of any aircraft.)

Falcon 50

Primary flight controls. Elevator, rudder, and ailerons are operated by a system of push-pull rods and hydraulic servo-actuators (FIG. 11-20 and FIG. 11-21). The removable hydraulic servo actuator consists of dual barrels sliding on two pistons attached to the structure. Each barrel is powered by one of two hydraulic systems. The loss of pressure in one hydraulic system has no effect on flying characteristics. Even with a loss of pressure in both hydraulic systems, the aircraft is capable of manual flight up to 260 K (.76 Mach).

On all three linkages, *artificial feel units* (AFU), eliminate control feedback. On the aileron and elevator linkages, hydraulically operated "Arthur-Q" unit bellcranks, slaved to airspeed, increase forces as airspeed increases. A nonlinear bellcrank "Amedee" reduces aileron deflection near the neutral position. Each servo actuator also includes a return spring (secondary AFU) to keep the actuator in neutral position if the linkage becomes disconnected.

Rudder and aileron trims are electric. Adjustment of the neutral point of the AFU on the respective linkages is provided by pedestal-mounted switches.

Pitch trim is obtained by adjustment of the horizontal stabilizer around a hinge point by means of a wormscrew jack driven by two electrical motors. The normal mode is operated by rocker switches on either control yoke. Emergency operation is controlled by a toggle switch on the pedestal. Stabilizer movement is indicated to the pilot by an aural signal. A warning light indicates when the aircraft is not in proper takeoff configuration. Ailerons, rudder, and pitch trim positions are indicated on the instrument panel. The rudder servo is supplemented by a gust damper.

An electrical, emergency aileron drive motor is tied to the aileron linkage on the left side. An automatic Mach trim system operates between Mach .78 and Mach .86 but is not required for flight.

Secondary flight controls. Airbrakes may be extended throughout the complete flight envelope of the airplane (FIG. 11-21). An AIRBRAKES light is lit when airbrakes are extended. Flaps are extended to 20° and 48° by screw jacks driven by a hydraulic motor. The leading-edge slats are deployed automatically when the flap control handle is selected for flaps. In addition to normal modes of operation, the leading-edge slats are automatically actuated in certain conditions—in a clean configuration at angles of attack of 17° and 19° when indicated airspeed is below 265 KIAS. Automatic extension and retraction of slats are controlled by two stall vanes, located on either side of the forward fuselage, that measure the angle of attack.

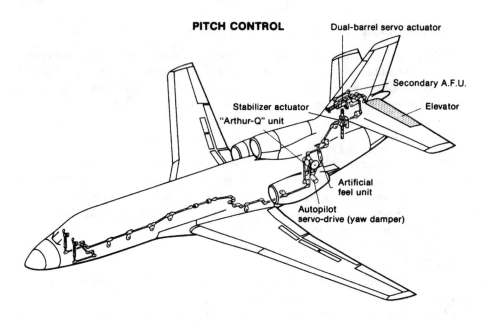

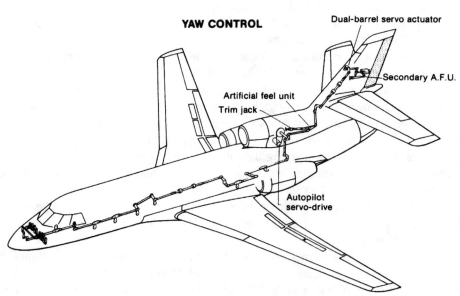

Fig. 11-20. *Aerodynamic control: pitch and yaw.* Falcon Jet Corporation

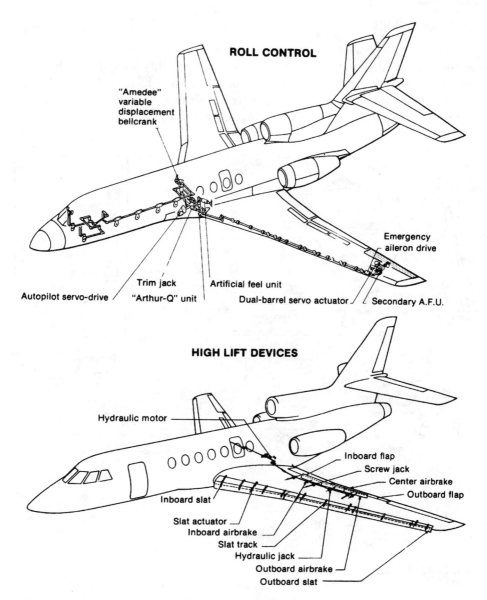

ROLL CONTROL

"Amedee" variable displacement bellcrank

Emergency aileron drive

Trim jack

Artificial feel unit

Autopilot servo-drive "Arthur-Q" unit Dual-barrel servo actuator Secondary A.F.U.

HIGH LIFT DEVICES

Hydraulic motor

Inboard flap
Screw jack
Center airbrake
Outboard flap

Inboard slat

Slat actuator
Inboard airbrake
Slat track
Hydraulic jack
Outboard airbrake
Outboard slat

Fig. 11-21. *Aerodynamic control: roll and high-lift devices.* Falcon Jet Corporation

Falcon 900B

Ice and rain protection. The Falcon 900 is certified with an ice protection system permitting safe flight in icing conditions. This capability is retained with one engine inoperative. Engine bleed air is used for thermal anti-icing of the wing leading edges, engine-air intakes, the heat exchanger ram air intake and center engine S-duct (FIG. 11-22). Airflow is monitored by pressure switches.

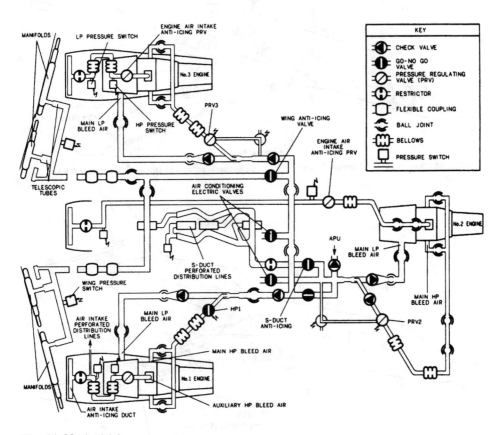

Fig. 11-22. *Anti-icing.*   Falcon Jet Corporation

The horizontal and vertical stabilizers do not require anti-ice protection. Pitot and static probes, angle-of-attack sensors, and the outside temperature probe are anti-iced by electrical resistors.

Electrical heating is used for windshield anti-icing, with each half of the center panel and the adjacent front panel heated by an independent system that incorporates a temperature regulator. Bird-impact strength requirements are met with the windshield heated or unheated.

Defogging of the inner surface of the windshield panes is by means of hot air from the flight deck conditioned air system and ventilation air from the instrument panel. The forward and aft side window panels are electrically defogged. The left and right windshield panels are each provided with an electrically operated, two-speed windshield wiper. They are stowed in the fairing below the windshields. Operation and storage is possible up to 215 knots. Passenger cabin windows will be defogged by ventilating the area between the panes.

Super King Air B200

Propeller electric deice system. The propeller electric deice system includes: electrically heated deice boots, slip rings and brush block assemblies, a timer for automatic operation, an ammeter, three circuit breakers located on the fuel control panel for left and right propeller and control circuit protection, and two switches located on the pilot's subpanel for automatic or manual control of the system (FIG. 11-23).

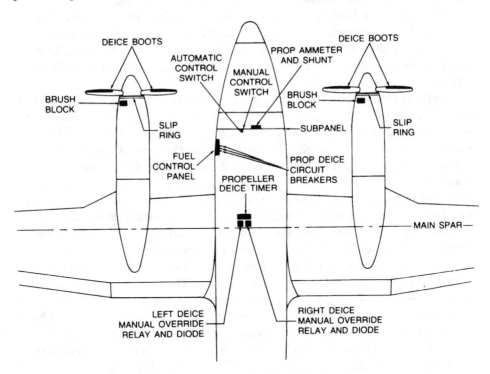

Fig. 11-23. *Propeller electric deice system schematic.* Beech Aircraft Corporation

A circuit breaker switch located on the pilot's subpanel, placarded PROP—AUTO—OFF, is provided to activate the automatic system. Upon placing the switch to the AUTO position, the timer diverts power through the brush block and slip ring to all heating elements on one propeller. Subsequently, the timer then diverts power to all heating elements on the other propeller for the same length of time. This cycle will continue as long as the switch is in the AUTO position. The system utilizes a metal foil type single heating element energized by dc voltage. The timer switches every 90 seconds, resulting in a complete cycle in approximately 3 minutes.

A manual prop deice system is provided as a backup to the automatic system. A control switch located on the left subpanel, placarded PROP—MANUAL—OFF, controls the manual override relay. Upon placing the switch in the MANUAL position, the automatic timer is overridden and power is then supplied to the heating elements of both propellers simultaneously. This switch is of the momentary type and must be held in position for approximately 90 seconds to dislodge ice from the propeller surface. Repeat this procedure as required to avoid significant buildup of ice that will result in loss of performance, vibration, and impingement upon the fuselage. The prop deice ammeter will not indicate a load while the propeller deice system is being utilized in the manual mode; however, the loadmeters will indicate an approximate .05 increase of load per meter while the manual prop deice system is in operation.

Pitot mast heat. Heating elements are installed in the pitot masts located on the nose. Each heating element is controlled by an individual circuit breaker switch placarded PITOT—LEFT—RIGHT, located on the pilot's subpanel. It is not advisable to operate the pitot heat system on the ground except for testing or for short intervals of time to remove ice or snow from the mast.

Surface deice system. The surface deice system removes ice accumulations from the leading edges of the wings and horizontal stabilizers. Ice removal is accomplished by alternately inflating and deflating the deice boots (FIG. 11-24). Pressure-regulated bleed air from the engines supplies pressure to inflate the boots. A venturi ejector, operated by bleed air, creates vacuum to deflate the boots and hold them down while not in use. To assure operation of the system in the event of failure of one engine, a check valve is incorporated in the bleed air line from each engine to prevent loss of pressure through the compressor of the inoperative engine. Inflation and deflation phases are controlled by a distributor valve. **Caution:** Operation of the surface deice system in ambient temperatures below –40°C can cause permanent damage to the deice boots.

A three-position switch in the ICE group on the pilot's subpanel identified: DEICE CYCLE—SINGLE—OFF—MANUAL, controls the deicing operation. The switch is spring-loaded to return to the OFF position from SINGLE or MANUAL. When the SINGLE position is selected, the distributor valve opens to inflate the wing boots. After an inflation period of approximately 6 seconds, an electronic timer switches the distributor to deflate the wing boots, and a 4-second inflation begins in the horizontal stabilizer boots. When these boots have inflated and deflated, the cycle is complete.

When the switch is held in the MANUAL position, all the boots will inflate simultaneously and remain inflated until the switch is released. The switch will return to the OFF position when released. After the cycle, the boots will remain in the vacuum holddown condition until again actuated by the switch.

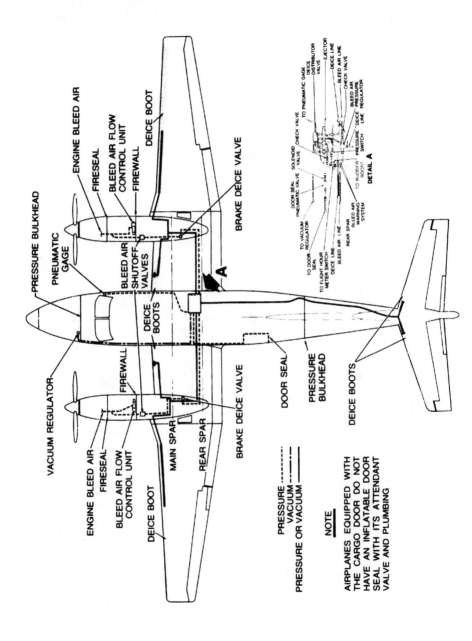

Fig. 11-24. *Pneumatic bleed air system and surface deice system schematic.* Beech Aircraft Corporation

For the most effective deicing operation, allow at least ½ inch of ice to form before attempting ice removal. Very thin ice might crack and cling to the boots instead of shedding. Subsequent cycling of the boots will then have a tendency to build up a shell of ice outside the contour of the leading edge, thus making ice removal efforts ineffective.

Bibliography

Airframe and Powerplant Mechanics Airframe Handbook. 1972, 1976. Washington, DC. U.S. Department of Transportation. Advisory Circular 65-15A.

Airframe and Powerplant Mechanics General Handbook. 1970, 1976. Washington, DC. U.S. Department of Transportation. Advisory Circular 65-9A.

Airframe and Powerplant Mechanics Powerplant Handbook. 1976. Department of Transportation, Federal Aviation Administration. Advisory Circular 65-12A.

Applied Science for the Aviation Technician. 1985. Casper, WY. IAP, Inc.

Beechcraft Beechjet 400A Pilot's Operating Manual. Beech Aircraft Corporation. 1991.

Beechcraft King Air C90B Pilot's Operating Handbook and FAA Approved Airplane Flight Manual. Beech Aircraft Corporation. 1991.

Beechcraft Super King Air B200 & B200C Pilot's Operating Handbook and FAA Approved Airplane Flight Manual. Beech Aircraft Corporation. 1991.

Crane, Dale 1975. Aircraft Hydraulic Systems. Basin, WY. Aviation Maintenance Publishers, Inc.

BIBLIOGRAPHY

Crane, Dale. 1975. Aircraft Fuel Metering Systems. Basin, WY. Aviation Maintenance Publishers, Inc.

Delp, Frank. 1979. Aircraft Propellers and Controls. Basin, WY. Aviation Maintenance Publishers, Inc.

Dole, Charles E. 1989. Flight Theory for Pilots - 3rd ed. Casper, WY. IAP, Inc.

Edwards, Mary and Elwyn Edwards. (1990). The Aircraft Cabin - Managing the Human Factors. Brookfield, VT: Gower Publishing Company.

Falcon 50 Aircraft and Interior Technical Description. Falcon Jet Corporation. 1991.

Falcon 900B Aircraft and Interior Technical Specifications. Falcon Jet Corporation. 1991.

Federal Aviation Regulations Part 91 General Operating and Flight Rules. U.S. Department of Transportation, Federal Aviation Administration.

Gamgram Bulletin #40 *Bonding vs Grounding*. Gammon Technical Products, Inc. Manasquan, NJ, May 1992.

Green, Roger G, et al. (1991). Human Factors For Pilots. Brookfield, VT: Gower Publishing Company.

Otis, Charles E. 1989. Aircraft Gas Turbine Powerplants. Casper, WY: IAP, Inc.

Rosen, Stan. 1980. DC Circuits. IAP, Inc. Casper, WY.

Shevell, Richard S. 1983. Fundamentals of Flight - 2nd ed. Englewood Cliffs, NJ: Prenctice-Hall, Inc.

Sullivan, Kenneth H. and Larry Milberry. 1989. Power - The Pratt & Whitney Canada Story. Toronto, Canada: CANAV Books.

The Aircraft Gas Turbine Engine and Its Operation. 1974. Pratt & Whitney Aircraft Group. Part No. PWA 182408.

TPE331 Turboprop Pilots Brief & Operational Tips. 1991. Garrett Engine Division. Part No. MC1418.

Treager, Irwin E. 1970. Aircraft Gas Turbine Engine Technology. New York, NY: McGraw-Hill Book Company.

Waite, Rick (Editor). 1989. Manual of Aviation Fuel Quality Control Procedures. Baltimore, MD: American Society for Testing and Materials. Manual MNL 5.

Wild, Thomas W. 1990. Transport Category Aircraft Systems. Casper, WY: IAP, Inc.

Index

A

absolute pressure, 225
absolute temperature scale, 225
acceleration, 67-68
accumulators, hydraulic systems, 168-171
  bladder type, 168
  diaphragm type, 168, **169**
  functions of accumulators, 171
  piston accumulators, 169, **170**, 171
  pressure-bump reduction, 171
  pressure-storage unit function, 171
  servicing hydraulic systems, warning, 169
  supplemental system pressure supply, 171
  system-pressure maintenance, 171
actuators, hydraulic systems, 173-176, **175**
  double-acting balanced, 175
  double-acting unbalanced, 175-176
  linear actuators, 175
  rotary actuators, 176
  single-acting actuators, 175
adiabatic process, 225
aeolipile, early turbine engine, 66, **66**
aerodynamic balance, 317
aerodynamic systems, 315-346
  Beech Super King Air B200 documentation, 343-346
  boundary-layer control devices, 320-321
  deicers, 324-334
  documentation, manufacturers', 339-346
  Falcon 50, documentation, 339
  Falcon 900B, documentation, 342-343
  flight control systems, 323-324
  high-lift devices, 319-320, **341**
  pitch control, **340**
  pneumatic bleed air system and surface deicer, **345**
  precipitation protection systems, 324-338
  rain elimination systems, 335-338
  roll control, **341**
  tabs, 315-318
  vortex generators, 321-322, **323**
  yaw control, **340**
aerodynamic twisting force, propellers, 135, **135**, 136
air inlet duct, turbine engines, 81-82
air-conditioning systems (*see also* air-distribution systems; pressurization systems, 225-226, 240-247

absolute temperature scale, 225
adiabatic process, 225
air distribution rates, 239-240, **240**
air-cycle machine (ACM) systems, 241-242
altitude, aircraft altitude, 225
altitude, cabin altitude, 225
ambient temperature, 225
basic requirements, 226
Beechjet 400A, 253-262
desiccants, alumina desiccant, 246
flow schematics, **243**
heat sinks, 244
latent heat, 245
premature vaporization (flashoff), 246
ram-air temperature rise, 225
refrigeration cycle, 244-247, **244**
refrigeration effect, 245
scales to measure temperature, 226
sensible heat, 245
superheating, 245
terminology of air-conditioning systems, 225
typical system, **227**
vapor-cycle (Freon) systems, 242, 244
air-cycle machine (ACM) cooling systems, 241-242
  heat exchangers, 241
air-distribution systems (*see also* air-conditioning systems), 239-240, **240**
  air ducts, 240, **241**
  Beechjet 400A, 253-262
  expansion bellows, 240, **241**
alternating current (ac), 7, 27
  alternators, 29
  capacitance, 29-31
  capacitive reactance, 31
  capacitors in ac circuits, 31, **32**
  counter-electromotive force (CEMF), 29
  cycles of ac, 29
  dc generator produces ac current, 37
  direct current (dc) compared to ac, 27
  frequency of ac, 29
  generation of alternating current, 29-32
  inductance, 29
  inductive reactance, 29
  phase of current and voltage, 31-32, **32**
  sine waves of ac, 29
  sine waves of ac, generation of sine wave, **30**
  transformer use, 27
  voltage curves, ac vs. dc, **27**

alternators, 29, 44-45
  brushless alternators, 45
  combined ac/dc systems, 46
  constant-speed drive (CSD) units, 46, **47**
  direct-connected, direct-current generator, 44
  integrated brushless alternators, 45
  maintenance for alternators, 47
  rectification, 45
  rectifier, alternator-rectifier units, 45
  revolving-arm vs. revolving-field alternators, 45
  rotator, 45
  rotors, 45
  stator, 45
  transformation, 45
  types of alternators, 44
  voltage regulation, 46
altitude
  aircraft altitude, 225
  cabin altitude, 225
ambient pressure, 225
ambient temperatures, 225
Ampere, Andre M., 8
ampere-hours rate, battery life, 20
ampere-load, dc generators, 43
amperes, 8
angle of attack, propellers, 135
area, 154
armatures, 37, 40, **40**
  dc motors, 50-51
  drum-type, **41**
asymmetric thrust, propellers, 140
atmosphere
  carbon dioxide , 220-222
  composition of atmosphere, 220-222, **221**
  ozone, 220-222
  pressure of atmosphere, atmospheric pressure, 222
  water vapor, 220-222
  weight of atmosphere vs. altitude, **222**
atmospheric pressure, 222
  absolute pressure, 225
  adiabatic process, 225
  ambient pressure, 225
  barometric pressure, 225
  differential pressure, 225
  gauge pressure, 225
atomic theory and electricity, 2-4
  atoms, 2-4
  atoms, hydrogen, **3**

Illustration is on page in **boldface**

atomic theory and electricity (cont)
  atoms, oxygen, **3**
  electrons, 3-4
  free electrons, 4
  matter and the universe, 2
  molecules, 2
  molecules, water molecules, **2**
  negative ions, negative charge, 4
  neutrons, 3-4
  Periodic Table of Elements, 4
  positive ions, positive charge, 4
  protons, 3-4
  shells, 4
auxiliary power unit (APU), 107-108
  battery vs., 19

B

balance
  aerodynamic balance, tabs, 317
  static balance, tabs, 318
balance tabs, 316-317, **316**
barometric pressure, 225
batteries, storage, 19-24
  ampere-hours ratings, 20
  auxiliary power units, ground power
    units vs., 19
  capacity of batteries, 20
  closed-circuit conditions, lead-acid
    batteries, 20
  hydrometer, 21
  lead-acid batteries, 19-21, **20**
  nickel-cadmium batteries, 21-24
  overcharge runaway, ni-cad batteries,
    23-24
  ratings for batteries, 20
  sulphation, lead-acid batteries, 20
  testing, lead-acid batteries, 21
  thermal runaway, ni-cad batteries, 23-
    24
beads, tires, 291
bearings, main bearings, turbine engines,
  102
Beech Aircraft Corporation, 58
Beech King Air 200
  pneumatic/vacuum system schematic,
    182
Beech King Air C90
  landing gear and brake systems, 296-
    299
  lubrication documentation, 127-128
  pitot cowling oil cooler, **128**
  propellers documentation, 149-151
Beech Super King Air B200
  aerodynamic systems documentation,
    343-346
  battery charge indicators, 61
  condition levers, 111
  electrical systems documentation, 59-61
  electrical systems schematic, **60**

engines, 109-111
external power for electrical systems,
  61
fire protection system documentation,
  310-313
fuel system schematic , **213**
fuel systems documentation, 212-216
power levers, 111
Pratt & Whitney PT6A-42 turboprop,
  110
propeller levers, 111
propeller reversing, 111-112
propulsion system controls, 111
turbine engines documentation, 109-
  112
Beechjet 400A
  environmental systems documentation,
    253-262
  hydraulic systems documentation, 183-
    185
  lubrication documentation, 128-130
  lubrication system, 128
  lubrication system schematic, **129**
  Pratt & Whitney JT15D-5 turbofan en-
    gine, **113**
  turbine engines documentation, 112-
    113
Bernoulli's principle, 158, **160**
bimetallic strips, 25
blade angle, propellers, 133, 135
bleed air, turbine engines compressors,
  thrust vs., 78, 82, 107-108
blowout, rich blowout, fuel systems, 197
boundary-layer control devices, 320-
  321, **321**, **322**
brake systems (*see also* landing gear systems), 281-289
  antiskid systems, 286, 288-289
  Beech King Air C90, 298-299
  care of brakes, 292-293
  expander tube brakes, 286, **287**
  multiple-disk brakes, 284-285, **285**
  power boost brake systems, 283-284,
    284
  power brake systems, 282-283, **282**
  propeller brakes, 143
  segmented rotor brakes, 285-286, **286**
breakdown diodes (*see* diodes, zener)
breakers, circuit (*see* circuit breakers)
breathers, oil system breathers, 122
bypass valve, theromostatic, oil system,
  123, **123**

C

cabin outflow valve, pressurization systems, 223
camber, propellers, 133, **134**
capacitance, 29-31
  farads to measure capacitance, 29

  measuring capacitance, 29
capacitive reactance, 31
capacitor-discharge ignition systems,
  turbine engines, 94-95, **96**
capacitors, 29-31
  ac circuit use, 31, **32**
  dc circuit use, **31**
  dielectrics, 29-31
  fixed, 31
  variable, 31
carbon dioxide in atmosphere, 220-222
cavitation, hydraulic-system pumps, 164
centrifugal cabin compressors, 226-229,
  228
centrifugal cabin supercharger, **229**
centrifugal force, propellers, 135, **135**, 136
centrifugal twisting force, propellers,
  135, **135**, 137
chafers, tires, 291
check valve, oil system, 122
chord lines, propellers, 133, **134**
circuit breakers, 25
coils, 17, 28-29
  dc motor, 51
combustion chambers, turbine engines,
  91-94
  annular-type, 92, 93, **94**
  can-annular type, 92, 93-94
  can-type, 92-93, **93**
  fuel-air ratio, 92
  types of combustion chambers, 92
commutators, 37-38, 40, **41**, 42
  dc motors, 51-52
compasses, magnetic compass, 15-16
compensator, nosewheel steering, 278
compound dc motors, 53-54, **54**
compressors, pneumatic systems
  two-stage compressor, **177**
compressors, pressurization systems
  centrifugal cabin compressor, 226-229,
    228
  centrifugal cabin supercharger, **229**
  positive-displacement compressor, 226
compressors, turbine engines, 82-91
  axial-flow compressors, 82, **83**, 86-91,
    86
  advantages, 88
  disadvantages, 89
  dual vs. triple spool system, 88, 89
  outer case temps, **126**
  rotors, 86, **87**, 88
  rotors, bulb vs. fir tree rotor assem-
    bly, 86, 88
  stators, 86, **87**, 88
  stators, inlet guide vanes, 88
  stators, straightening vane, 88
  bleed air vs. thrust, 78, 82, 107-108
  centrifugal-flow compressors, 82, 83-
    86, **83**, **84**

Illustration is on page in **boldface**

advantages, 85-86
components, **84**
diffuser, 83, **84**, 85
disadvantages, 86
impeller, 83, **84, 85**
manifold, 83, **84**
foreign object damage, 89-91, **90**
stalling compressors, 91
conductors, 7
cross section of conductor and resistance, 10
direct current (dc) circuits, 10
electromagnetism, magnetic field around conductor, **17**
length of conductor and resistance, 9, **10**
material of conductors and resistance, 9
parallel conductors, force between, 48-49, **49**
resistance, conductor components determining resistance, 9
temperature of conductor vs. resistance, 10
wire conductors, 7
constant-speed drive (CSD) alternator units, 46, **47**
constant-speed propellers (*see* propellers, constant-speed)
continuity, 26
coolers, oil coolers, 123-124
air-cooled heat exchangers, 123-124
fuel-cooled heat exchangers, 123-124, **124**
pitot cowling oil cooler, Beech C90, **128**
cooling systems, turbine engines (*see* lubrication and cooling systems)
core, transformer, 32
Coulomb, Charles A., 8
coulombs, 8
counter-electromotive force (CEMF), 29
dc motors, 57
current, 7-8
alternating current (ac), 7, 27
alternating current produced by dc generator, 37
amperes for measuring current, 8
conductors, 7
coulombs for measuring current, 8
dc motors, current-carrying wire, **48**
direct current (dc), 7, 27-29
electron movement, 8, **8**
intensity (I) of current flow, 8
measuring current flow, 8
phase in ac circuits, 31-32, **32**
speed of electricity vs. speed of light, 8
cycles of alternating current, 29

D

deicers, 324-334
boot construction, 327
boot construction, cross section, **327**
boot inflation cycle, **326**
boot preventive maintenance, 329
location of ice vs. method of control, 325
pneumatic bleed air system, **345**
pneumatic deicing systems, 325-329
schematic, **328**
propeller ice control systems, 329-331, **331, 343**
thermal anti-icing systems, 329
heated leading edge, **330**
typical system, **342**
windshield deicing systems, 331-334
electrical, **332**
laminated windshield, **332**
schematic, **334**
desiccants, air-conditioning systems, 246
diamagnetic materials, 15
die-out, lean die-out, fuel systems, 197
dielectrics, capacitors, 29-31
differential pressure, 225
differential relay switches, 42-43
diffuser, centrifugal-flow turbine engines compressor, 83, **84**, 85
diodes, 33-34
breakdown diodes (*see* diodes, zener)
zener diodes, 33-34
direct current (dc) circuits, 7, 10-14
alternating current (ac) compared to dc, 27
basic circuit, **11**
capacitor in dc circuit, **31**
circuit configurations, 13-14
coils, 28-29
conductors, 10
galvanometer, electromagnetism, 27-28, **28**
generation of direct current, 27-29, **28**
parallel circuit, 13-14, **14**
sum of resistance, 14
power measurements, 10
resistance, 10
resistors, 11
series circuit, 13, **13**
series-parallel circuit, 14, **15**
source, 10
symbols used in dc circuits, **11**
voltage curves, ac vs. dc, **27**
discontinuity, 26
drag, 132
drag struts, 267, **268**
dry-sump lubrication/cooling (*see* lubrication and cooling systems)
ductwork, air-distribution systems, 240, **241**

dump valves
fuel systems, 209-210
pressurization systems, 223-224, 231
dynamic electricity, 4

E

electrical systems, 35-63
alternators, 44-45
Beech Aircraft Super King Air B200, 59, **60**
combined ac/dc systems, fixed vs. variable frequency, 46
documentation, manufacturers', 58-63
Falcon Jet Falcon 50, 61-62, **61**
fixed frequency systems, 46
generators, dc, 35-44
motors, ac (*see* motors, ac; motors, dc)
motors, dc (*see* motors, ac; motors, dc)
variable frequency systems, 46
electricity, principles of electricity, 1-34
aircraft systems using electricity, 1-2
alternating current (ac), 29
alternators, 44-45
atomic theory and electricity, 2-4
batteries, storage, 19-24
capacitance and capacitors, 29-31
capacitive reactance, 31
circuit breakers, 25
counter-electromotive force (CEMF), 29
current flow, 7-8
diodes, 33-34
direct current (dc) circuits, 10
dynamic electricity, 4
electromagnetism, 17-18
electromotive force (EMF), 6-7
failure of circuits, primary causes, 25-26
fuses, 25
generators, dc, 35-44
inductance, 29
inductive reactance, 29
insulators, 9
inverters, 34
limiters, current or isolation limiters, 25
magnetism, 14-17
Ohm's law, 9
open circuits, 25, **26**
power, 10
protecting circuits, circuit protection, 24-25
rectifiers, 34
resistance, 8-10
semiconductors, 33
short circuits, 24-25, **26**
sine waves, ac, 29
generation of sine wave, **30**

electricity, principles of electricity (cont)
  speed of electricity vs. speed of light, 8
  static electricity, 4-6
  thermal protectors, 25
  transformers, 32-33
  transistors, 33
  volts, 7
electromagnetism (*see also* magnetism), 17-18
  coils, 17, 28-29
  electromagnets, 17, **18**
  galvanometers, 27-28, **28**
  magnetic field around conductor, **17**
  solenoids, 18, **18**
electromagnets, 17, **18**
electromotive force (EMF), 6-7
  counter-electromotive force (CEMF), 29, 57
  positive and negative charges, 6-7
  water pressure compared to EMF, **7**
electronic ignition systems, turbine engines, 95-97, **97**
electrons, atomic theory and electricity, 3-4
  free electrons, 4
  movement of electrons, **8**
electrostatic fields, 5-6, **6**
energy, 71-72
  conservation of energy (hydraulic systems), 156-159
  kinetic energy, 71-72, 156-157
  potential energy, 71, 156-157
environmental systems, 219-262
  air-conditioning (*see* air-conditioning systems)
  Beechjet 400A, environmental systems documentation, 253-262
  documentation, manufacturers' documentation, 253-262
  heating systems (*see* heating systems)
  oxygen-supplement systems (*see* oxygen-supplement systems)
  pressurization systems (*see* pressurization systems)
  temperature-control systems (*see* temperature-control systems)
  typical system, **254**
  Beechjet 400A, 253-262
equalizer circuit, dc generator, 43, **44**
exhaust system, turbine engines, 102-103
expansion bellows, air-distribution systems, 240, **241**
extinguishers (*see* fire protection systems)

F

failure of electrical circuits, 25-26
  continuity, 26
  discontinuity, 26
  open circuits, 25, **26**
  short circuits, 25, **26**
Falcon 50
  aerodynamic systems documentation, 339
  electrical systems documentation, 61-62
  electrical systems schematic, **61-62**
  fuel systems documentation, 216-217
Falcon 900B
  aerodynamic systems documentation, 342-343
  fire protection system documentation, 313-314
  Garrett TFE31-5BR-1C turbofan engine, **114**
  hydraulic systems documentation, 185-186
  landing gear and brake systems, 299-300
  turbine engines documentation, 113-114
Falcon Jet Corporation, 58
Faraday, Michael, 27
farads, 29
feathering, propellers, 140, 146
  auto-feather systems, 140-141
  thrust-sensitive signals (TSS), 140-141
ferrous materials, magnetism, 14
field frame, 40, **40**
filters
  fuel filters, 206-208, **207**
    wafer-screen type, **208**
  hydraulic systems, 172-173
    micron-type, **174**
  oil filters, 120-121, **121**
fire point, hydraulic fluid, 161
fire protection systems, 301-314, **313**
  Beech Super King Air B200, 310-313
  Class A fires, 301
  Class A zones, 302
  Class B fires, 301
  Class B zones, 303
  Class C fires, 301
  Class C zones, 303
  Class D zones, 303
  Class X zones, 303
  classification of fires, 301
  detection systems, 303-304
  detectors, 304-308
    continuous-element detectors, 308
    Fenwal spot detector, 305, **305**, **308**
    Lindberg system, 308, **309**
    thermal switch, 304-305, **305**
    thermocouple detectors, 305-306
  documentation, manufacturers', 310-314
  extinguisher systems, agent characteristics, 310
  extinguishers, 309-310
    agent characteristics, 310
    carbon dioxide, 302
    dry chemical, 302
    halogenated hydrocarbon, 302
    high-rate-of-discharge, 309
    portable, 302
    system schematic, **312**
  Falcon 900B, documentation, 313-314
  fire panel, **314**
  overheat warnings, 309
  zone classification, 302-303
fire protection systems, 313
fittings, hydraulic systems, 172
flaps, 319-320, **321**
flash point, hydraulic fluids, 161
flashoff, 246
flight-control systems, 323-324
flippers, tires, 291
flow control valves, hydraulic systems, 167
fluid power systems (*see* hydraulic systems)
follow-up linkage, nosewheel steering, 277-278
force, 67
free electrons, atomic theory and electricity, 4
frequency, alternating current (ac), 29
friction, 74
fuel gauges, 210-211, **211**, 216
fuel systems, 187-217
  additives, contamination-control, 193-194
  auxiliary fuel transfer systems, Beech, 214
  Beech Super King Air B200, fuel systems documentation, 212-216
  components of fuel systems, 204-211
  contamination, 189-195
    controlling contamination, 193-194
    detecting contamination, 193
    foreign particles, 191, **192**
    microbial growth, 191-192
    minimizing contamination, 194-195
    sediment, 192-193
    sediment, settling rates, **193**
    water, 190-191
    wrong fuels, 191
  controls, 198-204
  documentation, manufacturers' documentation, 211-217
  draining fuel, 215
  dump valves, 209-210
  Falcon 50, fuel systems documentation, 216-217
  filters, 206-208, **207**, **208**
  firewall shutoff, 215
  fuels, aviation gasoline, 214-215

Illustration is on page in **boldface**

gaseous, 187
Jet A, Jet A-1, Jet B, 188
liquid, 188
safe-handling tips, 195-197
solid, 187
specific gravity, 188
turbine engine fuels, 188-189
vaporization at atmospheric pressure, **189**
volatility, 189
wrong fuels as contaminants, 191
gauge, fuel-quantity indicating units, 210-211, **211**, 216
governor, speed-set governor, 202
grounding during refueling, 195
heaters, fuel heaters, 206, **207**
hydromechanical (JFC) controls, 198-200, **199**
Jet A, Jet A-1, Jet B fuels, 188
lean die-out, 197
maintaining the FCU control system, 204
maintaining the fuel systems, 211
manifolds, 208
measuring fuel use, mass-flow method, 210
metering system operations, 200, **201**
minimum-pressure valve, 202
overspeed protection for engine, 204
pressure regulation valves, 200
pressurization valves, 209-210
pumps, fuel pumps, 204-206
purging system, 215-216
rich blowout, 197
routing of fuel, 215
safety in fuel-handling, 195-197
scheduling system, 202-204, **203**
schematic of Beechcraft system, **213**
schematic of Falcon system, **217**
shutoff valve, 202
specific gravity of fuels, 188
spray nozzles, simplex vs. duplex, 208-209, **209**, **210**
static electricity hazards during refueling, 195-197
theory of fuel system operations, 197
throttle valve, 202
volatility of fuels, 189
water injection reset system, 204
fuses, 25
hydraulic fuses, 167

G

galvanometers, 27-28, **28**
Garrett TFE31-5BR-1C turbofan engine, **114**
Garrett TPE331 turboprop engine, **101**
gauge pressure, 225
generators, dc, 35-44

alternating current produced by dc generator, 44
ampere-load, 43
armature, 37, 40, **40**
drum-type, **41**
basic dc generator, **37**
commutators, 37-38, 40, **41**, 42
current, alternating current produced by dc generator, 37
differential relay switches, 42-43
equalizer circuit, 43, **44**
field frame, 40, **40**
maintenance for dc generators, 43
neutral plane, 39
off-line state in generators, 43
operation of basic dc generator, 38-39, **38**
operational theory of generators, 36
output of elementary generator, **37**
paralleling generators, 43
parts of dc generator, 40, **40**
ripple effect, 39, **39**
support components for generators, 42-43
voltage regulators, 42
rheostats, 42, **42**
voltage
inducing maximum voltage, **36**
output voltage factors, 39
yoke (see field frame)
gerotor oil pumps, 119-120, **120**
governors, fuel systems, 202
governors, propellers, 137, 144-146, **145**
pitot valve control, 145-146
speeder spring pressure, 145
ground power unit (GPU), 107-108
battery vs., 19
grounding during refueling, 195-197

H

heat (see also temperature)
resistance vs. temperature, 10
resistance-produced heat, 9
thermal protectors: intermittent switches/bimetallic strips, 25
heat exchanger, 123-124
air-cycle (ACM) cooling systems, 241
heat sinks, air-conditioning systems, 244
heaters, fuel heaters, 206, **207**
heating systems
Beechjet 400A, 253-262
compressed-air heating, 247
electric heaters, 247
radiant panel heat, 247
Hero, development of turbine engine, 66
high-lift devices, 319-320, **320**, **341**
flaps, 319-320, **321**
horsepower, 69-70, **70**
thrust horsepower, 132

hoses, hydraulic systems, 172
hydraulic fluids, 160-162
chemical stability, 161
fire point, 161
flash point, 161
mineral base fluids, 162
phosphate ester base fluids, 162
Skydrol fluid, 162
types of hydraulic fluid, 161-162
vegetable base fluids, 162
viscosity, 160-161
hydraulic systems, 153-176, 181-186
accumulators, 168-171
actuating cylinders, 173-176
area, 154-155
backup systems, 164
Beechjet 400A, hydraulic systems documentation, 183-185
Bernoulli's principle, 158, **160**
cavitation in pumps, 164
check valves, 167
components of hydraulic systems, 163-164
contaminants, **174**
abrasive vs. nonabrasive, 172-173
distance, 155-156
documentation, manufacturers' documentation, 183-186
dynamic fluids, 157
failures in hydraulic systems, 153-154
Falcon 900B, hydraulic systems documentation, 185-186
filters, 172-173
micron-type, **174**
flow control valves, 167
fluid power and conservation of energy, 156-159
fluid power systems, 154, **186**
fluid (see hydraulic fluids)
fuses, hydraulic fuses, 167
hydrostatic paradox, 158, **159**
kinetic energy, 156-157
lines, hoses, and fittings, 163, 172
maintenance of hydraulic systems, 181
mechanical advantage of hydraulic systems, 158, **158**
open vs. closed systems, 154
Pascal's law, 157-158, **157**
pneumatics vs. hydraulic systems (see also pneumatic systems), 153-154
potential energy, 156-157
power, 156
pressure, 155
pressure control valves, 167-168, **168**
pressure regulators, 168
pressurization of reservoir, 163
pumps, 165-166
can-type pump, **167**
cavitation in pumps, 164

hydraulic systems (cont)
  constant-displacement, 165
    gear-type, **165**
    piston-type angular pump, **166**
    variable-displacement, 166
  reservoirs, 163-164, **164**
    pressurization, 163
  servicing hydraulic systems, warning, 169
  static fluids, 157
  theory of operation, 154-160
  typical system, **184**
  valves, 167-168
    check valves, 167
    flow control, 167
    hydraulic fuses, 167
    pressure control, 167-168, **168**
    pressure regulators, 168
  volume, 154
  work, 156
hydrometers, battery testing, 21
hydroplaning, 295
hydrostatic paradox, 158, **159**
hypoxia, 219-220

I

ignition systems, turbine engines, 94-98
  capacitor-discharge type, 94-95, **96**
  electronic ignition, 95-97, **97**
  igniter plugs, 97-98, **98**
  typical system, **95**
impeller, centrifugal-flow turbine engines compressor, 83, **84, 85**
inductance, 29
induction ac motors, 58
inductive reactance, 29
inlet guide vanes (IGV), stators, axial-flow compressors, 88
innerliner, tires, 291
insulators, 9
intermittent switches, 25
inverters, 34
ions, atomic theory and electricity, 4

J

jet engines (*see* turbine engines)
JFC12-11 fuel systems, **199**
joules, 70

K

kilowatts, 70

L

landing gear systems, 263-300, **300**
  alignment, 266
  Beech King Air C90, 296
  Bogie truck landing gear assembly, **265**
  brake systems (*see* brake systems)

documentation, manufacturers's documentation, 295-300
drag struts, 267, **268**
dual main landing gear arrangement, **264**
emergency extension, 271, 273-275
Falcon 900B, 299
gear indicators, 273-274, **274**
ground locks, 273
main landing gear, 266-271
nose gear vs. tailwheel, 263
nosewheels (*see also* nosewheels), 275-281
  centering cams, 274-275, **275**
plumbing schematic, **297**
  electrical, 267, **269**
  hydraulic, 268-271, **270**
safety switches, 273, **273**
safety systems, 271, 273-275
shock struts, 264-265
  metering pin type, **266**
support, 266-267
torque links, **267**
trunnion and bracket arrangement, **268**
wheels and tires (*see* wheels and tires)
wing landing gear operation, 271, **272**
lead-acid batteries, 19-21, **20**
leading edge, propellers, 133, **134**
lean die-out, fuel systems, 197
limiter, current or isolation limiters (*see also* fuses), 25
lubrication and cooling systems, 115-130
  Beech King Air C90, documentation, 127-128
  Beechjet 400A, documentation, 128-130
  check valve, 122
  compressor temperature, axial-flow compressor, **126**
  coolers, 123-124
  cooling systems, 126-127
  documentation, manufacturers', 127-130
  dry-sump systems, 115, 117-126, **118, 125**
  exhaust turbine bearing, 115
  filters, 120-121, **121**
  gauge connections, 122
  jets or nozzles, 122
  lubrication system schematic, Beechjet 400A, **129**
  magnetic chip detectors, 128
  methods used, 115-116
  nacelle cooling arrangement, **127**
  pitot cowling oil cooler, Beech C90, **128**
  pressure relief valve, 121-122
  pump, oil pump, 117
    gerotor vs. piston type pumps, 119-120, **120**

tank, oil tank, **119**
thermostatic bypass valve, 123, **123**
vents or breathers, 122
wet-sump systems, 115, 116-117, **116**

M

magnetism, 14-17
  arrangement of molecules in magnetic materials, **16**
  compasses, 15-16
  diamagnetic materials, 15
  electromagnetism (*see* electromagnetism)
  ferrous materials, 14
  magnetic field of earth, **16**
  magnets
    permanent, 17
    residual magnetism, 17
    temporary, 17
  nonferrous materials, 14
  nonmagnetic materials, 15
  residual magnetism, 17
main bearings, turbine engines, 102
manifolds, centrifugal-flow turbine engines compressor, 83, **84**
manifolds, fuel systems, 208
mass, 73
matter, atomic theory and electricity, 2
mechanical advantage, hydraulic systems, 158, **158**
metering systems, fuel systems, 200, **201**
minimum-pressure valve, fuel systems, 202
molecules, atomic theory and electricity, 2
  water molecules, **2**
momentum, 73
motors, ac, 58
  induction motors, 58
  maintenance of ac motors, 58
  synchronous motors, 58
  types of ac motors, 58
motors, dc, 46-58
  armature, 50-51
  basic operations, **52**
  coil, 51
  commutator or slip-ring, 51-52
  compound motors, 53-54, **54**
  counter-electromotive force (CEMF), 57
  current-carrying wire, force on, 48, **48**
  cutaway view, **51**
  duty, types of duty for dc motors, 54
  load characteristics, **55**
  maintenance of dc motors, 58
  parallel conductors, force between, 48-49, **49**
  parts of dc motor, 50-52
  reversing motor direction, 55-56, **56**

Illustration is on page in **boldface**

series motors, 52-53, **53**
shunt motors, 53, **54**
  variable speed control, **57**
speed of motor, 56-57
split-field motors, 55, **56**
theory of operation in dc motors, 48
torque, 49-50, **50**
types of dc motors, 52-54
variable speed control, 56, **57**

N

negative charges
  atomic theory and electricity, 4
  electromotive force (EMF), 6-7
neutral plane, dc generators, 39
neutrons, atomic theory and electricity,
  3-4
Newton's laws of motion, 74-76
  friction, 74
  thrust (*see also* thrust), 74
nickel-cadmium batteries, 21-24
noise
  engine noise in turbine engines, 105-
  106
  static electricity-caused noise, 5
nonferrous materials, magnetism, 14
nosewheels, 275-281
  centering cams, 274-275, **275**
  compensator, 278
  follow-up linkage, 277-278
  mechanical and hydraulic units, **276**
  shimmy dampers, 278-281
  steer dampers, 281
  steering operation, 277
    compensator, 278
    follow-up linkage, 277-278
    hydraulic flow diagram, **277**

O

Oersted, Hans Christian, 17
Ohm's law, 9
Ohm, George S., 9
ohms, 9
oil filters, 120-121, **121**
oil jets or nozzles, 122
oil pumps, 117
  cutaway view, **120**
  gerotor type, 119-120, **120**
  gerotor vs. piston type pumps, 119-120
on-speed in propellers, 146
open circuits, 25, **26**
outflow valves, pressurization system,
  230, **230**
overcharge runaway, ni-cad batteries,
  23-24
overspeed in propellers, 146
overspeed protection, fuel systems, 204
oxygen-supplement systems, 219-222,
  247-253

altitude vs. oxygen available, 220
composition of atmosphere, 220-222,
  221
continuous-flow oxygen, 248-249, **249**
dangers associated with oxygen use,
  252-253
depressurization, rapid or explosive,
  248
diluter-demand regulators, **249**, 250,
  251
flow indicator, **252**
human's need for oxygen and hypoxia,
  219-220
hypoxia, 219-220
pressure gauge, **252**
pressure-demand oxygen, 249-252,
  250
regulator, narrow panel type, 252,
  252
typical system, **260**
ozone in atmosphere, 220-222

P

parallel circuit, dc, **14**
paralleling generators, 43
Pascal's law, 157-158, **157**
phase, alternating current (ac), 31-32, **32**
physics
  acceleration, 67-68
  area, 154-155
  Bernoulli's principle, 158, **160**
  drag, 132
  dynamic fluids, 157
  energy, 71-72
    conservation of energy (hydraulic
    systems), 156-159
    kinetic energy, 71-72, 156-157
    potential energy, 71, 156-157
  force, 67
  forward motion, propellers, 137
  friction, 74
  hydraulic and pneumatic systems, 154-
  160
  hydrostatic paradox, 158, **159**
  mass, 73
  mechanical advantage, hydraulic systems, 158, **158**
  momentum, 73
  Newton's laws of motion, 74-76
  Pascal's law, 157-158, **157**
  power, 69-70, 156
    horsepower, 69-70, **70**
    joule per second, watts, kilowatts, 70
  pressure, 155
  pressure, distance, 155-156
  rotational motions, propellers, 137
  speed, 72
  static fluids, 157
  thrust, 74, 132

factors affecting thrust, 76-78
thrust horsepower, 132
velocity, 72-73
  uniform motion, 72-73
  variable motion, 73
volume, 154
work, 67, 68, 156
pitch control, aerodynamic systems, **340**
pitch of propellers, 135
  effective pitch, 133, **133**
  geometric pitch, 133, **133**
  pitch-change mechanism, **145**
pitot cowling oil cooler, Beech C90, **128**
pitot valve, propeller-governor control,
  145-146
plugs, igniter plugs, turbine engines ignition systems, 97-98, **98**
pneumatic systems (*see also* hydraulic
  systems), 154, 176-186
  applications and uses, 176, 181
  Beech King Air 200 pneumatic/vac-
  uum system schematic, **182**
  compressors, two-stage compressor,
  177
  high-pressure systems, 177-180
  low-pressure systems, 180-181
  maintenance of pneumatic systems,
  181
  moisture separator, 178
  power system schematic, **179**
  types of hydraulic systems, 177-181
positive charges
  atomic theory and electricity, 4
  electromotive force (EMF), 6-7
potentiometers (*see also* resistors, variable; rheostats), 11, **12**
power, 10, 69-70
  horsepower, 69-70, **70**
  hydraulic systems, 156
  joules, 70
  kilowatts, 70
  watts, 70
Pratt & Whitney JT15D-5 turbofan engine, **113**
Pratt & Whitney PT6A-42 turboprop engine, **110**
precipitation protection systems, 324-
  338
  deicers, 324-334
pressure control valves, hydraulic systems, 167-168, **168**
pressure regulation valves, fuel systems,
  200
pressure relief valve, lubrication, 121-
  122
pressure valves, fuel systems, 209-210
pressurization systems (*see also* air-conditioning system; air-distribution
  systems), 222-239

Illustration is on page in **boldface**

pressurization systems (cont)
  absolute pressure, 225
  adiabatic process, 225
  air-conditioning and pressure systems, 225-226
  altitude, aircraft altitude, 225
  altitude, cabin altitude, 225
  ambient pressure, 225
  atmospheric pressure, 222
  barometric pressure, 225
  basic requirements, 226
  Beechjet 400A, 253-262
  cabin outflow valves, 223
  cabin pressure, 226
    centrifugal cabin compressor, 226-229, **228**
    centrifugal cabin supercharger, **229**
    positive-displacement compressor, 226
  controls
    cabin pressure, 233-239
    pressurization controllers, 231-232, **232**
  regulator, 233-237, **234, 235, 236**
  safety valve, 237-239, **238**
  typical control system, **256**
  depressurization, rapid or explosive, 248
  differential mode of operation, 235
  differential pressure, 225
  dump valves, 223-224, 231
  flow schematics, **243**
  gauge pressure, 225
  goals of aircraft pressurization, 223
  internal fuselage stresses, design concern, 224-225
  isobaric mode of operation, 235
  outflow valve, 230, **230**
  positive-displacement compressors, 226
  pressure vessel areas, 224
  regulator, cabin pressure, 233-237, **234, 235, 236**
  safety valve, cabin pressure, 237-239, **238**
  squat switch, 230
  terminology of pressurization systems, 225
  typical system, **227**
  unpressurized mode of operation, 235
  valves, 230-231
    safety valve, 237-239, **238**
propellers, 131-151
  aerodynamic factors, 132, **134**, 137
  aerodynamic twisting force, 135, **135**, 136
  afterbody assembly, turboprops, 143
  angle of attack, 135
  asymmetric thrust, 140

balancing propellers, 148-149
Beech King Air C90, propellers documentation, 149-151
blade angle, 133, 135
  setting angle, 148
blades, 132
  butt or base, 133, **134**
  cleaning, 149
  cross sections, 133, **134**
  shank, 133, **134**
brakes, 143
cambered edge, 133, **134**
centrifugal force, 135, **135**, 136
centrifugal twisting force, 135, **135**, 137
chord lines, 133, **134**
cleaning propeller blades, 149
constant-speed props, 137-142
  asymmetric thrust, 140
  auto-feather systems, 140-141
  feathering, 140
  fixed force, 144
  flight operation mode, 138-141
  governor, 137
  ground operation mode, 141-142
  negative torque signal, 139-140
  power levels, 139
  power vs. engine speed, 139
  reverse thrust, 137
  thrust-sensitive signals (TSS), 140-141
  torque, 139
  turboprops, 143-144
  variable force, 144
deicers, 329-331, **331, 343**
documentation, manufacturers' documentation, 149-151
drag, 132
feathering, 140, 146
  auto-feather systems, 140-141
  thrust-sensitive signals (TSS), 140-141
forces acting on propellers, 135-136, **135, 138**
forward motion, 137
governors, 137, 144-146, **145**
hub, 132
leading edge, 133, **134**
negative torque signal, 139-140
on-speed conditions, 146
operational systems, 142-146
overspeed conditions, 146
parts and components of propeller assembly, **144**
pitch, 135
  effective pitch, 133, **133**
  geometric pitch, 133, **133**
  pitch change mechanism, **145**
pitot valve control of governors, 145-146

power levels, 139
power vs. engine speed, 139
preflight inspection/service, 147-148
reduction gear assembly, 142, **142**
relative angle, 135
reverse thrust, 137, 141-142
rotational motion, 137
slippage, 132-133
speed vs. length of blade, 131
speeder spring pressure in governors, 145
synchrophasers, 146-147
terminology associated with propellers, 132
theory of propeller operation, 132-137
thrust, 132
thrust bending force, 135, **135**, 136
thrust horsepower, 132
thrust-sensitive signals (TSS), 140-141
torque, 139
torque bending force, 135, **135**, 136
tracking adjustment for blades, 148
trailing edge, 133, **134**
turboprop engines, 132
turbopropeller assembly, 143-146
underspeed conditions, 146
vibration in propellers, 148-149
protecting circuits, circuit protection, 24-25
  bimetallic strips, 25
  circuit breakers, 25
  fuses, 25
  intermittent switches, 25
  limiter, current or isolation, 25
  short circuits, 24-25
  thermal protectors, 25
protons, atomic theory and electricity, 3-4
pumps, fuel pumps, 204-206
  constant-displacement pumps, 204, 205-206
  engine-driven pumps, **205**
  variable-displacement pumps, 204, 206
pumps, hydraulic pumps, 165-166
  can-type pump, **167**
  constant-displacement pumps, 165
  gear-type pump, **165**
  piston-type angular pump, **166**
  variable-displacement pumps, 166
pumps, oil pumps, 117
  cutaway view, **120**
  gerotor vs. piston type pumps, 119-120, **120**

R

radios, noise, static electricity, 5
rain elimination systems (*see* windshield wipers)
ram-air temperature rise, 225

Illustration is on page in **boldface**

reactance, capacitive, 31
reactance, inductive, 29
rectifiers, 34
    alternator-rectifier units, 45
    alternators, rectification processes, 45
reduction gear assembly, propellers, 142,
    142
refrigeration cycle, air-conditioning,
    244-247, **244**
refrigeration effect, 245
refueling
    grounding during refueling, 195-197
    safety in fuel-handling, 195-197
    static electricity hazards, 5, 195-197
regulator, cabin pressure regulator, 233-
    237, **234, 235, 236**
relative angle, propellers, 135
relays, differential relay switches, 42-43
reservoirs, hydraulic systems, 163-164
    in-line reservoir, **164**
resistance, 8-10
    conductor components determining re-
        sistance, 9
    conductor materials contributing to re-
        sistance, 9
    cross section of conductor contributing
        to resistance, 10
    direct current (dc) circuits, 10
    heat produced by resistance, 9
    insulators, 9
    length of conductor contributing to re-
        sistance, 9, **10**
    Ohm's law, 9
    ohms to measure resistance, 9
    parallel circuit, 14
    resistors, 11
    temperature of conductor vs. resis-
        tance, 10
resistors, 11
    fixed, 11
    variable (*see also* potentiometers;
        rheostats), 11
reverse thrust
    propellers, 137, 141-142
    thrust reversers in turbine engines,
        104-105, **104**
rheostats (*see also* potentiometers; resis-
    tors, variable), 11, **12**
    voltage regulation using rheostat, 42,
        42
rich blowout, fuel systems, 197
ripple effect, dc generators, 39, **39**
roll control, aerodynamic systems, **341**
rotors, alternators, 45
rotors, axial-flow turbine engines com-
    pressors, 86, **87**, 88
    bulb root, 86, **88**
    fir tree root, 86, **88**
rotors, turbine engines, 99, **99**

S

scheduling systems, fuel systems, 202-
    204, **203**
semiconductors, 33
sensible heat, 245
series circuit, dc, **13**
series dc motors, 52-53, **53**
series-parallel circuit, dc, **15**
servo tabs, 316, **316**
shank of propeller blade, 133, **134**
shells, atomic theory and electricity, 4
shimmy dampers, nosewheels, 278-281
    piston-type damper, 279, **280**
    steer dampers, 281
    vane-type damper, 279-281, **281**
shock hazards
    batteries, overcharge or thermal run-
        away, 23-24
    static electricity, 5
shock struts, 264-265
    metering pin type, **266**
short circuits, 24-25, **26**
shunt dc motors, 53, **54**
    variable speed control, **57**
shutoff valve, fuel systems, 202
sidewall, tires, 291
sine waves, ac, 29
    generating sine waves, **30**
slippage of propellers, 132-133
solenoids, 18, **18**
specific gravity, fuels, 188
speed, 72
speed of electricity vs. speed of light, 8
split-field dc motors, 55, **56**
spools, turbine engines, 77
spray nozzles, fuel systems, simplex vs.
    duplex, 208-209, **209-210**
spring tags, **316**, 317
squat switch, pressurization systems, 230
starter systems, turbine engines, 106-108
    air-turbine starter systems, 107
    auxiliary power unit (APU), 107-108
    bleed air systems, 107-108
    combustion starter systems, 108
    cutaway view, **107**
    direct-cranking electric motors, 106
    electric motor starter systems, 106-107
    ground power unit (GPU), 107-108
    starter-generator electric motors, 106
static balance, 318
static electricity, 4-6
    electrostatic fields, 5-6, **6**
    noise, avionic/radio noise, 5
    refueling hazards, 5, 195-197
    shock hazard, 5
stators, alternators, 45
stators, axial-flow turbine engines com-
    pressors, 86, **87**, 88
    inlet guide vanes, 88

straightening vanes, 88
stators, turbine engines, 99, **99**
steering operation, nosewheel, 277
    hydraulic flow diagram, **277**
straightening vanes (IGV), stators, axial-
    flow compressors, 88
sulphation, lead-acid batteries, 20
Super King Air (*see* Beech Super King
    Air B200)
superchargers, centrifugal cabin super-
    charger, **229**
superheating, 245
synchronous ac motors, 58
synchrophasers, propellers, 146-147

T

tabs, 315-318
    aerodynamic balance, 317
    balance tabs, 316-317, **316**
    servo tabs, 316, **316**
    spring tabs, **316**, 317
    static balance, 318
    three forms of aerodynamic balance,
        317
    trim tabs, 316, **316**
temperature
    absolute temperature scale, 225
    adiabatic process, 225
    altitude vs. temperature, 223
    ambient temperature, 225
    latent heat, 245
    ram-air temperature rise, 225
    refrigeration effect, 245
    resistance vs. temperature, 10
    scales to measure temperature, 226
    sensible heat, 245
    superheating, 245
    thermal protectors: intermittent
        switches/bimetallic strips, 25
thermal protectors, 25
thermal runaway, ni-cad batteries, 23-24
throttle valve, fuel systems, 202
thrust, 132
    asymmetric thrust, propellers, 140
    compressor bleed air in turbine engines
        vs. thrust, 76-78
    factors affecting thrust, 76-78
    operating speed of turbine engines, 77-78
    pounds of thrust measurement, 79
    reverse thrust, 137
        propellers, 141-142
    temperature of incoming air in turbine
        engines vs. thrust, 78
    thrust horsepower, 132
thrust bending force, propellers, 135,
    135, 136
tires (*see* wheels and tires)
torque
    dc motors, 49-50, **50**

torque (cont)
  negative torque, propellers, 139-140
  propellers, 139
torque bending force, propellers, 135,
  135, 136
torque links, landing gear, **267**
trailing edge, propellers, 133, **134**
transformers, 27, 32-33
  alternators, transformation processes,
    45
  core, 32
  primary winding, 32
  secondary winding, 32
  step-down/up transformers, 32, **33**
  turns ratio, 33
transistors, 33
trim tabs, 316, **316**
turbine engines, 65-114
  acceleration, 67-68
  accessory section, 106
  aeolipile, early turbine engine, 66, **66**
  aerodynamic-blockage reverser sys-
    tems, 105
  air inlet duct, 81-82
  aircraft using turbine engines, 65-66
  auxiliary power unit (APU), 107-108
  balloon analogy, Newton's third law of
    motion, **75**
  Beechcraft Beechjet 400A, turbine en-
    gines documentation, 112-113
  Beechcraft Super King Air B200, tur
    bine engines documentation,
    109-112
  classification of turbine engines in air-
    craft use, 78-79
  cold-section maintenance, 109
  combustion chambers, 91-94
  components of turbine engines, 81-94
  compressors, 82-91
    bleed air vs. thrust, 78, 82
    centrifugal- vs. axial-flow compres
      sors, 82, 83-91, **83, 84**
  controlled descents, 101
  cooling (see lubrication and cooling
    systems)
  dive brakes, 105
  documentation, manufacturers' docu-
    mentation, 109-114
  energy, 71-72
  exhaust collector, 103, **103**
  exhaust cone, 102
  exhaust gas energy output, 98-99
  exhaust nozzle, 102
  exhaust section, 102-103
  Falcon 900B, turbine engines docu-
    mentation, 113-114
  fixed-shaft turbine configurations,
    100-101
  force, 67

foreign object damage to compressors,
  89-91, **90**
free-shaft turbine configurations, 100,
  101
friction, 74
fuel-air ratio mixtures, 92
fuels for turbine engines, 188-189
Garrett TFE31-5BR-1C turbofan engine, **114**
ground power unit (GPU), 107-108
high-pressure turbine, 101-102
history of turbine-engine development,
  66-67
hot-section maintenance, 109
ignition systems, 94-98
  capacitor-discharge type, 94-95, **96**
  electronic ignition, 95-97, **97**
  igniter plugs, 97-98, **98**
  typical system, **95**
interstage turbine temperature (ITT),
  100
low-pressure turbine, 102
lubrication (see lubrication and cooling
  systems)
main bearings, 102
maintenance for turbine engines, 108-
  109
mass, 73
mechanical-blockage reverser systems,
  105
momentum, 73
Newton's laws of motion (see Newton's laws; physics)
noise, engine noise, 105-106
operating speed, designed engine operating speed and thrust, 77-78
physics of turbine-engine operation
  (see also physics), 67-76
pounds of thrust measurement, 79
power, 69-70
  horsepower, 69-70, **70**
  joule per second, watts, kilowatts, 70
Pratt & Whitney JT15D-5 turbofan engine, **113**
Pratt & Whitney PT6A-42 turboprop,
  110
reverse thrust, rapid reverse, 101
reversers, thrust reversers, 104-105,
  104
rotors, 99, **99**
speed, 72
speed brakes, 105
spools, 77
stalled compressors, 91
starter systems, 106-108
stators, 99, **99**
temperature of incoming air vs. thrust,
  78
temperature vs. performance, 99-100

thrust, 74
  compressor bleed air vs. thrust, 78,
    82, 107-108
  designed engine operating speed, 77-
    78
  factors affecting thrust, 76-78
  pounds of thrust measurement, 79
  reverse thrust, 101, 104-105
  temperature of incoming air vs.
    thrust, 78
total turbine temperature (TTT), 100
turbine assembly, 98-102
turbine inlet temperature (TIT), 100
turbine outlet temperature (TOT), 100
turbofans, 78, 80
  Garrett TFE31-5BR-1C turbofan en
    gine, **114**
  Pratt & Whitney JT15D-5 turbofan
    engine, **113**
turbojets, 78, 79-80
turboprops, 78, 80, 132, 143-146
  Garrett TPE331 engine, **101**
  Pratt & Whitney PT6A-42, **110**
turboshafts, 78, 80-81
velocity, 72-73
water injection systems, 108
wing spoilers, 105
work, 67, 68
turbofan engines (see also turbine engines), 80
  forward-fan engines, 80, **81**
  Garrett TFE31-5BR-1C turbofan en-
    gine, **114**
  Pratt & Whitney JT15D-5 turbofan en-
    gine, **113**
turbojet engines (see also turbine engines), 79-80
turboprop engines (see also turbine engines), 80
  Garrett TPE331 engines, **101**
  Pratt & Whitney PT6A-42, **110**
  propellers assembly (see also pro-
    pellers), 132, 143-146
turboshaft engines (see also turbine engines), 80-81
Turns ratio, 33

U
underspeed in propellers, 146

V
vaporization of fuels, **189**
  premature vaporization (flashoff), 246
velocity, 72-73
  uniform motion, 72-73
  variable motion, 73
vents, oil system vents, 122
viscosity, hydraulic fluids, 160-161

Illustration is on page in **boldface**

volatility of fuels, 189
Volta, Alessandro, 7
voltage, 7
  dc generators, output voltage factors, 39
  generators, dc, inducing maximum voltage, **36**
  phase in ac circuits, 31-32, **32**
  regulators (*see* voltage regulators)
  voltage curves, ac vs. dc, **27**
voltage regulators, 42
  alternators, 46
  rheostats, 42, **42**
volume, 154
vortex generators, 321-322, **323**

W

water injection reset system, fuel systems, 204
water injection systems, turbine engines, 108
water vapor in atmosphere, 220-222
Watt, James, 70

watts, 10, 70
wet-sump lubrication/cooling (*see* lubrication)
wheels and tires (*see also* landing gear systems), 289-295
  applications and uses, 289-290
  split wheel, **290**
  tires
    beads, 291
    braking, 294
    care of tires and brakes, 292-293
    casing plies/cord body, 291
    chafers, 291
    condition of landing field, 295
    construction, cutaway view, **291**
    flippers, 291
    hydroplaning, 295
    innerliner, 291
    pivoting, 294
    preflight inspection, 293
    sidewall, 291
    takeoff and landing, 294-295
    taxiing, 293-294

    tips for extending tire life, 293-295
    tread, 290
    tread reinforcement, 290
windings, transformers, 32
windshield wipers, 335-338
  circuit diagram, **337**
  electrical system, **336**
  hydraulic systems, 335, **337**
  pneumatic systems, 335, **338**
  rain repellants, 336-338
wing landing gear operation, 271, **272**
wire
  conductive wire, 7
  cross section vs. resistance, 10
  length of wire vs. resistance, 9, **10**
  material of wire vs. resistance, 9
  temperature of wire vs. resistance, 10
work, 67, 68
  hydraulic systems, 156

Y

yaw control, aerodynamic systems, **340**
yoke, generator (*see* field frame)

Illustration is on page in **boldface**

About the author

David A. Lombardo is an aviation consultant specializing in organizational development, training program design, and staff recruitment and training. He also provides assistance regarding obtaining aircraft, parts, equipment, and contract training. Lombardo produces and directs videos for aviation training and corporate promotionals.

He was formerly associate professor and associate dean of the Lewis University division of aviation. He has also been director of training for Frasca International, Incorporated; director of program development for Airmanship, Incorporated; and a King Air instructor at FlightSafety International.

Lombardo is the author of *Aircraft Systems: Understanding Your Airplane* (TAB book no. 2423).

Other Bestsellers of Related Interest

AIM/FAR 1993: Airman's Information Manual/Federal Aviation Regulations—
TAB/AERO Staff

This is an essential reference for anyone involved in general aviation—offering fingertip access to the latest rules and regulations. It's the most current, accurate, and comprehensive pilot resource available. Nothing else compares in content or price; it's even less expensive than the government publications. Don't fly without it! 592 pages, illustrated. Book No. 24396, $12.95 paperback, $22.95 hardcover

AVIATOR'S GUIDE TO NAVIGATION—
2nd Edition—Donald J. Clausing

Navigate safely around the skies using the expert advice you'll find here. This second edition gives you the most up-to-date information available on Loran-C, inertial navigation systems (INS), NAVSTAR/GPS, electronic flight information systems (EFIS), and Omega/VLF navigation systems. It's designed to help you understand and use the system that will best satisfy your navigation needs—and stay within your budget. 296 pages, 124 illustrations. Book No. 3998, $18.95 paperback, $28.95 hardcover

"The classic you've been searching for . . .
STICK AND RUDDER: An Explanation of the Art of Flying—Wolfgang Langewiesche

Students, certificated pilots, and instructors alike have praised this book as *"the most useful guide to flying ever written"*. The book explains the important phases of the art of flying, in a way the learner can use. It shows precisely what the pilot does when he flies, just how he does it, and why. 400 pages, 88 illustrations. Book No. 3820, $19.95 hardcover

THE PILOT'S AIR TRAFFIC CONTROL HANDBOOK—2nd Edition—Paul E. Illman

Keep up with the most recent changes in rules and regulations and gain an understanding of why air traffic control is essential for both safe and legal flight with this handbook. It familiarized you with the national airspace system, the federal facilities that comprise the ATC system, and the operating procedures required to use the system properly—including a close-up look at the new airspace designations currently being implemented. 240 pages, 88 illustrations. Book No. 4232, $18.95 paperback, $28.95 hardcover

STANDARD AIRCRAFT HANDBOOK—5th Edition—Edited by Larry Reithmaier, originally compiled and edited by Stuart Leavell and Stanley Bungay

Now updated to cover the latest in aircraft parts, equipment, and construction techniques, this classic reference provides practical information on FAA-approved metal airplane hardware. Techniques are presented in step-by-step fashion and explained in shop terms without unnecessary theory and background. All data on materials and procedures is derived from current reports by the nation's largest aircraft manufacturers. 240 pages, 213 illustrations. Book No. 3634, $11.95 paperback only

**THE ILLUSTRATED GUIDE TO AERODY-
NAMICS**—2nd Edition—H. C. "Skip" Smith

Avoiding technical jargon and scientific explanations, this guide demonstrates how aerodynamic principles affect every aircraft in terms of
lift, thrust, drag, in-air performance, stability, and
control. It reviews airfoil development and design, accelerated climb performance, takeoff ve-
hicles, load and velocity-load factors, hypersonic
flight, area rules, laminar flow airfoils, planform
shapes, computer-aided design, and high-performance lightplanes. 352 pages, 269 illustrations.
Book No. 3786, $28.95 hardcover only.

Prices Subject to Change Without Notice.

Look for These and Other TAB Books at Your Local Bookstore

To Order Call Toll Free 1-800-822-8158
(24-hour telephone service available.)

or write to TAB Books, Blue Ridge Summit, PA 17294-0840.

Title	Product No.	Quantity	Price

☐ Check or money order made payable to TAB Books

Charge my ☐ VISA ☐ MasterCard ☐ American Express

Acct. No. _____ Exp. _____

Signature: _____

Name: _____

Address: _____

City: _____

State: _____ Zip: _____

Subtotal	$ _____
Postage and Handling ($3.00 in U.S., $5.00 outside U.S.)	$ _____
Add applicable state and local sales tax	$ _____
TOTAL	$ _____

TAB Books catalog free with purchase; otherwise send $1.00 in check or
money order and receive $1.00 credit on your next purchase.

*Orders outside U.S. must pay with international money in U.S. dollars
drawn on a U.S. bank.*

**TAB Guarantee: If for any reason you are not satisfied with the book(s)
you order, simply return it (them) within 15 days and receive a full
refund.**

BC